Graduate Texts in Mathematics 8

G. Takeuti W. M. Zaring

Axiomatic Set Theory

Springer-Verlag New York · Heidelberg · Berlin

Gaisi Takeuti
Professor of Mathematics, University of Illinois

Wilson M. Zaring
Associate Professor of Mathematics, University of Illinois

AMS Subject Classification (1970): 02 K 05, 02 K 15

Library of Congress Catalog Card Number 72–85950.

Printed in the United States of America.

ISBN 0–387–90050–0 Springer-Verlag New York Heidelberg Berlin (soft cover)
ISBN 0–387–90051–9 Springer-Verlag New York Heidelberg Berlin (hard cover)
ISBN 3–540–90050–0 Springer-Verlag Berlin Heidelberg New York (soft cover)

Preface

This text deals with three basic techniques for constructing models of Zermelo-Fraenkel set theory: relative constructibility, Cohen's forcing, and Scott-Solovay's method of Boolean valued models. Our main concern will be the development of a unified theory that encompasses these techniques in one comprehensive framework. Consequently we will focus on certain fundamental and intrinsic relations between these methods of model construction. Extensive applications will not be treated here.

This text is a continuation of our book, "Introduction to Axiomatic Set Theory," Springer-Verlag, 1971; indeed the two texts were originally planned as a single volume. The content of this volume is essentially that of a course taught by the first author at the University of Illinois in the spring of 1969. From the first author's lectures, a first draft was prepared by Klaus Gloede with the assistance of Donald Pelletier and the second author. This draft was then revised by the first author assisted by Hisao Tanaka.

The introductory material was prepared by the second author who was also responsible for the general style of exposition throughout the text. We have included in the introductory material all the results from Boolean algebra and topology that we need. When notation from our first volume is introduced, it is accompanied with a definition, usually in a footnote. Consequently a reader who is familiar with elementary set theory will find this text quite self-contained.

We again express our deep appreciation to Klaus Gloede and Hisao Tanaka for their interest, encouragement, and hours of patient hard work in making this volume a reality. We also thank our typist, Mrs. Carolyn Bloemker, for her care and concern in typing the final manuscript.

Urbana, Illinois
March 23, 1972

G. Takeuti
W. M. Zaring

Contents

Introduction

In this book, we present a useful technique for constructing models of Zermelo-Fraenkel set theory. Using the notion of Boolean valued relative constructibility, we will develop a theory of model construction. One feature of this theory is that it establishes a relationship between Cohen's method of forcing and Scott-Solovay's method of Boolean valued models.

The key to this theory is found in a rather simple correspondence between partial order structures and complete Boolean algebras. This correspondence is established from two basic facts; first, the regular open sets of any topological space form a complete Boolean algebra; and second, every Boolean algebra has a natural order. With each partial order structure $\mathbf{P}$, we associate the complete Boolean algebra of regular open sets determined by the order topology on $\mathbf{P}$. With each Boolean algebra $\mathbf{B}$, we associate the partial order structure whose universe is that of $\mathbf{B}$ minus the zero element and whose order is the natural order on $\mathbf{B}$.

If $\mathbf{B}_1$ is a complete Boolean algebra, if $\mathbf{P}$ is the associated partial order structure for $\mathbf{B}_1$, and if $\mathbf{B}_2$ is the associated Boolean algebra for $\mathbf{P}$, then it is not difficult to show that $\mathbf{B}_1$ is isomorphic to $\mathbf{B}_2$ (See Theorem 1.40). This establishes a kind of duality between partial order structures and complete Boolean algebras; a duality that relates partial order structures, which have broad and flexible applications, to the very beautiful theory of Boolean valued models. It is this duality that provides a connecting link between the theory of forcing and the theory of Boolean valued models.

Numerous background results are needed for our general theory. Many of those results are well known and can be found in standard textbooks. However, to assist the reader who may not know all that we require, we devote §1 to a development of those properties of Boolean algebras, partial order structures, and topologies that will be needed later.

Throughout this text, we will use the following variable conventions. Lower case letters $a, b, c, \ldots$ are used only as set variables. Capital letters $A, B, C, \ldots$ will be used both as set variables and as class variables; in any given context, capital letters should be assumed to be set variables unless we specifically state otherwise.

1. Boolean Algebra

In preparation for later work, we begin with a review of the elementary properties of Boolean algebras.

Definition 1.1. A structure $\langle B, +, \cdot, ^-, \mathbf{0}, \mathbf{1}\rangle$ is a Boolean algebra with universe B iff $\mathbf{0}$ and $\mathbf{1}$ are two (distinct) elements of B; $+$ and $\cdot$ are binary operations on B; $^-$ is a unary operation on B; and $\forall a, b, c \in B$.

1. $a + b = b + a$	$ab = ba$	Commutative Laws.
2. $a + (b + c) = (a + b) + c$	$a(bc) = (ab)c$	Associative Laws.
3. $a + bc = (a + b)(a + c)$	$a(b + c) = ab + ac$	Distributive Laws.
4. $\mathbf{0} + a = a$	$\mathbf{1}a = a$	Identity Laws.
5. $a + {}^-a = \mathbf{1}$	$a({}^-a) = \mathbf{0}$	Complementation Laws.

Remark. There are alternative definitions of a Boolean algebra. The reader might find it instructive to compare the definitions given in the standard texts.

Examples. 1. If $a \neq 0$ then $\langle \mathscr{P}(a)^*, \cup, \cap, ^-, 0, a\rangle$ is a Boolean algebra. If $a = 1$ we have a very special 2-element Boolean algebra that we denote by **2**. Every 2-element Boolean algebra is isomorphic to **2**.

2. If $a \neq 0$, $b \subseteq \mathscr{P}(a)$, $0 \in b$, $a \in b$, and if b is closed under set union, intersection, and relative complement then $\langle b, \cup, \cap, ^-, 0, a\rangle$ is a Boolean algebra. Such an algebra, i.e., one whose elements are sets and whose operations are union, intersection, and relative complement, we will call a *natural* Boolean algebra.

3. If for a first order logic whose language contains at least one predicate symbol we define an equivalence relation between sentences by

$$\phi \sim \psi \quad \text{iff} \quad \vdash [\phi \leftrightarrow \psi]$$

then the collection of equivalence classes is the universe for a Boolean algebra called the *Lindenbaum-Tarski algebra.* The operations are logical disjunction, conjunction, negation; $\vee$, $\wedge$, $\neg$, with the distinguished elements being truth and falsehood, i.e., **1** is the equivalence class of theorems and **0** is the equivalence class of contradictions.

Exercises. Prove the following for a Boolean algebra $\langle B, +, \cdot, ^-, \mathbf{0}, \mathbf{1}\rangle$:

1. $(\forall a)[a + b = a] \rightarrow b = \mathbf{0}$.
2. $(\forall a)[ab = a] \rightarrow b = \mathbf{1}$.

* $\mathscr{P}(a) = \{x | x \subseteq a\}$.

Notation: We will use the symbols $\mathbf{B}, \mathbf{B}', \mathbf{B}_1$ as variables on Boolean algebras. $|\mathbf{B}|$ is the universe of the Boolean algebra $\mathbf{B}$. When in a given context the symbols $\mathbf{0}$ and $\mathbf{1}$ appear it will be understood that they are the distinguished elements of whatever Boolean algebra is under discussion. If there are two or more Boolean algebras in the same discussion we will write $\mathbf{0_B}, \mathbf{1_B}, \mathbf{0_{B'}}, \mathbf{1_{B'}}$ to differentiate between the distinguished elements of the different spaces. If no confusion is likely the subscripts will be dropped. The same convention will be used in denoting Boolean operations.

Theorem 1.2. If $\langle B, +, \cdot, ^-, \mathbf{0}, \mathbf{1}\rangle$ is a Boolean algebra then $\forall a, b \in B$

1. $a + a = a \qquad aa = a$ Idempotent Laws.
2. $a + ab = a \qquad a(a + b) = a$ Absorption Laws.

Proof.

1. $a + a = (a + a)\mathbf{1} = (a + a)(a + {}^-a) = a + a({}^-a) = a + \mathbf{0} = a.$
2. $a + ab = a\mathbf{1} + ab = a(\mathbf{1} + b) = a({}^-b + b + b) = a({}^-b + b) = a\mathbf{1} = a.$

The proofs of the multiplicative properties are left to the reader.

Theorem 1.3. If $\langle B, +, \cdot, ^-, \mathbf{0}, \mathbf{1}\rangle$ is a Boolean algebra then

1. ${}^-\mathbf{0} = \mathbf{1}, {}^-\mathbf{1} = \mathbf{0}.$
2. $(\forall a \in B)[\mathbf{1} + a = \mathbf{1} \wedge \mathbf{0}a = \mathbf{0}].$

Proof.

1. ${}^-\mathbf{0} = \mathbf{0} + {}^-\mathbf{0} = \mathbf{1}.$
2. $\mathbf{1} + a = ({}^-a + a) + a = {}^-a + (a + a) = {}^-a + a = \mathbf{1}.$

The remaining proofs are left to the reader.

Theorem 1.4. If $\langle B, +, \cdot, ^-, \mathbf{0}, \mathbf{1}\rangle$ is a Boolean algebra then $\forall a, b \in B$

1. $a + b = \mathbf{1} \wedge ab = \mathbf{0} \rightarrow b = {}^-a.$
2. ${}^-({}^-a) = a.$
3. ${}^-(a + b) = ({}^-a)({}^-b), {}^-(ab) = {}^-a + {}^-b.$
4. $ab = a \leftrightarrow a + b = b.$

Proof.

1. $$\begin{aligned} b &= b\mathbf{1} = b(a + {}^-a) = ba + b({}^-a) \\ &= \mathbf{0} + b({}^-a) = a({}^-a) + b({}^-a) \\ &= (a + b)({}^-a) = \mathbf{1}({}^-a) = {}^-a. \end{aligned}$$
2. Since ${}^-a + a = \mathbf{1}$ and $({}^-a)a = \mathbf{0}$, we have from 1, ${}^-({}^-a) = a.$
3. $$\begin{aligned} (a + b) + ({}^-a)({}^-b) &= a + (b + {}^-a)(b + {}^-b) \\ &= a + (b + {}^-a) = \mathbf{1} + b = \mathbf{1} \\ (a + b)({}^-a)({}^-b) &= [a({}^-a) + b({}^-a)]({}^-b) \\ &= b({}^-a)({}^-b) = \mathbf{0}. \end{aligned}$$

Hence by 1, ${}^-(a + b) = ({}^-a)({}^-b).$

4. If $ab = a$ then $a + b = ab + b = b$. If $a + b = b$ then $ab = a(a + b) = a.$

The proof of the other half of 3 we leave as an exercise for the reader.

Definition 1.5. If $\langle B, +, \cdot, {}^-, \mathbf{0}, \mathbf{1}\rangle$ is a Boolean algebra then $\forall a, b \in B$

1. $(a - b) \stackrel{\Delta}{=} a({}^-b)$.
2. $(a \Rightarrow b) \stackrel{\Delta}{=} {}^-a + b$.
3. $(a \Leftrightarrow b) \stackrel{\Delta}{=} (a \Rightarrow b)(b \Rightarrow a)$.
4. $(a \leq b) \stackrel{\Delta}{\leftrightarrow} ab = a$.

Remark. We will refer to $\leq$ as the natural order on the Boolean algebra.

Theorem 1.6. If $\langle B, +, \cdot, {}^-, \mathbf{0}, \mathbf{1}\rangle$ is a Boolean algebra with natural order $\leq$ then $\forall a, b, c \in B$

1. $a \leq a$.
2. $a \leq b \wedge b \leq a \rightarrow a = b$.
3. $a \leq b \wedge b \leq c \rightarrow a \leq c$.

Proof.

1. $aa = a$.
2. $a = ab = ba = b$.
3. If $a = ab \wedge b = bc$ then $a = ab = a(bc) = (ab)c = ac$.

Theorem 1.7. If $\langle B, +, \cdot, {}^-, \mathbf{0}, \mathbf{1}\rangle$ is a Boolean algebra with natural order $\leq$ then $\forall a, b \in B$

1. $a \leq b \leftrightarrow {}^-b \leq {}^-a$.
2. $a \leq b \leftrightarrow a - b = \mathbf{0}$.
3. $a \leq b \leftrightarrow (a \Rightarrow b) = \mathbf{1}$.

Proof. 1. If $a \leq b$ then $a = ab$. Therefore ${}^-a = {}^-(ab) = {}^-a + {}^-b$. Then by Theorem 1.4.4 $({}^-b)({}^-a) = {}^-b$, i.e., ${}^-b \leq {}^-a$. Conversely if ${}^-b \leq {}^-a$ then

$${}^-({}^-a) \leq {}^-({}^-b) \quad \text{i.e.,} \quad a \leq b.$$

2. If $a \leq b$ then $a = ab$. Therefore $a({}^-b) = (ab)({}^-b) = \mathbf{0}$. Conversely if $a({}^-b) = \mathbf{0}$ then $a = a\mathbf{1} = a(b + {}^-b) = ab + a({}^-b) = ab$ i.e., $a \leq b$.

3. If $a \leq b$ then $a = ab$ and ${}^-a = {}^-a + {}^-b$. Therefore $(a \Rightarrow b) = {}^-a + b = ({}^-a + {}^-b) + b = {}^-a + \mathbf{1} = \mathbf{1}$. Conversely if $(a \Rightarrow b) = \mathbf{1}$ then $a = a\mathbf{1} = a({}^-a + b) = ab$ i.e., $a \leq b$.

Theorem 1.8. If $\langle B, +, \cdot, {}^-, \mathbf{0}, \mathbf{1}\rangle$ is a Boolean algebra with natural order $\leq$ then $\forall a, b, c, d \in B$

1. $\mathbf{0} \leq b \leq \mathbf{1}$.
2. $[a \leq b] \wedge [c \leq d] \rightarrow [ac \leq bd] \wedge [a + c \leq b + d]$.

Proof. 1. $\mathbf{0} = \mathbf{0}b \wedge b = b\mathbf{1}$.
2. If $a = ab$ and $c = cd$ then $(ac)(bd) = (ab)(cd) = ac$ and

$$(a + c)(b + d) = ab + ad + cb + cd = a + ad + cb + c = a + c.$$

Exercises. Prove the following for a Boolean algebra $\langle B, +, \cdot, {}^-, \mathbf{0}, \mathbf{1}\rangle$:

1. $a \leq {}^-b \leftrightarrow ab = \mathbf{0}$.
2. $a \leq (a + b) \wedge b \leq (a + b)$.

3. $ab \leq a \wedge ab \leq b$.
4. $[a \leq c \wedge b \leq c] \rightarrow (a + b) \leq c$.
5. $[c \leq a \wedge c \leq b] \rightarrow c \leq ab$.

Definition 1.9. If $\langle B, +, \cdot, ^-, \mathbf{0}, \mathbf{1}\rangle$ is a Boolean algebra with natural order $\leq$, if $A \subseteq B$ and $b \in B$ then

1. $b = \sum_{a \in A} a \overset{\Delta}{\leftrightarrow} (\forall a \in A)[a \leq b] \wedge (\forall b' \in B)[(\forall a \in A)[a \leq b'] \rightarrow b \leq b']$.

2. $b = \prod_{a \in A} a \overset{\Delta}{\leftrightarrow} (\forall a \in A)[b \leq a] \wedge (\forall b' \in B)[(\forall a \in A)[b' \leq a] \rightarrow b' \leq b]$.

Definition 1.10. A Boolean algebra $\langle B, +, \cdot, ^-, \mathbf{0}, \mathbf{1}\rangle$ is complete iff

$$(\forall A \subseteq B)(\exists b, b' \in B)\left[b = \sum_{a \in A} a \wedge b' = \prod_{a \in A} a\right].$$

Example. If $a \neq 0$ then the Boolean algebra $\langle \mathscr{P}(a), \cup, \cap, ^-, 0, a\rangle$ is complete. Indeed if $A \subseteq \mathscr{P}(a)$ and $A \neq 0$, then

$$\sum_{b \in A} b = \bigcup (A) \wedge \prod_{b \in A} b = \bigcap (A).$$

Theorem 1.11. If $\langle B, +, \cdot, ^-, \mathbf{0}, \mathbf{1}\rangle$ is a Boolean algebra and $A \subseteq B$ then

1. $^-\sum_{a \in A} a = \prod_{a \in A} (^-a)$.

2. $^-\prod_{a \in A} a = \sum_{a \in A} (^-a)$.

Proof. 1. Since $(\forall b \in A)[b \leq \sum_{a \in A} a]$ we have $^-\sum_{a \in A} a \leq {}^-b$ and hence

$$^-\sum_{a \in A} a \leq \prod_{a \in A} (^-a).$$

Also $(\forall b \in A)[\prod_{a \in A} (^-a) \leq {}^-b]$. Therefore $b \leq {}^-\prod_{a \in A} (^-a)$, hence

$$\sum_{a \in A} a \leq {}^-\prod_{a \in A} (^-a)$$

i.e.,

$$\prod_{a \in A} (^-a) \leq {}^-\sum_{a \in A} a.$$

2. Left to the reader.

Theorem 1.12. If $\langle B, +, \cdot, ^-, \mathbf{0}, \mathbf{1}\rangle$ is a Boolean algebra, if $b, c \in B$, $A \subseteq B$, and

$$b = \sum_{a \in A} a$$

then

$$cb = \sum_{a \in A} ca.$$

Proof. If $a \in A$ then by Definition 1.9, $a \leq b$ and hence $ca \leq cb$. If for each $a \in A$, $ca \leq d$ then since $a = (^-c + c)a = {}^-ca + ca \leq {}^-c + d$ it follows from Definition 1.9 that $b \leq {}^-c + d$. Hence $cb \leq d$ and again from Definition 1.9 $\sum_{a \in A} ca = cb$.

Remark. Having now reviewed the basic properties of Boolean algebras we turn to the problem of characterizing complete Boolean algebras. As a first step in this direction we will show that the collection of regular open sets of a topological space is the universe of a Boolean algebra that is almost a natural algebra.

Definition 1.13. The structure $\langle X, T \rangle$ is a topological space iff $X \neq 0$,

1. $T \subseteq \mathscr{P}(X) \wedge 0 \in T \wedge X \in T$.
2. $A \subseteq T \rightarrow \bigcup (A) \in T$.
3. $(\forall N, N' \in T)[N \cap N' \in T]$.

T is a topology on X iff $\langle X, T \rangle$ is a topological space. If $a \in X$ and $N \in T$ then N is a neighborhood of a iff $a \in N$. If N is a neighborhood of a we write $N(a)$.

Theorem 1.14. $\mathscr{P}(X)$ is a topology on X.

Proof. Left to the reader.

Definition 1.15. T is the discrete topology on X iff $T = \mathscr{P}(X)$.

Definition 1.16. If T is a topology on X and $A \subseteq X$ then

1. $A^0 \triangleq \{x \in A \mid (\exists N(x))[N(x) \subseteq A]\}$.
2. $A^- \triangleq \{x \in X \mid (\forall N(x))[N(x) \cap A \neq 0]\}$.

Theorem 1.17. If T is a topology on X and $A \subseteq X$ then $A^0 \in T$.

Proof. If $B = \{N \in T \mid N \subseteq A\}$ then $B \subseteq T$. Furthermore

$$\begin{aligned} x \in A^0 &\leftrightarrow \exists N(x) \subseteq A \\ &\leftrightarrow \exists N(x) \in B \\ &\leftrightarrow x \in \bigcup (B). \end{aligned}$$

Then $A^0 = \bigcup (B) \in T$.

Definition 1.18. T' is a base for the topology T on X iff

1. $T' \subseteq T$.
2. $(\forall A \subseteq X)[A = A^0 \rightarrow (\exists B \subseteq T')[A = \bigcup (B)]]$.

Theorem 1.19. If $X \neq 0$, if T' is a collection of subsets of X with the properties

1. $(\forall a \in X)(\exists A \in T')[a \in A]$.
2. $(\forall a \in X)(\forall A_1, A_2 \in T')[a \in A_1 \cap A_2 \rightarrow$
$(\exists A_3 \in T')[a \in A_3 \wedge A_3 \subseteq A_1 \cap A_2]]$.

Then T' is a base for a topology on X.

Proof. If $T = \{B \subseteq X \mid (\exists C \subseteq T')[B = \bigcup(C)]\}$ then $0 = \bigcup(0) \in T$ and from property 1, $X = \bigcup(T') \in T$. This establishes property 1 of Definition 1.13.

To prove 2 of Definition 1.13 we wish to show that $\bigcup(S) \in T$ whenever $S \subseteq T$. From the definition of T it is clear that if $S \subseteq T$ then $\forall B \in S$, $\exists C \subseteq T'$

$$B = \bigcup(C).$$

If

$$C_B = \{A \in T' \mid A \subseteq B\}$$

then

$$B = \bigcup(C_B)$$

and

$$\bigcup_{B\in S} B = \bigcup_{B\in S} \cup(C_B)$$

$$= \bigcup\left(\bigcup_{B\in S} C_B\right).$$

Since $\bigcup_{B\in S} C_B \subseteq T'$, $\bigcup(S) \in T$.

If $B_1, B_2 \in T$ then $\exists C_1, C_2 \subseteq T'$

$$B_1 = \bigcup(C_1) \wedge B_2 = \bigcup(C_2).$$

Therefore

$$B_1 \cap B_2 = \left(\bigcup_{A_1\in C_1} A_1\right) \cap \left(\bigcup_{A_2\in C_2} A_2\right)$$

$$= \bigcup_{\substack{A_1\in C_1\\ A_2\in C_2}} (A_1 \cap A_2)$$

$$= \bigcup_{\substack{A_1\in C_1\\ A_2\in C_2\\ A_3\subseteq A_1\cap A_2}} A_3 \qquad \text{(By 2).}$$

Then $B_1 \cap B_2 \in T$; hence T is a topology on X. Clearly T' is a base for T.

Definition 1.20. If T is a topology on X and $A \subseteq X$ then

1. A is open iff $A = A^0$.
2. A is regular open iff $A = A^{-0}$.
3. A is closed iff $A = A^-$.
4. A is clopen iff A is both open and closed.
5. A is dense in X iff $A^- = X$.

Remark. From Theorem 1.17 we see that if T is a topology on X then T is the collection of open sets in that topology. A base for a topology is simply a collection of open sets from which all other open sets can be generated by unions.

For the set of real numbers R the intervals $(a, b) \triangleq \{x \in R \mid a < x < b\}$ form a base for what is called the natural topology on R. In this topology $(0, 1)$, and indeed every interval (a, b), is not only open but regular open. $[a, b] \triangleq \{x \in R \mid a \leq x \leq b\} = (a, b)^-$. Thus for example $[1, 2]$ is closed.

Furthermore $(0, 1) \cup (1, 2)$ is open but not regular open. The set of all rationals is dense in R. In this topology there are exactly two clopen sets 0 and R.

Theorem 1.21. 1. In any topology on X both 0 and X are clopen.

2. In the discrete topology on X every set is clopen and the collection of singleton sets is a base.

Proof. Left to the reader.

Remark. The next few theorems deal with properties that are true in every topological space $\langle X, T\rangle$. In discussing properties that depend upon X but are independent of the topology T, it is conventional to suppress reference to T and to speak simply of a topological space X. Hereafter we will use this convention.

Theorem 1.22. If $A \subseteq X$ and if $B \subseteq X$ then

1. $A^0 \subseteq A \subseteq A^-$.
2. $A^{00} = A^0 \wedge A^{--} = A^-$.
3. $A \subseteq B \rightarrow A^0 \subseteq B^0 \wedge A^- \subseteq B^-$.
4. $(X - A)^- = X - A^0 \wedge (X - A)^0 = X - A^-$.

Proof.

1.
$$\begin{aligned} x \in A^0 &\rightarrow \exists N(x) \subseteq A \\ &\rightarrow x \in A \\ x \in A &\rightarrow (\forall N(x))[N(x) \cap A \neq 0] \\ &\rightarrow x \in A^-. \end{aligned}$$

2.
$$\begin{aligned} x \in A^0 &\rightarrow \exists N(x) \subseteq A \\ &\rightarrow (\exists N(x))[x \in (N(x) \cap A^0) \wedge (N(x) \cap A^0) \in T \\ &\qquad\qquad \wedge (N(x) \cap A^0) \subseteq A^0] \\ &\rightarrow \exists N(x) \subseteq A^0 \\ &\rightarrow x \in A^{00}. \end{aligned}$$

Since by 1, $A^{00} \subseteq A^0$ we conclude that $A^{00} = A^0$.

$$\begin{aligned} x \in A^{--} &\rightarrow (\forall N(x))[N(x) \cap A^- \neq 0] \\ &\rightarrow (\forall N(x))(\exists y)[y \in N(x) \wedge y \in A^-] \\ &\rightarrow (\forall N(x))(\exists y)(\exists N'(y))[N'(y) \cap A \neq 0 \wedge N'(y) \subseteq N(x)] \\ &\rightarrow (\forall N(x))[N(x) \cap A \neq 0] \\ &\rightarrow x \in A^-. \end{aligned}$$

Since by 1, $A^- \subseteq A^{--}$ it follows that $A^{--} = A^-$.

3. If $A \subseteq B$ then

$$\begin{aligned} x \in A^0 &\rightarrow \exists N(x) \subseteq A \\ &\rightarrow \exists N(x) \subseteq B \\ &\rightarrow x \in B^0 \\ x \in A^- &\rightarrow (\forall N(x))[N(x) \cap A \neq 0] \\ &\rightarrow (\forall N(x))[N(x) \cap B \neq 0] \\ &\rightarrow x \in B^-. \end{aligned}$$

4. $x \in (X - A)^- \leftrightarrow (\forall N(x))[N(x) \cap (X - A) \neq 0]$
$\leftrightarrow (\forall N(x))[N(x) \nsubseteq A]$
$\leftrightarrow x \notin A^0$
$\leftrightarrow x \in X - A^0.$

$x \in (X - A)^0 \leftrightarrow \exists N(x) \subseteq (X - A)$
$\leftrightarrow (\exists N(x))[N(x) \cap A = 0]$
$\leftrightarrow x \notin A^-$
$\leftrightarrow x \in X - A^-.$

Theorem 1.23. If $A \subseteq X$ and if $B \subseteq X$ then

1. A regular open implies A open.
2. A is open iff $X - A$ is closed.
3. A is closed iff $X - A$ is open.
4. $A \subseteq B$ and A dense in X implies B dense in X.

Proof.

1. If $A = A^{-0}$ then $A^0 = A^{-00} = A^{-0} = A$.
2. $A = A^0 \leftrightarrow (X - A) = (X - A^0)$
$\leftrightarrow (X - A) = (X - A)^-.$
3. Left to the reader.
4. $A \subseteq B \rightarrow A^- \subseteq B^-$. But A dense in X implies $A^- = X$. Hence $B^- = X$.

Theorem 1.24. If C is a clopen set in the topological space X and $B^- - B^0 \subseteq C$ then $B^- - C$ is clopen.

Proof. If $x \in B^- - C$ then since $B^- - B^0 \subseteq C$

$$x \in B^0 \wedge x \notin C.$$

Since C is closed $X - C$ is open. Therefore $B^0 \cap (X - C)$ is open. Then $x \in B^0 \cap (X - C)$ implies $\exists N(x) \subseteq B^0 \cap (X - C) \subseteq B^- - C$. Thus $B^- - C$ is open.

If $(\forall N(x))[N(x) \cap (B^- - C) \neq 0]$ then

$$(\forall N(x))[N(x) \cap B^- \neq 0 \wedge N(x) \cap (X - C) \neq 0].$$

Since B^- is closed $x \in B^-$; since C is open $X - C$ is closed, hence $x \in X - C$. Therefore $x \in B^- - C$ and $B^- - C$ is closed.

Theorem 1.25. If C is a clopen set in the topological space X then $X - C$ is clopen.

Theorem 1.26. The clopen sets of a topological space form a natural Boolean algebra.

Proof. Left to the reader.

Theorem 1.27. If $A \subseteq X$ and $B \subseteq X$ then
1. $(A \cup B)^- = A^- \cup B^-$, $(A \cap B)^0 = A^0 \cap B^0$,
2. $(A \cap B)^- \subseteq A^- \cap B^-$, $A^0 \cup B^0 \subseteq (A \cup B)^0$.

Proof. 1. Since $A \subseteq A \cup B$ and $B \subseteq A \cup B$ we have $A^- \subseteq (A \cup B)^-$ and $B^- \subseteq (A \cup B)^-$. Therefore $(A^- \cup B^-) \subseteq (A \cup B)^-$.
$x \in (A \cup B)^- \wedge x \notin A^-$

$$\begin{aligned}&\to (\forall N(x))[N(x) \cap (A \cup B) \neq 0] \wedge (\exists N'(x))[N'(x) \cap A = 0]\\ &\to (\forall N(x))(\exists N'(x))[N(x) \cap N'(x) \cap A = 0 \wedge\\ &\quad (N(x) \cap N'(x)) \cap (A \cup B) \neq 0]\\ &\to (\forall N(x))[N(x) \cap B \neq 0]\\ &\to x \in B^-.\end{aligned}$$

Thus $(A \cup B)^- \subseteq (A^- \cup B^-)$ and hence $(A \cup B)^- = A^- \cup B^-$.

$$\begin{aligned}x \in (A^0 \cap B^0) &\leftrightarrow [\exists N(x) \subseteq A] \wedge [\exists N'(x) \subseteq B]\\ &\leftrightarrow \exists N(x) \subseteq A \cap B\\ &\leftrightarrow x \in (A \cap B)^0.\end{aligned}$$

2. Left to the reader.

Theorem 1.28. If $A \subseteq X$, and if $B \subseteq X$, then

1. $A = A^0 \to A \subseteq A^{-0}$.
2. $(A^{-0})^{-0} = A^{-0}$.
3. $(A^0 \cap B^-) \subseteq (A \cap B)^-$.
4. $(A \cup B^{-0})^{-0} \subseteq (A^- \cup B^{-0})$.

Proof. 1. If $A = A^0$ then since $A \subseteq A^-$ we have $A = A^0 \subseteq A^{-0}$.

2. Since $A^{-0} \subseteq A^-$ and $A^{--} = A^-$ we have $(A^{-0})^{-0} \subseteq A^{--0} = A^{-0}$. Since A^{-0} is open we have from 1, $A^{-0} \subseteq (A^{-0})^{-0}$. Therefore

$$(A^{-0})^{-0} = A^{-0}.$$

3. $$\begin{aligned}x \in (A^0 \cap B^-) &\to [(\exists N'(x) \subseteq A) \wedge (\forall N(x))[N(x) \cap B \neq 0]\\ &\to (\exists N'(x))(\forall N(x))[(N(x) \cap N'(x)) \cap (A \cap B) \neq 0]\\ &\to (\forall N(x))[N(x) \cap A \cap B \neq 0]\\ &\to x \in (A \cap B)^-.\end{aligned}$$

4. $$\begin{aligned}X - (A^- \cup B^{-0}) &= (X - A^-) \cap (X - B^{-0})\\ &= (X - A^-) \cap (X - B^{-0-0}) &&\text{(By 2)}\\ &\subseteq [(X - A^-) \cap (X - B^{-0-})]^- &&\text{(By 3)}\\ &= [X - (A^- \cup B^{-0-})]^-\\ &= X - (A^- \cup B^{-0-})^0\\ &= X - (A \cup B^{-0})^{-0}.\end{aligned}$$

Therefore $(A \cup B^{-0})^{-0} \subseteq (A^- \cup B^{-0})$.

Theorem 1.29. If $A \subseteq X$ and if $B \subseteq X$ then

1. A and B are regular open implies $A \cap B$ is regular open.
2. A and B are open implies $[A \cap B = 0 \leftrightarrow A^{-0} \cap B^{-0} = 0]$.

Proof. 1. If A and B are regular open then since $A \cap B \subseteq A \wedge A \cap B \subseteq B$ we have

$$(A \cap B)^{-0} \subseteq A^{-0} = A$$

$$(A \cap B)^{-0} \subseteq B^{-0} = B.$$

Therefore $(A \cap B)^{-0} \subseteq A \cap B$. But also

$$(A \cap B) = (A^0 \cap B^0) = (A \cap B)^0 \subseteq (A \cap B)^{-0}.$$

Therefore $(A \cap B) = (A \cap B)^{-0}$.

2. If $A = A^0$ and $B = B^0$ then $A = A^0 \subseteq A^{-0} \wedge B = B^0 \subseteq B^{-0}$. Thus

$$(A \cap B) \subseteq (A^{-0} \cap B^{-0}).$$

Consequently $(A^{-0} \cap B^{-0}) = 0$ implies $A \cap B = 0$.

Conversely

$$\begin{aligned} A \cap B = 0 &\rightarrow A \subseteq X - B \\ &\rightarrow A^- \subseteq X - B \\ &\rightarrow A^{-0} \subseteq (X - B)^0 \subseteq (X - B)^{0-} \\ &\rightarrow A^{-0} \subseteq X - B^{-0} \\ &\rightarrow A^{-0} \cap B^{-0} = 0. \end{aligned}$$

Theorem 1.30. The class F of all regular open sets of a nonempty topological space X is a complete Boolean algebra with operations $A + B \triangleq (A \cup B)^{-0}$, $AB \triangleq A \cap B$, $^-A \triangleq (X - A)^0$, and distinguished elements $\mathbf{0} \triangleq 0$ and $\mathbf{1} \triangleq X$.

Proof. Clearly addition as defined is a binary operation on regular open sets. From Theorem 1.29 multiplication as defined is a binary operation on regular open sets.

If A is regular open then

$$(^-A)^{-0} = (X - A^-)^{-0} = (X - A^{-0})^0 = {}^-A.$$

Thus complementation, as defined, is a unary operation on regular open sets. Both $\mathbf{0}$ and $\mathbf{1}$ are regular open and since $X \neq 0$ we have $\mathbf{0} \neq \mathbf{1}$. If A, B are regular open sets then

$$\begin{aligned} A + B &= (A \cup B)^{-0} = (B \cup A)^{-0} = B + A \\ AB &= A \cap B = B \cap A = BA. \end{aligned}$$

Thus addition and multiplication are commutative. For the proof of associativity we have from Theorems 1.22 and 1.27

$$\begin{aligned} (A \cup B)^{-0} &\subseteq (A \cup B)^- = A^- \cup B^- \\ A^{-0} \cup B^{-0} &\subseteq (A^- \cup B^-)^0 = (A \cup B)^{-0}. \end{aligned}$$

Thus if A, B, and C are regular open sets

$$\begin{aligned} (A + B) + C &= [(A \cup B)^{-0} \cup C]^{-0} \subseteq [(A^- \cup B^-) \cup C^-]^{-0} \\ &= [(A \cup B) \cup C]^{-0} \\ &= [(A^{-0} \cup B^{-0}) \cup C]^{-0} \subseteq [(A \cup B)^{-0} \cup C]^{-0} \\ &= (A + B) + C. \end{aligned}$$

Thus

$$\begin{aligned}(A + B) + C &= [(A \cup B) \cup C]^{-0} \\ &= [(B \cup C) \cup A]^{-0} \\ &= (B + C) + A \\ &= A + (B + C).\end{aligned}$$

Consequently addition is associative. Since multiplication is set intersection it too is associative.

For the proof of the distributive law we use both Theorem 1.27 and Theorem 1.28.

$$\begin{aligned}A + (BC) &= [A \cup (B \cap C)]^{-0} \\ &= [(A \cup B) \cap (A \cup C)]^{-0} \subseteq (A \cup B)^{-0} \cap (A \cup C)^{-0} \\ &= (A + B)(A + C).\end{aligned}$$

$$\begin{aligned}&(A + B)(A + C) \\ &\quad = (A \cup B)^{-0} \cap (A \cup C)^{-0} \\ &\quad = [(A \cup B^{-0})^{-0} \cap (A \cup C^{-0})^{-0}]^{-0} \subseteq [(A^{-} \cup B^{-0}) \cap (A^{-} \cup C^{-0})]^{-0} \\ &\quad = [A^{-} \cup (B \cap C)]^{-0} \subseteq [A^{-} \cup (B \cap C)^{-}]^{-0} \\ &\quad = [A \cup (B \cap C)]^{-0} = A + BC.\end{aligned}$$

$$\begin{aligned}A(B + C) &= A \cap (B \cup C)^{-0} \\ &= A^{0} \cap (B \cup C)^{-0} \subseteq [A^{0} \cap (B \cup C)^{-}]^{0} \subseteq [A \cap (B \cup C)]^{-0} \\ &= [(A \cap B) \cup (A \cap C)]^{-0} = AB + AC.\\ AB + AC &= [(A \cap B) \cup (A \cap C)]^{-0} \\ &= [A \cap (B \cup C)]^{-0} \subseteq A^{-0} \cap (B \cup C)^{-0} = A(B + C).\\ \mathbf{0} + A &= (0 \cup A)^{-0} = A^{-0} = A.\\ \mathbf{1}A &= X \cap A = A.\\ A + A^{-} &= [A \cup (X - A)^{0}]^{-0} = [A \cup (X - A^{-})]^{-0} \\ &= A^{-} \cup (X - A^{-})^{-}]^{0} = [A^{-} \cup (X - A)]^{0} \\ &= X^{0} = X.\\ A(^{-}A) &= A \cap (X - A)^{0} = A \cap (X - A^{-}) = \mathbf{0}.\end{aligned}$$

Thus the collection F of regular open sets forms a Boolean algebra. To prove that this algebra is complete we note, from the definition of multiplication and the definition of the natural order $\leq$ for a Boolean algebra, that for $A, B \in F$

$$\begin{aligned}A \leq B &\leftrightarrow A = AB \\ &\leftrightarrow A \subseteq B.\end{aligned}$$

If H is a set of regular open sets then $[\bigcup(H)]^{-0}$ is regular open. Also

$$\begin{aligned}A \in H &\rightarrow A \subseteq \bigcup(H) \\ &\rightarrow A = A^{-0} \subseteq \left[\bigcup(H)\right]^{-0}.\end{aligned}$$

Furthermore if $B \in F$ and $(\forall A \in H)[A \subseteq B]$ then $\bigcup(H) \subseteq B$ and

$$\left[\bigcup(H)\right]^{-0} \subseteq B^{-0} = B.$$

Thus

$$\sum_{A \in H} A = \left[\bigcup (H)\right]^{-0}.$$

Similarly $[\bigcap (H)]^{-0}$ is regular open and

$$A \in H \to \bigcap (H) \subseteq A$$
$$\to \left[\bigcap (H)\right]^{-0} \subseteq A^{-0} = A.$$

If $B \in F$ and $(\forall A \in H)[B \subseteq A]$ then $B \subseteq \bigcap (H)$ and hence

$$B = B^{-0} \subseteq \left[\bigcap (H)\right]^{-0}.$$

Therefore

$$\prod_{A \in H} A = \left[\bigcap (H)\right]^{-0}.$$

Remark. We have now shown that the regular open sets in any given topological space form a complete Boolean algebra. We next wish to show that every complete Boolean algebra is isomorphic to the complete Boolean algebra of regular open sets of some topological space. This topological space is determined by the given Boolean algebra in a way we must now make clear. The key is certain properties of partial orderings.

Definition 1.31. The structure $\langle P, \leq \rangle$ is a partial order structure iff $\leq$ is a subset of $P \times P$ and $\forall x, y \in P$

1. $x \leq x$.
2. $x \leq y \wedge y \leq x \to x = y$.
3. $x \leq y \wedge y \leq z \to x \leq z$.

The relation $\leq$ is a partial ordering of P iff $\langle P, \leq \rangle$ is a partial order structure.

Definition 1.32. If $\leq$ is a partial ordering of P then

$$[x] \triangleq \{y \in P \mid y \leq x\}, x \in P.$$

Theorem 1.33. If $\leq$ is a partial ordering of P then $T' \triangleq \{[x] \mid x \in P\}$ is a base for a topology on P. Furthermore if $A \subseteq P$ then in this topology

1. $x \in A^0 \leftrightarrow [x] \subseteq A$.
2. $x \in A^- \leftrightarrow [x] \cap A \neq 0$.
3. $x \in A^{-0} \leftrightarrow (\forall y \leq x)[[y] \cap A \neq 0]$.
4. $x \in (P - A)^0 \leftrightarrow (\forall y \leq x)[y \notin A]$.
5. A is dense in P iff $(\forall x \in P)[[x] \cap A \neq 0]$.
6. x is a $\leq$-minimal element of P iff $x \in P \wedge [x] = \{x\}$.

Proof. $(\forall x \in P)[x \in [x] \in T']$.

$$(\forall x \in P)(\forall [y], [z] \in T')[x \in [y] \cap [z] \to x \in [x] \subseteq [y] \cap [z]].$$

Therefore, by Theorem 1.19, T' is a base for a topology on P. Furthermore if $A \subseteq P$ then in this topology

1. $x \in A^0 \leftrightarrow (\exists [y])[x \in [y] \subseteq A]$
   $\leftrightarrow [x] \subseteq A.$
2. $x \in A^- \leftrightarrow (\forall [y])[x \in [y] \to [y] \cap A \neq 0]$
   $\leftrightarrow [x] \cap A \neq 0.$
3. $x \in A^{-0} \leftrightarrow [x] \subseteq A^-$
   $\leftrightarrow (\forall y \leq x)[[y] \cap A \neq 0].$
4. $x \in (P - A)^0 \leftrightarrow [x] \subseteq P - A$
   $\leftrightarrow (\forall y \leq x)[y \notin A].$
5. A is dense in $P \leftrightarrow A^- = P$
   $\leftrightarrow (\forall x \in P)[x \in A^-]$
   $\leftrightarrow (\forall x \in P)[[x] \cap A \neq 0].$

6. The proof is left to the reader.

Remark. If $\langle P, \leq \rangle$ is a partial order structure then the topology with base $\{[x] \mid x \in P\}$ we call the topology on P induced by the order relation $\leq$ or the order topology on P. Hereafter when we speak of an open subset of a partially ordered set P we will mean open in the topology on P induced by the partial ordering.

Theorem 1.34. If $\langle P, \leq \rangle$ is a partial order structure and A is a collection of open subsets of P then

$$\bigcap_{a \in A} a$$

is open.

Proof.

$$\begin{aligned} p \in \bigcap_{a \in A} a &\to (\forall a \in A)[p \in a] \\ &\to (\forall a \in A)[[p] \subseteq a] \qquad \text{since } a \text{ is open} \\ &\to [p] \subseteq \bigcap_{a \in A} a. \end{aligned}$$

Remark. It then follows, that for a Boolean algebra of regular open sets of a partial order structure, infimum coincides with set intersection:

Theorem 1.35. If $\langle P, \leq \rangle$ is a partial order structure and A is a collection of regular open subsets of P then

$$\prod_{a \in A} a = \bigcap_{a \in A} a.$$

Proof. By Theorem 1.34

$$\bigcap_{a \in A} a \subseteq \left(\bigcap_{a \in A} a\right)^{-0} = \prod_{a \in A} a.$$

On the other hand, if $a \in A$ then

$$\bigcap_{a \in A} a \subseteq a$$

$$\left(\bigcap_{a \in A} a\right)^{-0} \subseteq a^{-0} = a.$$

Therefore

$$\left(\bigcap_{a \in A} a\right)^{-0} \subseteq \bigcap_{a \in A} a.$$

Exercises.

1. A is a regular open subset of a Boolean algebra iff

   (i) $(\forall x \in A)[[x] \subseteq A]$ and
   (ii) $(\forall x \notin A)(\exists y \leq x)[[y] \cap A = 0]$.

Remark. From Theorem 1.6 we see that if $\mathbf{B}$ is a Boolean algebra with natural order $\leq$ then $\langle |\mathbf{B}|, \leq \rangle$ is a partial order structure hence $\leq$ induces a topology on $|\mathbf{B}|$. In this topology the collection of regular open sets forms a Boolean algebra $\mathbf{B}'$. We wish to show that if $\mathbf{B}$ is complete and if $B_0 = |\mathbf{B}| - \{\mathbf{0}\}$ then $\langle B_0, \leq \rangle$ is a partial order structure from which we obtain a complete Boolean algebra $\mathbf{B}_0'$ that is isomorphic to $\mathbf{B}$.

Definition 1.36. f is a Boolean homomorphism iff there are Boolean algebras $\mathbf{B}_1$ and $\mathbf{B}_2$ such that $f\colon |\mathbf{B}_1| \to |\mathbf{B}_2|$ and $\forall x, y \in |\mathbf{B}_1|$.

1. $f(x + y) = f(x) + f(y)$.
2. $f(xy) = f(x)f(y)$.
3. $f(^-x) = {}^-f(x)$.

Under these conditions we say that f is a homomorphism from $\mathbf{B}_1$ into $\mathbf{B}_2$. If

$$f\colon |\mathbf{B}_1| \xrightarrow[\text{onto}]{} |\mathbf{B}_2|$$

then f is an epimorphism from $\mathbf{B}_1$ onto $\mathbf{B}_2$. If

$$f\colon |\mathbf{B}_1| \xrightarrow[\text{onto}]{1-1} |\mathbf{B}_2|$$

then f is an isomorphism from $\mathbf{B}_1$ onto $\mathbf{B}_2$.

f is a homomorphism, or epimorphism, or isomorphism on $\mathbf{B}_1$ iff there exists a Boolean algebra $\mathbf{B}_2$ such that f is a homomorphism, epimorphism, or isomorphism from $\mathbf{B}_1$ into or onto $\mathbf{B}_2$.

Theorem 1.37. If $\mathbf{B}_1$ and $\mathbf{B}_2$ are Boolean algebras and $f\colon |\mathbf{B}_1| \to |\mathbf{B}_2|$ such that $\forall x, y \in |\mathbf{B}_1|$

1. $f(x + y) = f(x) + f(y)$.
2. $f(^-x) = {}^-f(x)$.

then f is a Boolean homomorphism.

Proof.

$$f(xy) = f(^-(^-x + ^-y)) = ^-f(^-x + ^-y) = ^-(f(^-x) + f(^-y))$$
$$= ^-(^-f(x) + ^-f(y)) = f(x)f(y).$$

Theorem 1.38. If f is a homomorphism from $\mathbf{B}_1$ into $\mathbf{B}_2$ then

1. $f(\mathbf{0}) = \mathbf{0}$.
2. $f(\mathbf{1}) = \mathbf{1}$.
3. $(\forall x, y \in |\mathbf{B}_1|)[x \leq y \rightarrow f(x) \leq f(y)]$.

Proof.

1. $f(\mathbf{0}) = f(\mathbf{0}(^-\mathbf{0})) = f(\mathbf{0})(^-f(\mathbf{0})) = \mathbf{0}$.
2. $f(\mathbf{1}) = f(\mathbf{1} + {}^-\mathbf{1}) = f(\mathbf{1}) + {}^-f(\mathbf{1}) = \mathbf{1}$.
3. $x \leq y \rightarrow x = xy$
   $\rightarrow f(x) = f(x)f(y)$
   $\rightarrow f(x) \leq f(y)$.

Theorem 1.39. If $\mathbf{B}$ is a Boolean algebra with natural order $\leq$ and if $B_0 = |\mathbf{B}| - \{\mathbf{0}\}$ then $\leq$ partially orders B_0 and

$$(\forall a, b \in B_0)[[a] \cap [b] = 0 \leftrightarrow ab = \mathbf{0}].$$

Proof. Left to the reader.

Theorem 1.40. If $\mathbf{B}$ is a complete Boolean algebra, if $B_0 = |\mathbf{B}| - \{\mathbf{0}\}$, if T is the topology on B_0 induced by the natural ordering $\leq$ and if $\mathbf{B}_0$ is the Boolean algebra of regular open subsets of B_0, as determined by T, then $\mathbf{B}$ and $\mathbf{B}_0$ are isomorphic.

Proof. If $(\forall b \in B)[F(b) = [b]]$ then clearly F is a function. Furthermore

$$\begin{aligned} a \in [b]^- &\leftrightarrow [a] \cap [b] \neq 0 \\ &\leftrightarrow ab \neq \mathbf{0} \\ &\leftrightarrow a \nleq {}^-b \\ &\leftrightarrow a \in B_0 - [^-b]. \end{aligned}$$

Thus $[b]^{-0} = (B_0 - [^-b])^0$.

$$\begin{aligned} a \in [b]^{-0} &\leftrightarrow [a] \subseteq B_0 - [^-b] \\ &\leftrightarrow [a] \cap [^-b] = 0 \\ &\leftrightarrow a(^-b) = \mathbf{0} \\ &\leftrightarrow a \leq b \\ &\leftrightarrow a \in [b]. \end{aligned}$$

Thus $[b] = [b]^{-0}$ and hence F maps B into the collection of regular open sets.

$$\begin{aligned} x \in [a] + [b] &\leftrightarrow x \in ([a] \cup [b])^{-0} \\ &\leftrightarrow [x] \subseteq ([a] \cup [b])^- \\ &\leftrightarrow (\forall y \leq x)[[y] \cap ([a] \cup [b]) \neq 0] \\ &\leftrightarrow (\forall y \leq x)[ya \neq \mathbf{0} \vee yb \neq \mathbf{0}] \\ &\leftrightarrow (\forall y \leq x)[y(a + b) \neq \mathbf{0}] \\ &\leftrightarrow (\forall y \leq x)[[y] \cap [a + b] \neq 0] \\ &\leftrightarrow [x] \subseteq [a + b]^- \\ &\leftrightarrow x \in [a + b]^{-0} = [a + b]. \end{aligned}$$

Thus $[a + b] = [a] + [b]$ and hence

$$F(a + b) = F(a) + F(b).$$

$$\begin{aligned}[a][b] &= \{x \in B_0 \mid x \leq a \wedge x \leq b\}\\ &= \{x \in B_0 \mid x \leq ab\}\\ &= [ab].\end{aligned}$$

Therefore $F(ab) = F(a)F(b)$. Furthermore

$$^-[b] = (B_0 - [b])^0 = B_0 - [b]^-.$$

Then

$$\begin{aligned} a \in {}^-[b] &\leftrightarrow a \neq \mathbf{0} \wedge [a] \cap [b] = 0\\ &\leftrightarrow a \neq \mathbf{0} \wedge ab = \mathbf{0}\\ &\leftrightarrow a \neq \mathbf{0} \wedge a \leq {}^-b\\ &\leftrightarrow a \in [{}^-b].\end{aligned}$$

Thus $^-[b] = [{}^-b]$ and hence $F({}^-b) = {}^-F(b)$.

We have proved that F is a homomorphism of $\mathbf{B}$ into $\mathbf{B}_0$. To prove that F is an epimorphism, i.e., onto, we note that if A is a regular open subset of B_0 then

$$A = \bigcup_{b \in A} [b].$$

Therefore

$$A = A^{-0} = \left(\bigcup_{b \in A} [b]\right)^{-0} = \sum_{b \in A} [b].$$

Since $\mathbf{B}$ is complete $(\exists a \in B)[a = \sum_{b \in A} b]$. Furthermore $\forall x \in A$

$$\begin{aligned} x \leq a &\leftrightarrow x = ax\\ &\leftrightarrow [x] = [a][x]\\ &\leftrightarrow [x] \leq [a].\end{aligned}$$

Also $x \leq a' \leftrightarrow [x] \leq [a']$. Consequently

$$[x \leq a' \rightarrow a \leq a'] \leftrightarrow [[x] \leq [a'] \rightarrow [a] \leq [a']].$$

Thus

$$a = \sum_{b \in A} b \leftrightarrow [a] = \sum_{b \in A} [b].$$

Since $F(a) = [a] = A$ it follows that F is an onto map.

Finally if $[a] = [b]$ then $a \leq b$ and $b \leq a$. Hence $a = b$ i.e.,

$$F(a) = F(b) \rightarrow a = b.$$

We then conclude that F is an isomorphism of $\mathbf{B}$ onto $\mathbf{B}_0$.

Exercises. Determine the complete Boolean algebra of regular open sets in the following partial order structure where $a \neq 0$.

1. $\langle \mathscr{P}(a), \subseteq \rangle$.
2. $\langle \mathscr{P}(a) - \{0\}, \subseteq \rangle$.

3. $\langle \mathscr{P}(a), \leq \rangle$ where $x \leq y \leftrightarrow y \subseteq x$.
4. $\langle \mathscr{P}(a) - \{0\}, \leq \rangle$ where $x \leq y \leftrightarrow y \subseteq x$.

Definition 1.41. I is an ideal in the Boolean algebra $\mathbf{B}$ iff $I \subseteq |\mathbf{B}|$ and

1. $\mathbf{0} \in I$.
2. $a, b \in I \rightarrow a + b \in I$.
3. $a \in I \wedge b \in |\mathbf{B}| \rightarrow ab \in I$.

An ideal I is

1. a proper ideal iff $\mathbf{1} \notin I$,
2. a principal ideal iff $(\exists b \in |\mathbf{B}|)[I = [b]]$,
3. a trivial ideal iff $I = \{\mathbf{0}\}$.

Theorem 1.42. If I is an ideal in the Boolean algebra $\mathbf{B}$ then

1. $(\forall a, b \in |\mathbf{B}|)[a \leq b \in I \rightarrow a \in I]$.
2. $\mathbf{1} \in I \rightarrow I = B$.

Proof. Left to the reader.

Definition 1.43. If f is a Boolean homomorphism of $\mathbf{B}_1$ into $\mathbf{B}_2$ then

$$\ker(f) \triangleq \{a \in |\mathbf{B}_1| \mid f(a) = \mathbf{0}\}.$$

Theorem 1.44. If f is a Boolean homomorphism on $\mathbf{B}$ then $\ker(f)$ is a proper ideal in $\mathbf{B}$. Furthermore if $\ker(f) = \{\mathbf{0}\}$ then f is an isomorphism.

Proof. Since $f(\mathbf{0}) = \mathbf{0}$, $\mathbf{0} \in \ker(f)$. Furthermore $\ker(f) \subseteq |\mathbf{B}|$. If $a, b \in \ker(f)$ then

$$f(a + b) = f(a) + f(b) = \mathbf{0} + \mathbf{0} = \mathbf{0}.$$

Therefore $a + b \in \ker(f)$. If $a \in \ker(f)$ and $b \in |\mathbf{B}|$ then

$$f(ab) = f(a)f(b) = \mathbf{0}f(b) = \mathbf{0}.$$

Therefore $ab \in \ker(f)$. Thus $\ker(f)$ is an ideal in $\mathbf{B}$. But

$$f(\mathbf{1}) = \mathbf{1} \neq \mathbf{0}.$$

Therefore $\mathbf{1} \notin \ker(f)$. Hence $\ker(f)$ is a proper ideal.

If $\ker(f) = \{\mathbf{0}\}$ and $f(x) = f(y)$ then $f(x - y) = \mathbf{0}$, consequently $x - y = \mathbf{0}$ and $x \leq y$. Similarly $y - x \in \ker(f)$ and hence $y \leq x$. Therefore $x = y$ and f is one-to-one.

Theorem 1.45. If I is an ideal in the Boolean algebra $\mathbf{B}$ then

$$(\forall a, b \in |\mathbf{B}|)[a + b \in I \rightarrow a \in I \wedge b \in I].$$

Proof.

$$a = a + ab = a(a + b) \in I$$
$$b = b + ab = b(a + b) \in I.$$

Definition 1.46. If I is a proper ideal in the Boolean algebra $\mathbf{B}$ then

1. $(\forall a \in |\mathbf{B}|)[a/I \triangleq \{x \in |\mathbf{B}| \mid a(^-x) + x(^-a) \in I\}$.
2. $|\mathbf{B}|/I \triangleq \{a/I \mid a \in |\mathbf{B}|\}$.

Theorem 1.47. If I is a proper ideal in the Boolean algebra $\mathbf{B}$ then

1. $a/I = b/I \leftrightarrow a(^-b) + b(^-a) \in I$.
2. $a/I = c/I \wedge b/I = d/I \to (a + b)/I = (c + d)/I$.
3. $a/I = c/I \wedge b/I = d/I \to ab/I = cd/I$.
4. $a/I = b/I \to {}^-a/I = {}^-b/I$.

Proof. 1. If $a/I = b/I$ then since $a(^-a) + a(^-a) = \mathbf{0} \in I$ we have $a \in a/I$ and hence $a \in b/I$. Therefore $a(^-b) + b(^-a) \in I$. Conversely if $a(^-b) + b(^-a) \in I$ and $x \in a/I$ then by Theorem 1.42 $a(^-b), b(^-a), x(^-a), a(^-x) \in I$. Therefore

$$a(^-b)x + b(^-a)(^-x) + x(^-a)(^-b) + a(^-x)b \in I$$
$$(x(^-b) + b(^-x))(a + {}^-a) \in I$$
$$x(^-b) + b(^-x) \in I$$

i.e., $x \in b/I$. Similarly $x \in b/I \to x \in a/I$.

2. If $a(^-c) + c(^-a), b(^-d) + d(^-b) \in I$ then by Theorem 1.45 $a(^-c), c(^-a), b(^-d), d(^-b) \in I$.

Therefore $a(^-c)(^-d) + c(^-a)(^-b) + b(^-d)(^-c) + d(^-b)(^-a) \in I$

$$(a + b)(^-c)(^-d) + (c + d)(^-a)(^-b) \in I$$
$$(a + b)^-(c + d) + (c + d)^-(a + b) \in I$$

i.e., $(a + b)/I = (c + d)/I$.

3. As in 2, $a(^-c), c(^-a), b(^-d), d(^-b) \in I$ implies

$$a(^-c)b + c(^-a)d + b(^-d)a + d(^-b)c \in I$$
$$(ab)^-(cd) + (cd)^-(ab) \in I$$

i.e., $ab/I = cd/I$.

4. If $a(^-b) + b(^-a) \in I$ then

$$(^-a)^-(^-b) + (^-b)^-(^-a) = a(^-b) + b(^-a) \in I$$

i.e., ${}^-a/I = {}^-b/I$.

Theorem 1.48. If I is a proper ideal in $\mathbf{B}$ then $|\mathbf{B}|/I$ is the universe of a Boolean algebra, $\mathbf{B}/I$, with operations

$$a/I + b/I = (a + b)/I, \qquad a/I \cdot b/I = ab/I, \qquad {}^-(a/I) = (^-a)/I$$

and distinguished elements $\mathbf{0}/I$ and $\mathbf{1}/I$.

Proof. Left to the reader.

Theorem 1.49. If I is a proper ideal in $\mathbf{B}$ and

$$(\forall a \in |\mathbf{B}|)[f(a) = a/I]$$

then f is a Boolean homomorphism of $\mathbf{B}$ onto $\mathbf{B}/I$ and $\ker(f) = I$.

Proof. Left to the reader.

Definition 1.50. If **B** is a complete Boolean algebra then
1. a homomorphism f on **B** is complete iff

$$(\forall A \subseteq |\mathbf{B}|)\left[f\left(\sum_{b\in A} b\right) = \sum_{b\in A} f(b) \wedge f\left(\prod_{b\in A} b\right) = \prod_{b\in A} f(b)\right],$$

2. an ideal I in **B** is complete iff

$$(\forall A \subseteq I)\left[\sum_{b\in A} b \in I\right].$$

Definition 1.51. 1. A Boolean algebra **B** is M-complete iff

$$(\forall A \subseteq |\mathbf{B}|)\left[A \in M \rightarrow \sum_{b\in A} b \in |\mathbf{B}|\right].$$

2. A homomorphism f on **B** is M-complete iff

$$(\forall A \subseteq |\mathbf{B}|)\left[A \in M \rightarrow f\left(\sum_{b\in A} b\right) = \sum_{b\in A} f(b) \wedge f\left(\prod_{b\in A} b\right) = \prod_{b\in A} f(b)\right].$$

3. An ideal I in **B** is M-complete iff

$$(\forall A \subseteq I)\left[A \in M \rightarrow \sum_{b\in A} b \in I\right].$$

Theorem 1.52. If $\mathbf{B} = \langle B, \cup, \cap, ^{-}, \mathbf{0}, \mathbf{1}\rangle$ is a complete Boolean algebra, if M is a standard transitive model of ZF, and if $\mathbf{1} \in M$ then $B^M \triangleq \langle B \cap M, \cup, \cap, ^{-}, \mathbf{0}, \mathbf{1}\rangle$ is an M-complete Boolean algebra.

Proof. If $a, b \in |\mathbf{B}| \cap M$ then

$a \cup b \in |\mathbf{B}| \cap M, \qquad a \cap b \in |\mathbf{B}| \cap M$ and $^{-}a = \mathbf{1} - a \in |\mathbf{B}| \cap M.$

Since $\mathbf{0}, \mathbf{1} \in |\mathbf{B}| \cap M$ it follows that $\mathbf{B}^M$ is a Boolean algebra. Furthermore if $A \subseteq (|\mathbf{B}| \cap M)$ and $A \in M$ then since **B** is complete and M satisfies the Axioms of Unions

$$\bigcup(A) \in |\mathbf{B}| \cap M.$$

Therefore $\mathbf{B}^M$ is M-complete.

Theorem 1.53. If $\langle P, \leq\rangle$ is a partial order structure, if $\langle P, \leq\rangle \in M$, M a standard transitive model of ZF, and if **B** is the Boolean algebra of regular open subsets of P, then $|\mathbf{B}| \cap M$ is the universe of an M-complete Boolean subalgebra, $\mathbf{B}^M$, of **B**.

Proof. Since $\langle P, \leq\rangle \in M$ and M is transitive

$$(\forall p \in P)[[p] \in M].$$

Since M satisfies the Axiom Schema of Replacement

$$\{\langle p, [p]\rangle \mid p \in P\} \in M.$$

Then $\forall b \in |\mathbf{B}| \cap M, b^0, b^-, b^{-0} \in M$. Consequently if $a, b \in |\mathbf{B}| \cap M$

$$a + b \in |\mathbf{B}| \cap M, \qquad ab \in |\mathbf{B}| \cap M, \qquad {}^-a \in |\mathbf{B}| \cap M.$$

Since $0, P \in M$, $\mathbf{B}^M$ is a subalgebra of $\mathbf{B}$.

If $A \subseteq |\mathbf{B}| \cap M$ and $A \in M$ then since $\mathbf{B}$ is complete and M is a model of ZF

$$\sum_{a \in A} a \in |\mathbf{B}| \cap M.$$

Thus $\mathbf{B}^M$ is M-complete.

Theorem 1.54. 1. The kernel of every complete Boolean homomorphism is a complete ideal.

2. The kernel of every M-complete Boolean homomorphism is an M-complete ideal.

Proof. 1. If f is a complete Boolean homomorphism on $\mathbf{B}$ and $A \subseteq \ker(f)$ then

$$f\left(\sum_{a \in A} a\right) = \sum_{a \in A} f(a) = \mathbf{0}.$$

Consequently $\sum_{a \in A} a \in \ker(f)$.

2. If $A \subseteq \ker(f)$ and $A \in M$ then

$$f\left(\sum_{a \in A} a\right) = \sum_{a \in A} f(a) = \mathbf{0}.$$

Hence $\sum_{a \in A} a \in \ker(f)$.

Theorem 1.55. Every complete Boolean ideal is principal.

Proof. If I is a complete ideal then, since $I \subseteq I$, $\sum_{a \in I} a \in I$ and $(\forall b \in I)[b \leq \sum_{a \in I} a]$. Furthermore

$$\left(\forall b \leq \sum_{a \in I} a\right)\left[b = b\left(\sum_{a \in I} a\right) \in I\right].$$

Therefore

$$I = \left[\sum_{a \in I} a\right].$$

Definition 1.56. I is a maximal ideal in the Boolean algebra $\mathbf{B}$ iff I is a proper ideal in $\mathbf{B}$ and for each ideal J in $\mathbf{B}$

$$I \subseteq J \rightarrow I = J \vee J = |\mathbf{B}|.$$

Theorem 1.57. I is a maximal ideal in the Boolean algebra $\mathbf{B}$ iff

$$(\forall b \in |\mathbf{B}|)[b \in I \leftrightarrow {}^-b \notin I].$$

Proof. If $b \in |\mathbf{B}| \wedge b \notin I \wedge {}^-b \notin I$ and

$$J = \{x + y \mid x \leq b \wedge y \in I\}$$

then $\mathbf{0} \in J$. If $[x_1 \leq b] \wedge [y_1 \in I] \wedge [x_2 \leq b] \wedge [y_2 \in I]$ then

$$[x_1 + x_2 \leq b] \wedge [y_1 + y_2 \in I].$$

Therefore $(x_1 + x_2) + (y_1 + y_2) \in J$. Furthermore if $c \in |\mathbf{B}|$ then $c \leq \mathbf{1}$ and hence

$$x_1 c \leq x_1 \leq b \wedge y_1 c \in I.$$

Consequently $(x_1 + y_1)c \in J$.

Thus J is an ideal and $I \subseteq J$. Since $b \in J \wedge b \notin I$ we have $I \subset J$. Furthermore $^-b \notin J$ for otherwise

$$(\exists x \leq b)(\exists y \in I)[^-b = x + y].$$

Then $^-b = (^-b)(^-b) = (x + y)(^-b) = \mathbf{0} + y(^-b) \in I$. This is a contradiction. Hence $J \neq |\mathbf{B}|$ and I is not maximal.

If $[b \in |\mathbf{B}|] \wedge [b \in I] \wedge [^-b \in I]$ then $\mathbf{1} = b + (^-b) \in I$. Hence I is not a proper ideal. Thus if I is maximal then $(\forall b \in |\mathbf{B}|)[b \in I \leftrightarrow {}^-b \notin I]$.

Conversely if $I \subset J$ then $(\exists b \in J)[b \notin I]$. Therefore $^-b \in I \subseteq J$ and hence

$$^-b + b = \mathbf{1} \in J$$

i.e., $J = |\mathbf{B}|$ and I is maximal.

Theorem 1.58. If $A \neq 0$ and $\forall a \in A$, I_a is an ideal in the Boolean algebra $\mathbf{B}$ then $\bigcap_{a \in A} I_a$ is an ideal in $\mathbf{B}$.

Proof. Left to the reader.

Definition 1.59. If $\mathbf{B}$ is a Boolean algebra and $A \subseteq |\mathbf{B}|$ then

$$\bigcap \{I \mid A \subseteq I \wedge I \text{ is an ideal in } \mathbf{B}\}$$

is the ideal generated by A.

Theorem 1.60. If I is the ideal in the Boolean algebra $\mathbf{B}$ that is generated by $A \subseteq |\mathbf{B}|$, $A \neq 0$ then $b \in I$ iff $b \in |\mathbf{B}|$ and

$$(\exists b_1, \ldots, b_n \in A)[b \leq b_1 + \cdots + b_n].$$

Proof. If $J = \{b \in |\mathbf{B}| \mid (\exists b_1 \cdots b_n \in A)[b \leq b_1 + \cdots + b_n]\}$ then J is an ideal in $\mathbf{B}$ (details are left to the reader) and $A \subseteq J$. Thus $I \subseteq J$. But

$$b \in |\mathbf{B}| \wedge b \leq b_1 + \cdots + b_n \in I \rightarrow b \in I.$$

Therefore $J \subseteq I$.

Theorem 1.61. If I is a proper ideal in the Boolean algebra $\mathbf{B}$, if $^-a \notin I$ and if I_a is the ideal in $\mathbf{B}$ generated by $I \cup \{a\}$ then I_a is a proper ideal.

Proof. If $\mathbf{1} \in I_a$ then from Theorem 1.60

$$(\exists b \in I)[\mathbf{1} \leq b + a].$$

Then

$$^-a = {}^-a\mathbf{1} \leq {}^-ab \in I.$$

This is a contradiction. Hence I_a is a proper ideal.

Theorem 1.62. Every proper ideal I in a Boolean algebra $\mathbf{B}$ can be extended to a maximal ideal i.e., there exists a maximal ideal J in $\mathbf{B}$ for which $I \subseteq J$.

Proof. For any well ordering R of $|\mathbf{B}|$ we define

$$\begin{aligned} I_0 &= I. \\ I_{\alpha+1} &= \text{ideal generated by } I_\alpha \cup \{a\}, \text{ where } a \text{ is the first element of } |\mathbf{B}| \text{ for} \\ &\quad \text{which } a \notin I_\alpha \wedge {}^-a \notin I_\alpha \text{ if such an } a \text{ exists,} \\ &= I_\alpha \text{ otherwise.} \\ I_\alpha &= \bigcup_{\beta<\alpha} I_\beta, \ \alpha \in K_{\text{II}}. \end{aligned}$$

If $J = \bigcup_{\alpha \in On} I_\alpha$ then $I \subseteq J \subseteq |\mathbf{B}|$. Furthermore if $x, y \in J$ then

$$(\exists \alpha, \beta)[x \in I_\alpha \wedge y \in I_\beta].$$

If $\gamma = \max(\alpha, \beta)$ then $x, y \in I_\gamma$ and hence $x + y \in I_\gamma$. Also if $a \in |\mathbf{B}|$ then $ax \in I_\gamma$. Therefore, since $\mathbf{0} \in I \subseteq J$, J is an ideal in $\mathbf{B}$. Indeed since $J = \bigcup_{\alpha \in On} I_\alpha$ and $(\forall a)[\mathbf{1} \notin I_\alpha]$ it follows that $\mathbf{1} \notin J$ i.e., J is a proper ideal.

If J were not maximal then there would be a first element $a \in |\mathbf{B}|$ such that $a \notin J$ and ${}^-a \notin J$. If for each x that precedes a, in the order R, we define

$$F(x) = \mu_\alpha(x \in I_\alpha \vee {}^-x \in I_\alpha)^*$$

then $F``\{x \mid x \, R \, a\}$ is a set of ordinals. If

$$\beta = \bigcup F``\{x \mid x \, R \, a\}$$

then I_β is a proper ideal and for each x that precedes a

$$x \in I_\beta \vee {}^-x \in I_\beta.$$

Since a is the first element in $|\mathbf{B}|$ such that $a \notin I_\beta \wedge {}^-a \in I_\beta$ it follows that $a \in I_{\beta+1}$ and hence $a \in J$.

Since this is a contradiction we conclude that J is maximal.

Exercise. 1. Can two different topological spaces lead to the same Boolean algebra?

* $\mu_\alpha(P(\alpha))$ denotes the smallest ordinal having the property P.

2. Generic Sets

In the material ahead we will be interested in standard transitive models M of ZF and in partial order structures $\mathbf{P} = \langle P, \leq \rangle$ for which $\mathbf{P} \in M$. Although some of the results hold under more general conditions we will assume hereafter that this is the case i.e., M is a standard transitive model of ZF, $\mathbf{P} = \langle P, \leq \rangle$ is a partial order structure and $\mathbf{P} \in M$.

Definition 2.1. If $\leq$ is a partial ordering of P then

1. $(\forall a, b \in P)[\text{Comp}(a, b) \overset{\Delta}{\leftrightarrow} (\exists c \in P)[c \leq a \wedge c \leq b]$.
2. $(\forall S \subseteq P)[\text{Comp}(S) \overset{\Delta}{\leftrightarrow} (\forall a, b \in S)[\text{Comp}(a, b)]]$.

Remark. The symbol "Comp (a, b)" is read "a and b are compatible." Similarly "Comp (S)" is read "S is compatible." By definition a subset S of a partially ordered set P is compatible if and only if its elements are pairwise compatible.

Later we will be interested in partially ordered sets P whose elements "code" certain non-contradictory information. The ordering will be so defined that $a \leq b$ means that a contains all of the information that b does and perhaps more. Then $c \leq a \wedge c \leq b$ means that c contains all of the information in both a and b. Consequently the information in a is compatible (consistent) with that in b.

Definition 2.2. Let A be a given class. If $\mathbf{P} = \langle P, \leq \rangle$ is a partial order structure and $G \subseteq P$, then G is $\mathbf{P}$-generic over A iff

1. G is compatible.
2. $p \in G \wedge q \in P \wedge p \leq q \rightarrow q \in G$.
3. $S \in A \wedge S \subseteq P \wedge S^- = P \rightarrow G \cap S \neq 0$.

G is $\mathbf{P}$-generic over A in the strong sense if in addition

4. $(\forall p, q \in G)(\exists r \in G)[r \leq p \wedge r \leq q]$.

Remark. In Definition 2.2 the topology is that induced on P by the partial ordering $\leq$. Consequently, condition 3 asserts that every element of A that is a dense subset of P, in the order topology, has a nonempty intersection with G.

Theorem 2.3. If $\mathbf{P} = \langle P, \leq \rangle$ is a partial order structure, if p is a minimal element of P and if $G = \{q \in P \mid p \leq q\}$ then G is $\mathbf{P}$-generic over A (in the strong sense).

Proof. If $q_1, q_2 \in G$ then $p \leq q_1$ and $p \leq q_2$. Therefore G is compatible. If $q_1 \in G$, $q_2 \in P$ and $q_1 \leq q_2$ then since $p \leq q_1$ we have $p \leq q_2$, hence $q_2 \in G$. If $S \in A$ and S is a dense subset of P then $q \in P$ implies $[q] \cap S \neq 0$. In particular $p \in P$. Therefore $[p] \cap S \neq 0$. But by hypothesis p is a minimal element of P i.e., $[p] = \{p\}$. Consequently $p \in S$. By definition of G, $p \in G$. Hence $p \in G \cap S$ i.e.,

$$G \cap S \neq 0.$$

Remark. Definition 2.2 is more general than is necessary for most of our purposes. For the most part, we will be interested in sets that are **P**-generic over a standard transitive model of *ZF*.

Theorem 2.4. If $\mathbf{P} \in M$, M a standard transitive model of *ZF*, and if G is **P**-generic over M then G is **P**-generic over M in the strong sense.

Proof. If $a, b \in G$ and

$$S = \{c \in P \mid [c \leq a \wedge c \leq b] \vee [\neg \operatorname{Comp}(c, a) \wedge \neg \operatorname{Comp}(c, b)]\}$$

then $\forall c \in P$

1. $(\exists x \leq c)[\neg \operatorname{Comp}(x, a) \vee \neg \operatorname{Comp}(x, b)]$ or
2. $(\forall x \leq c)[\operatorname{Comp}(x, a) \wedge \operatorname{Comp}(x, b)]$.

If 1 is the case then $[c] \cap S \neq 0$. If 2 is the case then c and a are compatible. Consequently

$$(\exists c_1)[c_1 \leq c \wedge c_1 \leq a].$$

Again from 2, c_1 and b are compatible, hence

$$(\exists c_2)[c_2 \leq c_1 \wedge c_2 \leq b].$$

Then $c_2 \leq c$, $c_2 \leq a$ and $c_2 \leq b$ i.e., $[c] \cap S \neq 0$.

Therefore S is dense in P. Since G is **P**-generic over M and since $S \in M$

$$S \cap G \neq 0$$

i.e., $\exists c \in S \cap G$. Then $a, b, c \in G$. Since G is compatible it follows that c and a are compatible and c and b are compatible. But $c \in S$. Therefore $c \leq a$ and $c \leq b$.

Theorem 2.5. Let $\mathbf{P} = \langle P, \leq \rangle$ be a partial order structure with $\mathbf{P} \in A$ and let $G \subseteq P$. Suppose that for all S

1. $S \in A \wedge S \subseteq P \rightarrow S \cup \{p \mid [p] \cap S = 0\} \in A$.

Then for all S

2. $S \in A \wedge S \subseteq P \wedge S^- = P \rightarrow G \cap S \neq 0$

iff for all S

3. $S \in A \wedge S \subseteq P \rightarrow (\exists p \in G)[p \in S \vee [p] \cap S = 0]$.

Proof. $(3 \to 2)$. If $S \in A$ and S is a dense subset of P then

$$(\forall p)[[p] \cap S \neq 0].$$

Therefore by 3, $(\exists p \in G)[p \in S]$ i.e., $G \cap S \neq 0$.

$(2 \to 3)$. Conversely suppose that $S \in A$ and $S \subseteq P$. If

$$S' = S \cup \{p \mid [p] \cap S = 0\}$$

then, by 1, $S' \in A$. Furthermore S' is dense in P for otherwise

$$(\exists p \in P)[[p] \cap S' = 0].$$

But then

$$[p] \cap S = 0$$

and hence $p \in S'$ contradicting $[p] \cap S' = 0$.

It then follows that $G \cap S' \neq 0$ i.e.,

$$(\exists p \in G)[p \in S \vee [p] \cap S = 0].$$

Theorem 2.6. If $\mathbf{P} = \langle P, \leq \rangle$ is a partial order structure and if A is countable then every member of P is contained in some subset of P that is $\mathbf{P}$-generic over A in the strong sense.

Proof. Since A is countable we can enumerate the elements of A that are dense subsets of P:

$$S_1, S_2, \ldots.$$

If $a \in P$ then

$$(\exists p_1 \in S_1)[p_1 \leq a], \qquad \text{(Since } S_1 \text{ is dense in } P\text{)}$$

then

$$(\exists p_2 \in S_2)[p_2 \leq p_1], \qquad \text{(Since } S_2 \text{ is dense in } P\text{)}$$

etc. If

$$G = \{q \mid (\exists p_i)[p_i \leq q]\}$$

then

$$(\forall q, q' \in G)(\exists p_i, p_j)[p_i \leq q \wedge p_j \leq q'].$$

Since $p_i \leq p_j$ or $p_j \leq p_i$ it follows that G is compatible. Furthermore if $p \in G$, $q \in P$ and $p \leq q$ then $(\exists p_i)[p_i \leq p \leq q]$. Therefore $q \in G$ and in particular $a \in G$.

If $S \in A$ and S is a dense subset of P then $(\exists i)[S = S_i]$. Since $p_i \in S_i$ it follows that $G \cap S \neq 0$.

If $p, q \in G$ then $(\exists p_i \in G)[p_i \leq p \wedge p_i \leq q]$. Thus G is $\mathbf{P}$-generic over A in the strong sense and $a \in G$.

Remark. In Theorem 2.6 it is not necessary for A to be countable. It is sufficient for A to contain only countably many elements that are dense subsets of P. This will be the case if $\mathscr{P}^A(P)$ is countable.

Definition 2.7. F is a filter for the Boolean algebra $\mathbf{B}$ iff $0 \neq F \subseteq |\mathbf{B}|$ and

1. $x \in F \wedge y \in F \rightarrow xy \in F$.
2. $x \in F \wedge y \in |\mathbf{B}| \wedge x \leq y \rightarrow y \in F$.

Examples. 1. $\{\mathbf{1}\}$ is a filter for $\mathbf{B}$.

2. If $A \neq 0$ and $a \subseteq A$ then $\{x \subseteq A \mid a \subseteq x\}$ is a filter for the natural algebra on $\mathscr{P}(A)$.

Definition 2.8. If F is a filter on the Boolean algebra $\mathbf{B}$ then

1. F is an ultrafilter iff $(\forall x \in |\mathbf{B}|)[x \in F \leftrightarrow {}^{-}x \notin F]$.
2. F is a principal filter iff $(\exists a \in F)[F = \{x \in |\mathbf{B}| \mid a \leq x\}]$.
3. F is M-complete iff $(\forall A \in M)\left[A \subseteq F \rightarrow \prod_{a \in A} a \in F\right]$.
4. F is a proper filter iff $\mathbf{0} \notin F$.
5. F is a trivial filter iff $F = \{\mathbf{1}\}$.

Theorem 2.9. If F is a filter on the Boolean algebra $\mathbf{B}$ then
1. $\mathbf{1} \in F$.
2. $x \in F \wedge y \in |\mathbf{B}| \rightarrow x + y \in F$.

Theorem 2.10. If $\mathbf{B}$ is a Boolean algebra and F and I are nonempty subsets of $|\mathbf{B}|$ with the property that $a \in F$ iff ${}^{-}a \in I$ then

1. I is an ideal in $\mathbf{B} \leftrightarrow F$ is a filter for $\mathbf{B}$.
2. I is a maximal ideal $\leftrightarrow F$ is an ultrafilter.
3. I is a principal ideal $\leftrightarrow F$ is a principal filter.
4. I is M-complete $\leftrightarrow F$ is M-complete.
5. I is a proper ideal $\leftrightarrow F$ is a proper filter.
6. I is a trivial ideal $\leftrightarrow F$ is a trivial filter.

Theorem 2.11. 1. If F is an ultrafilter for $\mathbf{B}$ and

$$\begin{aligned} f(x) &= \mathbf{1}, \qquad x \in F \\ &= \mathbf{0}, \qquad x \in |\mathbf{B}| - F \end{aligned}$$

then f is a homomorphism from $\mathbf{B}$ to $\mathbf{2}$. If F and $\mathbf{B}$ are M-complete so is f.

2. If f is a homomorphism from $\mathbf{B}$ to $\mathbf{2}$ and $F = \{x \in |\mathbf{B}| \mid f(x) = \mathbf{1}\}$ then F is an ultrafilter for $\mathbf{B}$. If f and $\mathbf{B}$ are M-complete so is F.

Theorem 2.12. Every proper filter on a Boolean algebra $\mathbf{B}$ can be extended to an ultrafilter on $\mathbf{B}$ i.e., if F is a proper filter on $\mathbf{B}$ there exists an ultrafilter F' on $\mathbf{B}$ such that $F \subseteq F'$.

Theorem 2.13. If F_1 and F_2 are ultrafilters on a Boolean algebra $\mathbf{B}$ then

1. $F_1 \subseteq F_2 \rightarrow F_1 = F_2$.
2. $F_1 \neq F_2 \rightarrow [F_1 - F_2 \neq 0] \wedge [F_2 - F_1 \neq 0]$.

Proofs. Left to the reader.

Remark. From Theorem 2.12 and the definition of a generic set we can prove a very important result known as the Rasiowa-Sikorski Theorem.

Theorem 2.14. (Rasiowa-Sikorski.) If $\mathbf{B}$ is a Boolean algebra, if

1. $a_0 \in |\mathbf{B}| - \{\mathbf{0}\}$, and
2. $A_n \subseteq |\mathbf{B}| \wedge b_n \in |\mathbf{B}| \wedge b_n = \sum_{a \in A_n} a,\ n \in \omega,$

then there exists a Boolean homomorphism $h\colon |\mathbf{B}| \to |\mathbf{2}|$ such that

1. $h(a_0) = \mathbf{1}$, and
2. $h(b_n) = \sum_{a \in A_n} h(a),\ n \in \omega,$

Proof. Since every homomorphism maps $\mathbf{0}$ onto $\mathbf{0}$ and $b_n = \mathbf{0}$ iff $(\forall a \in A_n)[a = \mathbf{0}]$ there is no loss in generality if we assume that

$$(\forall n \in \omega)[\mathbf{0} \notin A_n \wedge b_n \neq \mathbf{0}].$$

If $\leq$ is the natural order on $\mathbf{B}$ and $P = |\mathbf{B}| - \{\mathbf{0}\}$ then $\mathbf{P} = \langle P, \leq \rangle$ is a partial order structure. If

$$S_n = \{p \in P \mid p \leq {}^{-}b_n \vee (\exists a \in A_n)[p \leq a]\}$$

then $\forall p \in P$

$$p \leq {}^{-}b_n \vee p \nleq {}^{-}b_n.$$

If $p \leq {}^{-}b_n$ then $p \in S_n$ and $[p] \cap S_n \neq 0$. If $p \nleq {}^{-}b_n$ then $pb_n \neq \mathbf{0}$ i.e.,

$$p \sum_{a \in A_n} a = \sum_{a \in A_n} pa \neq 0.$$

Therefore $(\exists a \in A_n)[pa \neq \mathbf{0}]$. But this implies

$$(\exists a \in A_n)[[p] \cap [a] \neq 0]$$

i.e.,

$$[p] \cap S_n \neq 0.$$

Thus S_n is dense in P. Since $\{S_n \mid n \in \omega\}$ is countable it follows from Theorem 2.6 that there exists a $G \subseteq P$ that is $\mathbf{P}$-generic over $\{S_n \mid n \in \omega\}$ in the strong sense and such that $a_0 \in G$.

Since G is $\mathbf{P}$-generic in the strong sense

$$(\forall x, y \in G)(\exists p \in G)[p \leq x \wedge p \leq y].$$

Therefore

$$p = px \leq xy$$

i.e.,

$$xy \in G.$$

Thus G is a filter for $\mathbf{B}$ and indeed, since $\mathbf{0} \notin G$, a proper filter.

By Theorem 2.12 G can be extended to an ultrafilter F. If

$$\begin{aligned} h(b) &= \mathbf{1}, \qquad b \in F \\ &= \mathbf{0}, \qquad b \in |\mathbf{B}| - F \end{aligned}$$

then f is a homomorphism from $\mathbf{B}$ onto $\mathbf{2}$ (Theorem 2.11). Since $a_0 \in F$, $h(a_0) = \mathbf{1}$. Since G is $\mathbf{P}$-generic over $\{S_n \mid n \in \omega\}$ and for each n, S_n is dense in P

$$S_n \cap F \neq 0$$

i.e.,

$$(\exists p \in F)[p \leq {}^{-}b_n \vee (\exists a \in A_n)[p \leq a]].$$

If $p \leq {}^{-}b_n$ then since h is order preserving

$$\mathbf{1} = h(p) = h({}^{-}b_n) = {}^{-}h(b_n)$$

i.e., $h(b_n) = \mathbf{0}$. Also

$$(\forall a \in A_n)[a \leq b_n].$$

Therefore

$$(\forall a \in A_n)[h(a) \leq h(b_n) = \mathbf{0}].$$

Thus

$$\sum_{a \in A_n} h(a) = \mathbf{0} = h(b_n).$$

If $(\exists b \in A_n)[p \leq b]$ then

$$\mathbf{1} = h(p) \leq h(b) \leq \sum_{a \in A_n} h(a) \leq h(b_n).$$

Remark. If in Theorem 2.14 we allow a collection of sums of arbitrary cardinality then the conclusion is false. If, however, $\mathbf{B}$ satisfies the countable chain condition, to be discussed in the next section, then a new axiom by Martin gives a generalization of the result for sets of sums of cardinality less than the continuum.

Theorem 2.15. If $\mathbf{B}$ is an M-complete Boolean algebra and $\mathscr{P}(|\mathbf{B}|) \cap M$ is countable then for each $b \neq \mathbf{0}$ in $|\mathbf{B}|$ there exists an M-complete homomorphism f from $\mathbf{B}$ onto $\mathbf{2}$ such that $f(b) = \mathbf{1}$.

Proof. If $S \in \mathscr{P}(|\mathbf{B}|) \cap M$ and if $b \in |\mathbf{B}|$ with $b \neq \mathbf{0}$ then, from the Rasiowa-Sikorski Theorem, there exists a homomorphism f from $\mathbf{B}$ into $\mathbf{2}$ such that $f(b) = \mathbf{1}$ and f preserves

$$\sum_{a \in S} a.$$

Theorem 2.16. If $\mathbf{P} = \langle P, \leq \rangle \in M$, if G is $\mathbf{P}$-generic over M if $\mathbf{B}$ is the Boolean algebra of regular open subsets of P and

$$F = \{b \in |\mathbf{B}| \cap M \mid b = b^{-0} \wedge b \cap G \neq 0\}$$

then F is a proper M-complete ultrafilter for the Boolean algebra $\mathbf{B}^M$.

Proof. Clearly $F \subseteq |\mathbf{B}| \cap M$. Since P is dense in P, $P \cap G \neq 0$. Also P is regular open and $\mathbf{P} \in M$. Therefore $P \in F$ i.e., $F \neq 0$.

If $x, y \in F$ then $xy = x \cap y \in |\mathbf{B}| \cap M$. Furthermore $x \cap G \neq 0$ and $y \cap G \neq 0$ i.e.,

$$[\exists c_1 \in x \cap G] \wedge [\exists c_2 \in y \cap G].$$

Since x and y are open

$$[c_1] \subseteq x \wedge [c_2] \subseteq y.$$

From Theorem 2.4

$$(\exists c \in G)[c \leq c_1 \wedge c \leq c_2].$$

Then $c \in (x \cap y) \cap G$ i.e., $(x \cap y) \cap G \neq 0$. Therefore $xy \in F$.

If $x \in F$, $y \in |\mathbf{B}| \cap M$ and $x \leq y$ then since $x \cap G \neq 0$ it follows that $y \cap G \neq 0$. Consequently $y \in F$ and F is a filter. Furthermore $\mathbf{0} \cap G = 0$, hence $\mathbf{0} \notin F$ and F is a proper filter.

To prove that F is M-complete we note that if $A \subseteq F$ and $A \in M$ then since $\mathbf{B}^M$ is M-complete $\prod_{a \in A} a \in |\mathbf{B}| \cap M$. We then need only prove that $\prod_{a \in A} a \cap G \neq 0$. For this purpose we appeal to Theorem 2.5.

Since M is a transitive model of ZF and $\mathbf{P} \in M$ it follows that for each S if $S \in M$ and S is a subset of P then

$$S \cup \{p \mid [p] \cap S = 0\} \in M.$$

Since G is $\mathbf{P}$-generic over M we have property 2 of Theorem 2.5. Consequently, by Theorem 2.5

$$(\forall S \in M)(\exists p \in G)[p \in S \vee [p] \cap S = 0].$$

In particular if $A \subseteq F$ and $A \in M$ then $\prod_{a \in A} a \in M$; hence

1. $$(\exists p \in G)\left[p \in \prod_{a \in A} a \vee [p] \cap \prod_{a \in A} a = 0\right].$$

If $[p] \cap \prod_{a \in A} a = 0$ then since $[p]$ and $\prod_{a \in A} a$ are each open

$$[p]^{-0} \cap \left(\prod_{a \in A} a\right) = 0.$$

(See Theorem 1.29.2.) Thus

$$[p]^{-0} \prod_{a \in A} a = \mathbf{0}$$

$$\prod_{a \in A} a \leq {}^{-}[p]^{-0}.$$

If $A' = A \cup \{[p]^{-0}\}$, then

$$\prod_{a \in A'} a = [p]^{-0} \prod_{a \in A} a = \mathbf{0}$$

and hence $\sum_{a \in A'} {}^{-}a = \mathbf{1}$. But $\sum_{a \in A'} {}^{-}a = (\bigcup_{a \in A'} {}^{-}a)^{-0} \subseteq (\bigcup_{a \in A'} {}^{-}a)^{-}$. Consequently if $S = \bigcup_{a \in A'} {}^{-}a$, then S is dense in $\mathbf{P}$ and $S \in M$. Since G is $\mathbf{P}$-generic over M

$$(\exists q \in G)\left[q \in \bigcup_{a \in A'} {}^{-}a\right]$$

$$(\exists q \in G)(\exists a \in A')[q \in {}^{-}a].$$

Then ${}^{-}a \in F$. On the other hand, $p \in [p]^{-0} \cap G$. Therefore $[p]^{-0} \in F$ i.e., $A' \subseteq F$. Therefore $a \in F$. But since F is a filter a, ${}^{-}a \in F$ implies $a({}^{-}a) = \mathbf{0} \in F$, which is a contradiction since F is a proper filter.

From this contradiction and 1 above we conclude that

$$(\exists p \in G)\left[p \in \prod_{a \in A} a\right]$$

i.e.,

$$\prod_{a \in A} a \in F.$$

Thus F is a proper M-complete filter.

Finally, if $a \in |\mathbf{B}| \cap M$, $a + {}^{-}a = \mathbf{1}$. Therefore $(a \cup {}^{-}a)^{-0} = P$. Consequently $a \cup {}^{-}a \in M$ and $a \cup {}^{-}a$ is dense in P. Then

$$(a \cup {}^{-}a) \cap G \neq 0$$

i.e.,

$$a \cap G \neq 0 \vee {}^{-}a \cap G \neq 0.$$

Therefore $a \in F \vee {}^{-}a \in F$. But since $a({}^{-}a) = \mathbf{0} \notin F$ we have

$$a \in F \leftrightarrow {}^{-}a \notin F.$$

F is an ultrafilter.

Theorem 2.17. If $\mathbf{P} = \langle P, \leq \rangle$ is a partial order structure, if $\mathbf{B}$ is the Boolean algebra of regular open subsets of P, if F is a proper M-complete ultrafilter in $\mathbf{B}^M$, and if $G = \{p \in P \mid [p]^{-0} \in F\}$ then G is $\mathbf{P}$-generic over M.

Proof. Clearly $G \subseteq P$. If $p, q \in G$ then $[p]^{-0} \in F$ and $[q]^{-0} \in F$. But

$$[p]^{-0}[q]^{-0} = [p]^{-0} \cap [q]^{-0} \in F.$$

Since F is proper, $0 \notin F$. Therefore

$$[p]^{-0} \cap [q]^{-0} \neq 0$$

and hence by Theorem 1.29.2

$$[p] \cap [q] \neq 0$$

i.e., p and q are compatible. Thus G is compatible.

If $p \in G$ and $p \leq q$ then $[p] \subseteq [q]$ hence $[p]^{-0} \leq [q]^{-0}$. But since $p \in G$ implies $[p]^{-0} \in F$ and since F is a filter $[q]^{-0} \in F$, and $q \in G$.

If $S \in M$ and S is dense in P then

$$S \subseteq \bigcup_{p \in S} [p]^{-0}$$

and hence

$$\left(\bigcup_{p \in S} [p]^{-0}\right)^{-0} = P$$

i.e.,

$$\sum_{p \in S} [p]^{-0} = \mathbf{1}.$$

If $G \cap S = 0$ then $(\forall p \in S)[p \notin G]$ and hence $[p]^{-0} \notin F$. But F is an ultrafilter. Therefore $^{-}([p]^{-0}) \in F$. Consequently

$$\mathbf{0} = \prod_{p \in S} {}^{-}([p]^{-0}) \in F.$$

Since F is proper this is a contradiction from which we conclude that $G \cap S \neq 0$. Therefore G is $\mathbf{P}$-generic over M.

Remark. In Theorem 2.16 we established a procedure for obtaining a proper M-complete ultrafilter F from a given G that is $\mathbf{P}$-generic over M. In Theorem 2.17 we showed how to obtain a G that is $\mathbf{P}$-generic over M from a proper M-complete ultrafilter F. If from a $\mathbf{P}$-generic G we obtain an ultrafilter F from which we in turn obtain a $\mathbf{P}$-generic G', how are G and G' related? We will show that in fact $G = G'$. Similarly if we proceed from F to G to F' then $F = F'$.

Theorem 2.18. If G is $\mathbf{P}$-generic over M then G is a maximal compatible subset of P.

Proof. If there exists a $p \notin G$ such that $G \cup \{p\}$ is compatible and if

$$S = [p] \cup \{q \mid \neg \operatorname{Comp}(p, q)\}$$

it is easily established that S is dense in P. Indeed if $q \in P$ either q is compatible with p or it is not. If q is compatible with p then $[q] \cap [p] \neq 0$; if q is not compatible with p then $q \in S$. In either case $[q] \cap S \neq 0$.

Since S is dense in P, $S \cap G \neq 0$. On the other hand, since $G \cup \{p\}$ is compatible G contains no elements incompatible with p. Therefore $[p] \cap G \neq 0$ i.e. $(\exists q \leq p)[q \in G]$. Since G is $\mathbf{P}$-generic it follows that $p \in G$. This is a contradiction.

Theorem 2.19. If $\mathbf{P} = \langle P, \leq \rangle \in M$, if $\mathbf{B}$ is the Boolean algebra of regular open subsets of P, if G is $\mathbf{P}$-generic over M, and if

$$\begin{aligned} F &= \{b \in |\mathbf{B}| \cap M \mid b = b^{-0} \wedge b \cap G \neq 0\} \\ G' &= \{p \mid [p]^{-0} \in F\} \end{aligned}$$

then $G = G'$.

Proof.

$$\begin{aligned} p \in G &\to [p]^{-0} \cap G \neq 0 \\ &\to p \in G' \end{aligned}$$

i.e., $G \subseteq G'$. Since, by Theorems 2.16 and 2.17, G' is **P**-generic over M it follows from Theorem 2.18 that $G = G'$.

Theorem 2.20. If **B** is a natural Boolean algebra, if F is a proper M-complete ultrafilter for $\mathbf{B}^M$, and if

$$\begin{aligned} G &= \{p \mid [p]^{-0} \in F\} \\ F' &= \{b \in |\mathbf{B}| \cap M \mid b = b^{-0} \wedge b \cap G \neq 0\} \end{aligned}$$

then $F = F'$.

Proof.

$$\begin{aligned} b \in F' &\to b = b^{-0} \wedge b \cap G \neq 0 \\ &\to (\exists p \in G)[p \in b] \\ &\to (\exists p)[[p]^{-0} \in F \wedge [p]^{-0} \leq b] \\ &\to b \in F. \end{aligned}$$

Thus $F' \subseteq F$. On the other hand, by Theorem 2.17, G is **P**-generic over M, and hence, by Theorem 2.16, F' is a proper ultrafilter. Then, by Theorem 2.13, $F = F'$.

3. Boolean σ-Algebras

Definition 3.1.

1. A Boolean algebra **B** is a σ-algebra iff

$$(\forall A \subseteq |\mathbf{B}|)\left[\overline{\overline{A}} = \omega \rightarrow \sum_{a\in A} a \in |\mathbf{B}| \wedge \prod_{a\in A} a \in |\mathbf{B}|\right].$$

2. A Boolean ideal I is a σ-ideal iff

$$(\forall A \subseteq I)\left[\overline{\overline{A}} = \omega \rightarrow \sum_{a\in A} a \in I\right].$$

3. A homomorphism f on a Boolean σ-algebra **B** is a σ-homomorphism iff

$$(\forall A \subseteq |\mathbf{B}|)\left[\overline{\overline{A}} = \omega \rightarrow f\left(\sum_{a\in A} a\right) = \sum_{a\in A} f(a) \wedge f\left(\prod_{a\in A} a\right) = \prod_{a\in A} f(a)\right].$$

Definition 3.2. A Boolean algebra $\mathbf{B}'$ is a subalgebra of the Boolean algebra **B** iff $|\mathbf{B}'| \subseteq |\mathbf{B}|$, the operations in $\mathbf{B}'$ are restrictions of the operations in **B** to $|\mathbf{B}'|$, and the distinguished elements of $\mathbf{B}'$ are the same as in **B**.

Theorem 3.3. If $I \neq 0$ and $\mathbf{B}_a$ is a subalgebra of the Boolean algebra $\mathbf{B} = \langle B, +, \cdot, ^-, \mathbf{0}, \mathbf{1}\rangle$ for $a \in I$ then $\mathbf{B}' = \langle \bigcap_{a\in I} |\mathbf{B}_a|, +, \cdot, ^-, \mathbf{0}, \mathbf{1}\rangle$ is a subalgebra of **B**. If in addition each $\mathbf{B}_a$ is a σ-algebra then $\mathbf{B}'$ is a σ-algebra.

Proof. Left to the reader.

Definition 3.4. 1. If $I \neq 0$ and $\mathbf{B}_a$ is a subalgebra of $\mathbf{B} = \langle B, +, \cdot, ^-, \mathbf{0}, \mathbf{1}\rangle$ for $a \in I$ then

$$\bigcap_{a\in I} \mathbf{B}_a \triangleq \left\langle \bigcap_{a\in I} |\mathbf{B}_a|, +, \cdot, ^-, \mathbf{0}, \mathbf{1}\right\rangle.$$

2. Let $A \subseteq |\mathbf{B}|$. Then $\bigcap \{\mathbf{B}' \mid \mathbf{B}'$ is a subalgebra of **B** and $A \subseteq |\mathbf{B}'|\}$ is the sub-algebra of **B** generated by A.

Definition 3.5. If $\langle X, T\rangle$ is a topological space and $A \subseteq X$, then A is a Borel set iff A belongs to the σ-subalgebra, generated by T, of the natural Boolean algebra on $\mathscr{P}(X)$.

Theorem 3.6. If $\langle X, T\rangle$ is a topological space, if

$$A_0 = T \cup \{X - a \mid a \in T\}$$

$$A_{\alpha+1} = \left\{a \mid (\exists f \in (A_\alpha)^{(\omega)})\left[\left[a = \bigcup_{i<\omega} f(i)\right] \vee \left[a = \bigcap_{i<\omega} f(i)\right]\right]\right\}$$

$$A_\alpha = \bigcup_{\beta<\alpha} A_\beta, \qquad \alpha \in K_{\mathrm{II}},$$

then $A = \bigcup_{\alpha\in\aleph_1} A_\alpha$ is the set of all Borel sets in X.

Proof. Clearly each element in A_0 is a Borel set. If A_α is a collection of Borel sets then so is $A_{\alpha+1}$. If for $\beta < \alpha$, A_β is a collection of Borel sets then

$$\bigcup_{\beta<\alpha} A_\beta$$

is a collection of Borel sets. Therefore A is a collection of Borel sets. To prove that A contains all Borel sets it is sufficient to prove that A is a Boolean σ-algebra.

Since $A_0 \subseteq A$ and $\mathbf{0}, \mathbf{1} \in A_0$ we have $\mathbf{0}, \mathbf{1} \in A$. Since union and intersection are associative, commutative, and distributive we need only prove that A has the closure and σ-closure properties.

We first note that $\alpha < \beta$ implies $A_\alpha \subseteq A_\beta$. If

$$b_0, b_1, \ldots$$

is an ω-sequence of elements of A then there exists an ω-sequence of ordinals

$$\alpha_0, \alpha_1, \ldots$$

each less than $\aleph_1$ and such that $b_0 \in A_{\alpha_0}, b_1 \in A_{\alpha_1}, \ldots$.

Since $\{\alpha_0, \alpha_1, \ldots\}$ is a set it has a supremum that is also less than $\aleph_1$. Therefore

$$(\exists\alpha < \aleph_1)(\forall i < \omega)[b_i \in A_\alpha].$$

Then

$$\sum_{i<\omega} b_i \in A_{\alpha+1} \wedge \prod_{i<\omega} b_i \in A_{\alpha+1}.$$

Definition 3.7. If X is a topological space and $A \subseteq X$ then

1. A is nowhere dense iff $A^{-0} = 0$.
2. A is meager iff A is the union of countably many nowhere dense sets i.e., $A = \bigcup_{i<\omega} A_i$ where $\forall i < \omega$, A_i is nowhere dense.

Theorem 3.8. If X is a topological space and $A \subseteq X$ then

1. A is open implies $A^- - A$ is meager.
2. A is closed implies $A - A^0$ is meager.

Proof.

$$
\begin{aligned}
1.\ (A^- - A)^0 &= [A^- \cap (X - A)]^{-0} \subseteq A^{-0} \cap (X - A)^{-0} \\
&= A^{-0} \cap (X - A^{0-}) \qquad \text{since } A \text{ is open} \\
&\subseteq A^{-0} \cap (X - A^{-0}) \\
&= 0.
\end{aligned}
$$

$$
\begin{aligned}
2.\ (A - A^0)^{-0} &= [A \cap (X - A^0)]^{-0} \\
&\subseteq A^0 \cap (X - A^0)^{-0} \qquad \text{since } A \text{ is closed} \\
&= A^0 \cap (X - A^{0-}) \\
&= 0.
\end{aligned}
$$

Theorem 3.9. 1. The collection of all meager sets in a topological space X is a proper σ-ideal in the natural algebra on $\mathscr{P}(X)$.

2. The collection of all meager Borel sets in a topological space X is a proper σ-ideal in the Boolean σ-algebra of Borel sets.

Proof. Left to the reader.

Theorem 3.10. If B is a Borel set of the topological space X then there exists an open set G and meager sets N_1 and N_2 such that

$$B = (G + N_1) - N_2,$$

i.e. every Borel set has the property of Baire.

Proof. If B is open then $B = (B + 0) - 0$. If B is closed

$$B = [B^0 + (B - B^0)] - 0.$$

Thus in the notation of Theorem 3.6, the result holds for each element of A_0. If it holds for each element of A_α and if $B \in A_{\alpha+1}$ then there is an ω-sequence $B_0, B_1, \ldots,$ of elements in A_α, such that $B = \sum_{i<\omega} B_i$ or $B = {}^-\sum_{i<\omega} B_i$. From our induction hypothesis there exist open sets G_i and meager sets $N_1{}^i$ and $N_2{}^i$ such that

$$B_i = (G_i + N_1{}^i) - N_2{}^i.$$

If $G = \sum_{i<\omega} G_i$ then G is open. Furthermore if

$$N_1 = B - G \wedge N_2 = G - B$$

then

$$N_1 = B - G \subseteq \sum_{i<\omega} B_i - \sum_{i<\omega} G_i \subseteq \sum_{i<\omega} (B_i - G_i) \subseteq \sum_{i<\omega} N_1{}^i$$

$$N_2 = G - B \subseteq \sum_{i<\omega} (G_i - B_i) \subseteq \sum_{i<\omega} N_2{}^i.$$

Thus N_1 and N_2 are meager and $B = (G + N_1) - N_2$.

If $B = {}^-C$ and $C = (G + N_1) - N_2$ for G open and N_1 and N_2 meager, then

$$B = ({}^-G + N_2) - (N_1 - N_2).$$

Since ${}^{-}G$ is closed ${}^{-}G - ({}^{-}G)^0$ is meager and hence

$$B = [({}^{-}G)^0 + ({}^{-}G - ({}^{-}G)^0) + N_2] - (N_1 - N_2)$$

where $({}^{-}G - ({}^{-}G)^0) + N_2$ and $N_1 - N_2$ are meager.

Corollary 3.11. If B is a Borel set of the topological space $\langle X, T\rangle$ then there exists a regular open set G and meager sets N_1 and N_2 such that

$$B = (G + N_1) - N_2.$$

Proof. By Theorem 3.10 there exists an open set G and meager sets N_1 and N_2 such that $B = (G + N_1) - N_2$. But

$$G = G^{-0} - (G^{-0} - G).$$

Hence

$$B = (G^{-0} + N_1) - [(G^{-0} - G - N_1) + N_2].$$

Definition 3.12. A is a compact set in the topological space $\langle X, T\rangle$ iff $A \subseteq X$ and

$$(\forall S \subseteq T)\left[A \subseteq \bigcup(S) \rightarrow (\exists S' \subseteq S)\left[\text{Fin}^*(S') \wedge A \subseteq \bigcup(S')\right]\right].$$

Definition 3.13. A topological space $\langle X, T\rangle$ is
1. a Hausdorff space

$$\text{iff } (\forall a, b \in X)[a \neq b \rightarrow (\exists N(a))(\exists N'(b))[N(a) \cap N'(b) = 0]],$$

2. a compact space iff X is a compact set,
3. a locally compact space iff $\forall a \in X, \exists N(a), N(a)^-$ is a compact set.

Theorem 3.14. If the topological space $\langle X, T\rangle$ is a Hausdorff space then $(\forall a, b \in X)[a \neq b \rightarrow (\exists N(a))[b \notin N(a)^-]]$.

Proof. By definition of a Hausdorff space

$$(\exists N(a))(\exists N'(b))[N(a) \cap N'(b) = 0].$$

Therefore $b \notin N(a)^-$.

Theorem 3.15. 1. Every compact set in a Hausdorff space is closed.
2. Every closed set in a compact space is compact.

Proof. 1. Let A be a compact set in a Hausdorff space $\langle X, T\rangle$. If $b \in A^- - A$ then by Theorem 3.14

$$(\forall a \in A)(\exists N(a))[b \notin N(a)^-].$$

Since $A \subseteq \bigcup\{N(a) \mid a \in A \wedge b \notin N(a)^-\}$ and since A is compact, there exists a finite collection of elements of A

$$a_1, \ldots, a_n$$

* Fin (S) means "S is finite".

and a neighborhood of each such point

$$N(a_1), \ldots, N(a_n)$$

such that

$$A \subseteq N(a_1) \cup \cdots \cup N(a_n)$$

and $b \notin N(a_1)^-$, $b \notin N(a_2)^-$, $\cdots$, $b \notin N(a_n)^-$. Therefore

$$\begin{gathered}(\exists N_1(b))[N_1(b) \cap N(a_1) = 0]\\ (\exists N_2(b))[N_2(b) \cap N(a_2) = 0]\\ \vdots\\ (\exists N_n(b))[N_n(b) \cap N(a_n) = 0].\end{gathered}$$

If

$$N(b) = N_1(b) \cap \cdots \cap N_n(b)$$

then $N(b) \cap A = 0$. But this contradicts the fact that $b \in A^-$.

Therefore $A^- - A = 0$ i.e., A is closed.

2. If A is a closed set in the compact space $\langle X, T\rangle$ and if

$$[S \subseteq T] \wedge \left[A \subseteq \bigcup (S)\right]$$

then

$$(\forall a \in X - A)[\exists N(a) \subseteq X - A].$$

Consequently $X \subseteq \bigcup (S) \cup \bigcup \{N(a) \mid a \in X - A \wedge N(a) \subseteq X - A\}$. Since $\langle X, T\rangle$ is compact there exists a finite collection of sets in S

$$D_1, \ldots, D_n$$

and a finite collection of sets in $\{N(a) \mid [a \in X - A] \wedge [N(a) \subseteq X - A]\}$

$$N(a_1), \ldots, N(a_m)$$

such that

$$X \subseteq D_1 \cup \cdots \cup D_n \cup N(a_1) \cup \cdots \cup N(a_m).$$

Then

$$A \subseteq D_1 \cup \cdots \cup D_n.$$

Definition 3.16. A set S has the finite intersection property iff every finite subset of S has a nonempty intersection.

Theorem 3.17. The topological space $\langle X, T\rangle$ is compact iff for each collection S of closed sets with the finite intersection property

$$\bigcap (S) \neq 0.$$

Proof. (By contradiction.) Suppose that $\langle X, T\rangle$ is a compact topological space and there exists a collection of closed sets S with the finite intersection property but for which $\bigcap (S) = 0$. Then

$$X - 0 = X - \bigcap (S) = \bigcup_{A \in S} (X - A).$$

Since $X - A$ is open for each $A \in S$ and $\langle X, T\rangle$ is compact $\exists A_0, \ldots, A_n \in S$

$$X = \bigcup_{i \leq n} (X - A_i) = X - \bigcap_{i \leq n} A_i.$$

Therefore $\bigcap_{i \leq n} A_i = 0$. This is a contradiction.

Conversely suppose that every collection of closed sets S with the finite intersection property also has the property $\bigcap (S) \neq 0$. Suppose also that there exists a collection of open sets T' such that

$$X \subseteq \bigcup (T')$$

but $\forall A_1, \ldots, A_n \in T'$

$$X \nsubseteq \bigcup_{i \leq n} A_i.$$

Then

$$\bigcap_{i \leq n} (X - A_i) = X - \bigcup_{i \leq n} A_i \neq 0.$$

Thus $\{X - A \mid A \in T'\}$ is a collection of closed sets with the finite intersection property. Then

$$X - \bigcup_{A \in T'} A = \bigcap_{A \in T'} (X - A) \neq 0$$

i.e.,

$$X \nsubseteq \bigcup (T').$$

This is a contradiction.

Theorem 3.18. If $\langle X, T\rangle$ is a topological space, if $X' \subseteq X$ and $T' = \{X' \cap N \mid N \in T\}$ then $\langle X', T'\rangle$ is a topological space. Furthermore

1. $\langle X', T'\rangle$ is a compact space if X' is a compact set,
2. $\langle X', T'\rangle$ is a Hausdorff space if $\langle X, T\rangle$ is Hausdorff.

Proof. Left to the reader.

Definition 3.19. If $\langle X, T\rangle$ is a topological space, if $X' \subseteq X$ and $T' = \{X' \cap N \mid N \in T\}$ then T' is the relative topology on X' induced by T and $\langle X', T'\rangle$ is a subspace of $\langle X, T\rangle$.

Theorem 3.20. If $\langle X, T\rangle$ is a topological space, if $X' \subseteq X$, if T' is the relative topology on X' induced by T, if B is a base for T and

$$B' = \{X' \cap N \mid N \in B\}$$

then B' is a base for T'.

Proof. Left to the reader.

Theorem 3.21. If $\langle X, T\rangle$ is a topological space, if $X' \subseteq X$, and if T' is the relative topology on X' induced by T then

1. A is an open set in T implies $A \cap X'$ is an open set in T'
2. A is closed in T implies $A \cap X'$ is closed in T'
3. A is clopen in T implies $A \cap X'$ is clopen in T'.

Proof. 1. If A is open in T then

$$(\forall a \in A \cap X')(\exists N(a) \in T)[N(a) \subseteq A].$$

Then $N(a) \cap X' \in T'$ and $a \in N(a) \cap X' \subseteq A \cap X'$. Thus $A \cap X'$ is open in T'.

2. If $a \in X'$ and $(\forall N(a) \in T')[N(a) \cap (A \cap X') \neq 0]$ then

$$(\forall N(a) \in T)[N(a) \cap X' \cap A \neq 0].$$

Thus

$$(\forall N(a) \in T)[N(a) \cap A \neq 0].$$

Since A is closed $a \in A$ and hence $a \in A \cap X'$ i.e., $A \cap X'$ is closed in T'.

3. If A is both open and closed in T then by 1 and 2 above $A \cap X'$ is both open and closed in T'.

Theorem 3.22. If $\langle X, T\rangle$ is a locally compact Hausdorff space then for each open set A and each $a \in A$ there exists an open set B such that

$$a \in B \wedge B^- \subseteq A.$$

Proof. If A is an open set in X and $a \in A$ then since $\langle X, T\rangle$ is locally compact $\exists N(a)$, $N(a)^-$ is compact. If

$$M = (N(a)^- \cap A)^0$$

then M^- is also compact. If

$$T' = \{M^- \cap A \mid A \in T\}$$

then $\langle M^-, T'\rangle$ is a compact Hausdorff space. In this space $M^- - M$ is closed and hence compact. Moreover

$$(\forall y \in M^- - M)(\exists N(y) \in T')[a \notin N(y)^-].$$

Since $M^- - M$ is compact there is a finite collection of elements of T'

$$N(y_1), \ldots, N(y_n)$$

such that

$$M^- - M \subseteq N(y_1) \cup \cdots \cup N(y_n)$$

and

$$a \notin N(y_1)^- \wedge \cdots \wedge a \notin N(y_n)^-.$$

Therefore there exist neighborhoods in T'

$$M_1(a), \ldots, M_n(a)$$

such that

$$M_i(a) \cap N(y_i) = 0 \qquad i = 1, \ldots, n.$$

If $M(a) = \bigcap_{i \leq n} M_i(a)$ then

$$M(a) \cap [N(y_1) \cup \cdots \cup N(y_n)] = 0$$

Therefore

$$M(a) \subseteq M^- - [N(y_1) \cup \cdots \cup N(y_n)].$$

But since $N(y_1) \cup \cdots \cup N(y_n)$ is open $M^- - [N(y_1) \cup \cdots \cup N(y_n)]$ is closed. Hence

$$M(a)^- \subseteq M^- - [N(y_1) \cup \cdots \cup N(y_2)].$$

Therefore

$$M(a)^- \cap [N(y_1) \cup \cdots \cup N(y_n)] = 0$$

and

$$M(a)^- \subseteq M = (N(a)^- \cap A)^0.$$

But since $M(a) \in T'$,

$$(\exists N \in T)[M(a) = M^- \cap N].$$

And since $M(x) \subseteq M = M^0$

$$M(a) = M \cap N \in T.$$

Theorem 3.23. (The Baire Category Theorem.) Every open meager set in a locally compact Hausdorff space is empty.

Proof. If B is an open meager set in the locally compact Hausdorff space $\langle X, T\rangle$ then there exists an ω-sequence of nowhere dense sets

$$A_0, A_1, \ldots$$

such that

$$B = \bigcup_{\alpha<\omega} A_\alpha.$$

If $B \neq 0$ then by Theorem 3.22

$$(\exists N_1 \in T)[N_1{}^- \subseteq B]$$

and since A_1 is nowhere dense

$$(\exists N_2 \subseteq N_1)[N_2 \cap A_1 = 0]$$

for otherwise $N_1 \subseteq A_1{}^{-0}$. Then

$$(\exists N_3 \in T)[N_3{}^- \subseteq N_2].$$

Inductively we define a nested sequence of neighborhoods such that

$$N_{n+1} \subseteq N_{n+1}^- \subseteq N_n, \qquad n < \omega.$$

Consequently

$$\bigcap_{n<\omega} N_{2n+1} = \bigcap_{n<\omega} N_{2n+1}^- \neq 0$$

(Theorem 3.17). Therefore

$$\exists x \in \bigcap_{n<\omega} N_{2n+1} \subseteq B.$$

But then $\forall n \in \omega$

$$x \in N_{2n+1} \wedge N_{2n+1} \cap A_n = 0.$$

Therefore $x \notin \bigcup_{\alpha \in \omega} A_\alpha$. This is a contradiction that compels the conclusion $B = 0$.

Theorem 3.24. If $\mathbf{B}$ is the Boolean σ-algebra of all Borel sets in the locally compact Hausdorff space $\langle X, T \rangle$ and if I is the σ-ideal of all meager Borel sets then $\mathbf{B}/I$ is isomorphic to $\mathbf{B}'$, the complete Boolean algebra of all regular open sets in X.

Proof. If

$$F(G) = G/I, \qquad G \in |\mathbf{B}'|$$

then $F(G_1) = F(G_2) \leftrightarrow G_1 - G_2 \in I \wedge G_2 - G_1 \in I$. Then $G_1 - G_2^-$ is meager and open. Thus, by the Baire Category Theorem

$$G_1 - G_2^- = 0.$$

Similarly

$$G_2 - G_1^- = 0.$$

Then $G_1 \subseteq G_2^- \wedge G_2 \subseteq G_1^-$. Since G_1 and G_2 are each regular open

$$G_1 = G_1^0 \subseteq G_2^{-0} = G_2 \quad \text{and} \quad G_2 = G_2^0 \subseteq G_1^{-0} = G_1.$$

Therefore $G_1 = G_2$ and hence F is one-to-one.

If $G \in |\mathbf{B}|$ then by Corollary 3.11 there exists a regular open set G' and meager sets N_1, N_2 such that

$$G = (G' + N_1) - N_2.$$

Then $G - G' \subseteq N_1 - N_2$ i.e., $G - G' \in I$. Similarly $G' - G \in I$ and hence

$$G/I = G'/I.$$

Then

$$F(G') = G'/I = G/I.$$

That is F is onto.

That F has the morphism properties is clear from its definition.

Definition 3.25. A Boolean algebra $\mathbf{B}$ satisfies the countable chain condition (c.c.c.) iff

$$(\forall S \subseteq |\mathbf{B}|)[\forall a, b \in S)[a \neq b \rightarrow ab = \mathbf{0}] \rightarrow \bar{\bar{S}} \leq \omega].$$

Theorem 3.26. If X is a topological space with a countable base then the Boolean algebra of regular open sets in X satisfies the countable chain condition.

Proof. If $U_1, U_2, \ldots$ is a countable base and if S is a pairwise disjoint subset of $|\mathbf{B}|$ then since the elements of S are open it follows that

$$(\forall A \in S)(\exists n < \omega)[U_n \subseteq A].$$

Furthermore

$$(\forall A, B \in S)[[U_n \subseteq A] \wedge [U_n \subseteq B] \rightarrow A = B].$$

Therefore S is countable.

Theorem 3.27. If $\mathbf{B}$ satisfies the c.c.c. then for each subset E of $|\mathbf{B}|$ there exists a countable subset D of E such that D and E have the same set of upper bounds.

Proof. If I is the ideal generated by E then $E \subseteq I$. Consequently every upper bound for I is an upper bound for E. Conversely

$$(\forall b \in I)(\exists b_1, \ldots, b_n \in E)[b \leq b_1 + \cdots + b_n].$$

Therefore every upper bound for E is also an upper bound for I.

From Zorn's Lemma there exists a maximal set F of disjoint elements of I. By the c.c.c. $\overline{\overline{F}} \leq \omega$. Since $F \subseteq I$ every upper bound for I is an upper bound for F. If b_0 is an upper bound for F that is not an upper bound for I then

$$(\exists b_1 \in I)[b_1 \not\leq b_0].$$

Therefore $b_1 - b_0 \in I$. Furthermore, since $(\forall b \in F)[b \leq b_0]$

$$(\forall b \in F)[b \cap (b_1 - b_0) = 0].$$

Then $F \cup \{b_1 - b_0\}$ is a collection of pairwise disjoint elements of I. But this contradicts the definition of F. Thus every upper bound for F is also an upper bound for I.

We have established that F is countable. If

$$F = \{f_n \mid n < \omega\}$$

then since $F \subseteq I$ and I is generated by E

$$(\forall n < \omega)(\exists b_1{}^n \cdots b_{m_n}{}^n \in E)[f_n \leq b_1{}^n + \cdots + b_{m_n}{}^n].$$

From this existence property and with the aid of the AC, we define a set D thus:

$$D \triangleq \{b_i{}^n \mid n < \omega \wedge i = 1, 2, \ldots, m_n\}$$

then $D \subseteq E$, $\overline{\overline{D}} \leq \omega$ and D and E have the same set of upper bounds.

Theorem 3.28. Every Boolean σ-algebra $\mathbf{B}$ satisfying the c.c.c. is complete.

Proof. By Theorem 3.27 if $E \subseteq |\mathbf{B}|$ then there exists a countable subset D of E such that D and E have the same set of upper bounds. Since D is countable and $\mathbf{B}$ is a σ-algebra

$$\sum_{b \in D} b$$

exists. Since D and E have the same set of upper bounds

$$\sum_{b \in E} b$$

exists. Indeed

$$\sum_{b \in E} b = \sum_{b \in D} b.$$

Remark. In later sections we will need certain properties of product topologies which we will now prove. We begin by defining projection functions for cross products. Hereafter the symbol $\prod$ will be used to denote cross products and Boolean products. We will rely on the context to make the meaning clear.

Definition 3.29.

$$(\forall a \in A)\left(\forall C \subseteq \prod_{a \in A} X_a\right)[p_a(C) \triangleq \{f(a) \mid f \in C\}].$$

Theorem 3.30. If $A \neq 0$ and $\forall a \in A$, $\langle X_a, T_a \rangle$ is a topological space then

$$T' = \left\{B \subseteq \prod_{a \in A} X_a \mid (\exists n \in \omega)(\exists \sigma \in A^n)(\forall i < n)[[p_{\sigma(i)}(B) \in T_{\sigma(i)}] \right.$$
$$\left. \wedge\ (\forall a \in A)[a \neq \sigma(i) \to p_a(B) = X_a]]\right\}$$

is a base for a topology on $\prod_{a \in A} X_a$.

Proof. If $A \neq 0$ then $\exists b \in A$. If $f \in \prod_{a \in A} X_a$ then $f(b) \in X_b$ and hence $\exists N(f(b)) \in T_b$. Then

$$f \in \left\{g \in \prod_{a \in A} X_a \mid g(b) \in N(f(b))\right\} \in T'.$$

Clearly, if $B_1, B_2 \in T'$ then $B_1 \cap B_2 \in T'$. Therefore by Theorem 1.18 T' is a base for a topology on $\prod_{a \in A} X_a$.

Definition 3.31. The topology T of Theorem 3.30 we call the (weak) product topology on $\prod_{a \in A} X_a$ induced by the topologies T_a, $a \in A$. This topology we will denote by

$$\prod_{a \in A} T_a.$$

$\langle \prod_{a \in A} X_a, \prod_{a \in A} T_a \rangle$ we call a product topological space.

Theorem 3.32. If $A \neq 0$ and $\forall a \in A$, $\langle X_a, T_a \rangle$ is a topological space then

$$(\forall a \in A)\left(\forall C \subseteq \prod_{a \in A} T_a\right)[p_a(C) \in T_a].$$

Proof. Left to the reader.

Theorem 3.33. (Tychonoff's Theorem.) If $A \neq 0$ and $\forall a \in A$, $\langle X_a, T_a \rangle$ is a compact topological space then the product topological space $\prod_{a \in A} X_a$, is also compact.

Proof. Let S be a collection of closed subsets of $\prod_{a \in A} X_a$, with the finite intersection property and

$$T \triangleq \left\{B \subseteq \mathscr{P}\left(\prod_{a \in A} X_a\right) \mid S \subseteq B \wedge B \text{ has the finite intersection property}\right\}.$$

Let $\{B_b \mid b \in I_0\}$ be a subset of T, linearly ordered by inclusion. Then $S \subseteq \bigcup_{b \in I_0} B_b$ and for each finite subset of $\bigcup_{b \in I_0} B_b$ there is a $b \in I_0$ such that that subset is contained in B_b. Since B_b has the finite intersection property $\bigcup_{b \in I_0} B_b$ has the finite intersection property. Thus

$$\bigcup_{b \in I_0} B_b \in T.$$

Since every linearly ordered subset of T has an upper bound in T, with respect to inclusion, T contains a maximal element by Zorn's Lemma. Thus there is a B in T such that no proper extension of B has the finite intersection property.

Since B has the finite intersection property $\{p_a(C) \mid C \in B\}$ has the finite intersection property for all $a \in A$. Therefore if

$$C_a = \{p_a(C)^- \mid C \in B\}$$

then C_a is a collection of closed subsets of X_a and C_a has the finite intersection property. Since $\langle X_a, T_a \rangle$ is compact it follows from Theorem 3.17 that $\bigcap (C_a) \neq 0$ i.e.,

$$\exists b_a \in \bigcap (C_a).$$

If $b = \prod_{a \in A} \{b_a\}$ and if $N(b)$ is any neighborhood of b in the product topology then

$$b_a \in p_a(N(b)) \in T_a.$$

Since

$$(\forall C \in B)[b_a \in p_a(C)^-]$$
$$(\forall C \in B)[p_a(N(b)) \cap p_a(C) \neq 0].$$

Consequently

$$(\forall C \in B)[N(b) \cap C \neq 0].$$

In particular since $S \subseteq B$

$$(\forall A \in S)[N(b) \cap A \neq 0].$$

Since S is a collection of closed sets, $b \in A$ for each A in S i.e.,

$$\bigcap (S) \neq 0.$$

Therefore by Theorem 3.17 the product topology is compact.

Theorem 3.34. If $A \neq 0$ and $\forall a \in A$, $\langle X_a, T_a \rangle$ is a Hausdorff space then the product topology on $\prod_{a \in A} X_a$ is Hausdorff.

Proof. If $f, g \in \prod_{a \in A} X_a$ and $f \neq g$ then $(\exists b \in A)[f(b) \neq g(b)]$. Since $\langle X_b, T_b \rangle$ is a Hausdorff space $\exists N(f(b)), N'(g(b)) \in T_b$

$$N(f(b)) \cap N'(g(b)) = 0.$$

Then $M = \{h \in \prod_{a \in A} X_a \mid h(b) \in N(f(b))\}$ is a neighborhood of f and $M' = \{h \in \prod_{a \in A} X_a \mid h(b) \in N'(g(b))\}$ is a neighborhood of g. But $M \cap M' = 0$. Therefore the product space is Hausdorff.

4. Distributive Laws

In this section we wish to discuss several generalized distributive laws for Boolean algebras that will be of importance in the work to follow.

Definition 4.1. If α and β are cardinal numbers then a complete Boolean algebra $\mathbf{B}$ satisfies the (α, β)-distributive law ((α, β)-DL) iff for each

$$\{b_{ij} \in |\mathbf{B}| \mid i \in I \wedge j \in J\} \quad \text{with} \quad \bar{\bar{I}} \leq \alpha \wedge \bar{\bar{J}} \leq \beta$$

$$\prod_{i \in I} \sum_{j \in J} b_{ij} = \sum_{f \in J^I} \prod_{i \in I} b_{i, f(i)}$$

$\mathbf{B}$ satisfies the complete distributive law iff it satisfies the (α, β)-distributive law for all α and β.

Remark. From Theorem 1.12 every complete Boolean algebra satisfies the $(2, \beta)$-DL. We can easily provide an example of a complete Boolean algebra that does not satisfy the $(\omega, 2)$-DL.

Example. If $\mathbf{B}$ is the complete Boolean algebra of all regular open sets of the product space $\mathbf{2}^{\omega}$ then $\mathbf{B}$ does not satisfy the $(\omega, 2)$-DL: If

$$b_{i0} \triangleq \{f \in \mathbf{2}^{\omega} \mid f(i) = \mathbf{0}\}, \qquad b_{i1} \triangleq \{f \in \mathbf{2}^{\omega} \mid f(i) = \mathbf{1}\}$$

then b_{i0}, and b_{i1} are each clopen and hence regular open. Then

$$\prod_{i \in \omega} (b_{i0} + b_{i1}) = \mathbf{1}.$$

But $\forall f \in \mathbf{2}^{\omega}$

$$\prod_{i \in \omega} b_{i, f(i)} = \left(\bigcap_{i \in \omega} b_{i, f(i)}\right)^{-0} = \{f\}^{-0} = \{f\}^{0} = \mathbf{0}.$$

Therefore

$$\sum_{f \in \mathbf{2}^{\omega}} \prod_{i \in \omega} b_{i, f(i)} = \mathbf{0}.$$

Theorem 4.2. If $\mathbf{B}$ is a complete Boolean algebra and

$$\{b_{ij} \mid i \in I \wedge j \in J\} \subseteq |\mathbf{B}|$$

then

$$\sum_{f \in J^I} \prod_{i \in I} b_{i, f(i)} \leq \prod_{i \in I} \sum_{j \in J} b_{ij}.$$

Proof.

$$(\forall f \in J^I)\Big[b_{i,f(i)} \leq \sum_{j \in J} b_{ij}\Big]$$

$$(\forall f \in J^I)\Big[\prod_{i \in I} b_{i,f(i)} \leq \prod_{i \in I} \sum_{j \in J} b_{ij}\Big]$$

$$\sum_{f \in J^I} \prod_{i \in I} b_{i,f(i)} \leq \prod_{i \in I} \sum_{j \in J} b_{ij}.$$

Theorem 4.3. If **B** is the complete natural Boolean algebra of all subsets of $A \neq 0$ then **B** satisfies the complete distributive law.

Proof. If $b_{ij} \subseteq A$ for $i \in I$ and $j \in J$ then

$$\begin{aligned} b \in \bigcap_{i \in I} \bigcup_{j \in J} b_{ij} &\leftrightarrow (\forall i \in I)(\exists j \in J)[b \in b_{ij}] \\ &\leftrightarrow (\exists f \in J^I)(\forall i \in I)[b \in b_{ij}] \\ &\leftrightarrow b \in \bigcup_{f \in I^I} \bigcap_{i \in I} b_{i,f(i)}. \end{aligned}$$

Remark. We next show that to within isomorphism the complete natural Boolean algebras are the only completely distributive complete Boolean algebras.

Theorem 4.4. For each completely distributive complete Boolean algebra **B** there exists a nonempty set A for which the natural algebra on $\mathscr{P}(A)$ is isomorphic to **B**.

Proof. If $\forall b \in |\mathbf{B}|$

$$\begin{aligned} a_{b\mathbf{0}} &= -b \\ a_{b\mathbf{1}} &= b \end{aligned}$$

and

$$A = \Big\{\prod_{b \in |\mathbf{B}|} a_{b,f(b)} \mid f \in \mathbf{2}^{|\mathbf{B}|} \wedge \prod_{b \in |\mathbf{B}|} a_{b,f(b)} \neq \mathbf{0}\Big\}$$

then since **B** is completely distributive

$$\mathbf{1} = \prod_{b \in |\mathbf{B}|} (a_{b\mathbf{0}} + a_{b\mathbf{1}}) = \sum_{f \in \mathbf{2}^{|\mathbf{B}|}} \prod_{b \in |\mathbf{B}|} a_{b,f(b)}.$$

Consequently

$$(\exists f \in \mathbf{2}^{|\mathbf{B}|})\Big[\prod_{b \in |\mathbf{B}|} a_{b,f(b)} \neq \mathbf{0}\Big]$$

i.e., $A \neq 0$. Furthermore

$$c \in |\mathbf{B}| \to c = c \prod_{b \in |\mathbf{B}|} (a_{b\mathbf{0}} + a_{b\mathbf{1}}) = \sum_{f \in \mathbf{2}^{|\mathbf{B}|}} \Big(c \prod_{b \in |\mathbf{B}|} a_{b,f(b)}\Big).$$

If $c \neq \mathbf{0}$ then $\exists f \in \mathbf{2}^{|\mathbf{B}|}$

$$c \prod_{b \in |\mathbf{B}|} a_{b,f(b)} \neq \mathbf{0}.$$

But $\forall f \in \mathbf{2}^{|\mathbf{B}|}$ if $f(c) = \mathbf{0}$ then $a_{c,f(c)} = {}^{-}c$ and

$$\prod_{b \in |\mathbf{B}|} a_{b,f(b)} \leq {}^{-}c$$

i.e.,

$$c \prod_{b \in |\mathbf{B}|} a_{b,f(b)} = \mathbf{0}.$$

If $f(c) = \mathbf{1}$ then $a_{c,f(c)} = c$ and

$$\prod_{b \in |\mathbf{B}|} a_{b,f(b)} \leq c$$

i.e.,

$$c \prod_{b \in |\mathbf{B}|} a_{b,f(b)} = \prod_{b \in |\mathbf{B}|} a_{b,f(b)}.$$

Thus, if $S = \{\prod_{b \in |\mathbf{B}|} a_{b,f(b)} \in A \mid f(c) = \mathbf{1}\}$ then $c = \sum_{b \in S} b$. Therefore, if we define F on $\mathscr{P}(A)$ by

$$F(S) = \sum_{b \in S} b,\ S \subseteq A$$

then F maps $\mathscr{P}(A)$ onto $|\mathbf{B}|$.

To prove that F is one-to-one we note that if $b_1, b_2 \in A \wedge b_1 \neq b_2$ then $\exists f_1, f_2 \in \mathbf{2}^{|\mathbf{B}|}$

$$b_1 = \prod_{b \in |\mathbf{B}|} a_{b,f_1(b)} \neq \prod_{b \in |\mathbf{B}|} a_{b,f_2(b)} = b_2.$$

Therefore $\exists b \in |\mathbf{B}|, f_1(b) \neq f_2(b)$ and hence

$$a_{b,f_1(b)} a_{b,f_2(b)} = \mathbf{0}.$$

Consequently $b_1 b_2 = \mathbf{0}$.

If $b \in A \wedge S \subseteq A \wedge b \leq F(S)$ then

$$bF(S) = \sum_{c \in S} bc \neq \mathbf{0}.$$

Hence $\exists c \in S, bc \neq 0$. But $bc \neq 0$ iff $b = c$. Therefore $b \in S$. Consequently if $S \subseteq A$ then

$$(\forall b \in A)[b \in S \leftrightarrow b \leq F(S)].$$

From this fact it follows that if $S \subseteq A$ and $S' \subseteq A$ then

$$F(S) = F(S') \rightarrow S = S',$$

i.e., F is one-to-one.

Furthermore

$$F(S \cup S') = \sum_{b \in S \cup S'} b.$$

But

$$\begin{aligned} b \in S &\rightarrow b \leq F(S) \leq F(S) + F(S') \\ b \in S' &\rightarrow b \leq F(S') \leq F(S) + F(S'). \end{aligned}$$

Therefore

$$F(S \cup S') \leq F(S) + F(S').$$

If $(\forall b \in S \cup S')[b \leq c]$ then

$$F(S) \leq c \wedge F(S') \leq c$$

hence

$$F(S) + F(S') \leq c$$

i.e.,

$$F(S \cup S') = F(S) + F(S').$$

Finally $\forall c \in A$

$$\begin{aligned} c \leq F(A - S) &\leftrightarrow c \in A - S \\ &\leftrightarrow c \notin S \\ &\leftrightarrow c \not\leq F(S) \\ &\leftrightarrow c \leq {}^{-}F(S). \end{aligned}$$

Therefore

$$F(A - S) = {}^{-}F(S).$$

Thus F is an isomorphism of $\mathscr{P}(A)$ onto $|\mathbf{B}|$.

5. Partial Order Structures and Topological Spaces

In the work ahead we will be interested in Boolean algebras that are associated with certain partial order structures (Definition 5.4) and Boolean algebras of regular open sets of certain topological spaces. Quite often we find that the Boolean algebra associated with a particular partial order structure is the same algebra as that of the regular open sets of a certain topological space even though there appears to be no connection between the partial order structure and the topological space. In this section we will establish such a connection. For a given partial order structure we will define a topological space of ultrafilters for the partial order structure (Definitions 5.2, 5.3, and 5.6). We will show that in general this topological space is a T_1-space (Theorem 5.7). If, however, the partial order structure is one associated with a Boolean algebra, then the topological space is in fact Hausdorff (Theorem 5.8).

Definition 5.1. A topological space $\langle X, T \rangle$ is a T_1-space iff it satisfies the T_1-axiom of separation: $\forall x, y \in X$

$$x \neq y \rightarrow (\exists N(x))[y \notin N(x)] \wedge (\exists N(y))[x \notin N(y)].$$

Remark. For the results we wish to prove we first define filter and ultrafilter for partial order structures.

Definition 5.2. Let $\mathbf{P} = \langle P, \leq \rangle$ be a partial order structure and let F be a nonempty subset of P. Then F is a *filter* for $\mathbf{P}$ iff

1. F is strongly compatible i.e., $(\forall x, y \in F)(\exists z \in F)[z \leq x \wedge z \leq y]$.
2. F is upward hereditary i.e., $(\forall x \in F)(\forall y \in P)[x \leq y \rightarrow y \in F]$.

Remark. From Definitions 2.2 and 5.2 and from Theorem 2.4 we see that if G is $\mathbf{P}$-generic over M, with M a standard transitive model of ZF, then G is a filter for $\mathbf{P}$. In fact G is an ultrafilter in the following sense.

Definition 5.3. F is an ultrafilter for the partial order structure $\mathbf{P}$ iff F is a maximal filter i.e., F is a filter for $\mathbf{P}$ and for each filter F'

$$F \subseteq F' \rightarrow F = F'.$$

Remark. Note that an ultrafilter for a partial order structure $\mathbf{P} = \langle P, \leq \rangle$ need not be a proper filter, i.e., P could be an ultrafilter. Indeed if P is compatible P is an ultrafilter.

We next establish a connection between filters for Boolean algebras and filters for partial order structures.

Definition 5.4. Let $\mathbf{B} = \langle B, +, \cdot, ^{-}, \mathbf{0}, \mathbf{1} \rangle$ be a Boolean algebra with natural order $\leq$ (see Definition 1.5). Let $P = B - \{\mathbf{0}\}$ and $\mathbf{P} = \langle P, \leq \rangle$. Then $\mathbf{P}$ is the partial order structure associated with $\mathbf{B}$.

Theorem 5.5. Let $\mathbf{B} = \langle B, +, \cdot, ^{-}, \mathbf{0}, \mathbf{1} \rangle$ be a Boolean algebra and $\mathbf{P} = \langle P, \leq \rangle$ be its associated partial order structure. If F is a nonempty subset of $|\mathbf{B}|$ then F is a proper filter for the Boolean algebra $\mathbf{B}$ iff F is a filter for the partial order structure $\mathbf{P}$.

Proof. Let F be a proper filter for $\mathbf{B}$. Then $\mathbf{0} \notin F$ i.e., $F \subseteq B - \{\mathbf{0}\}$. If $x, y \in F$ then $xy \in F$ and hence there exists a $z \in F$, namely xy, such that $z \leq x$ and $z \leq y$. Thus F is a filter for $\mathbf{P}$.

Conversely let F be a filter for $\mathbf{P}$. If $x, y \in F$ then there is a $z \in F$ such that $z \leq x$ and $z \leq y$. Therefore $z \leq xy$ and since F is upward hereditary $xy \in F$. Furthermore since $\mathbf{0} \notin P$ it follows that $\mathbf{0} \notin F$ i.e., F is a proper filter for $\mathbf{B}$.

Definition 5.6. Let $\mathbf{P} = \langle P, \leq \rangle$ be a partial order structure and let $\mathbf{F}$ be the set of all ultrafilters for $\mathbf{P}$. Then

$$N(p) \triangleq \{F \in \mathbf{F} \mid p \in F\}, p \in P$$
$$\mathbf{T} \triangleq \{G \subseteq \mathbf{F} \mid (\forall F \in G)(\exists p \in P)[F \in N(p) \subseteq G]\}.$$

Theorem 5.7. $\langle \mathbf{F}, \mathbf{T} \rangle$ is a T_1-space.

Proof. First of all we shall show that $\langle \mathbf{F}, \mathbf{T} \rangle$ is a topological space. From Definition 5.6 it is clear that 0 and $\mathbf{F}$ are each open. Let G_1 and G_2 be open sets and let $F \in G_1 \cap G_2$. Then there exist p and p' such that

$$F \in N(p) \subseteq G_1$$

and

$$F \in N(p') \subseteq G_2.$$

Then $p \in F$, $p' \in F$ and hence there exists a $z \in F$ such that $z \leq p$ and $z \leq p'$. Therefore, since every ultrafilter is upward hereditary

$$F \in N(z) \subseteq N(p) \cap N(p') \subseteq G_1 \cap G_2,$$

and hence $G_1 \cap G_2$ is open.

It is clear that if each G_a, $a \in A$, is open then $\bigcup \{G_a \mid a \in A\}$ is also open. Thus $\langle \mathbf{F}, \mathbf{T} \rangle$ is a topological space.

Next we will show that $\langle \mathbf{F}, \mathbf{T} \rangle$ satisfies the T_1-axiom of separation. Let F_1 and F_2 be different elements of $\mathbf{F}$. From the maximality of F_1 and of F_2, there is a $p \in F_1 - F_2$ and a $p' \in F_2 - F_1$. Then $F_2 \notin N(p)$ and $F_1 \notin N(p')$ i.e. $\langle \mathbf{F}, \mathbf{T} \rangle$ is a T_1-space.

Remark. There exist examples of partial order structures such that the corresponding topological space $\langle \mathbf{F}', \mathbf{T}' \rangle$ is not Hausdorff.

Theorem 5.8. Let $\mathbf{P} = \langle P, \leq \rangle$ be the partial order structure associated with the Boolean algebra $\mathbf{B}$. Then $\langle \mathbf{F}, \mathbf{T} \rangle$ is a Hausdorff space.

Proof. Suppose not. Then there would exist distinct $F_1, F_2 \in \mathbf{F}$ such that

$$(\forall p_1 \in F_1)(\forall p_2 \in F_2)(\exists F \in \mathbf{F})[p_1 \in F \wedge p_2 \in F].$$

Then $(\exists r \in F)[r \leq p_1 \wedge r \leq p_2]$ i.e., $p_1 p_2 \neq \mathbf{0}$.

If $G = \{p \in P \mid (\exists p_1 \in F)(\exists p_2 \in F)[p_1 p_2 \leqslant p]\}$ then G is a filter for $\mathbf{P}$. For, if $p, q \in G$ then

$$(\exists p_1, q_1 \in F_1)(\exists p_2, q_2 \in F_2)[p_1 p_2 \leq p \wedge q_1 q_2 \leq q].$$

Since F_1 and F_2 are filters $p_1 q_1 \in F_1$ and $p_2 q_2 \in F_2$. Furthermore $p_1 q_1 p_2 q_2 \leq pq$. Then $pq \in G$ i.e., for each $p, q \in G$ there is an $r \in G$, namely pq, such that $r \leq p$ and $r \leq q$. Clearly G is upward hereditary.

Since G is a filter, since $F_1 \subseteq G$ and $F_2 \subseteq G$, and since F_1 and F_2 are distinct ultrafilters we have a contradiction. Therefore $\langle \mathbf{F}, \mathbf{T} \rangle$ is Hausdorff.

Remark. In order for F to be a filter for a partial order structure $\mathbf{P} = \langle P, \leq \rangle$ we require that F be strongly compatible (Definition 5.2). This raises a very natural question. Why do we not define a more general notion by requiring that F only be compatible? That is, instead of requiring F to satisfy 1 of Definition 5.2 why do we not require instead that F satisfy the weaker requirement

1′. $$(\forall x, y \in F)(\exists z \in P)[z \leq x \wedge z \in y]?$$

For purposes of discussion let us call filters as originally defined strong filters and filters as newly proposed, weak filters. The change from strong filter to weak filter also changes the notion of ultrafilter for being maximal among weak filters is a stronger restriction than being maximal among strong filters. There are two interesting consequences of this fact. If ultrafilters are maximal among weak filters then the sets $N(p)$ of Definition 5.6 form only a subbase for the topological space $\langle \mathbf{F}, \mathbf{T} \rangle$. Furthermore this space is Hausdorff. The fact that $\langle \mathbf{F}, \mathbf{T} \rangle$ satisfies the T_2-axiom of separation was first pointed out by H. Tanaka.

Nevertheless, for the work that comes later we need strong filters and we want ultrafilter to mean a strong filter that is maximal among strong filters. Thus, later use brings us back to the definition as given.

We do not know whether every T_1-space is homeomorphic to a topological space $\langle \mathbf{F}, \mathbf{T} \rangle$ associated with some partial order structure or whether every Hausdorff space is homeomorphic to a topological space $\langle \mathbf{F}', \mathbf{T}' \rangle$ associated with the partial order structure associated with a Boolean algebra.

Let $\mathbf{P} = \langle P, \leq \rangle$ be a partial order structure and let $\mathbf{F}$ the set of all ultrafilters for $\mathbf{P}$. In order to investigate some relations between the topologies on $\mathbf{P}$ and $\mathbf{F}$ we introduce the following notation.

Definition 5.9. Let G_1 be an open subset of P and G_2 be an open subset of $\mathbf{F}$. Then

$$G_1^* \triangleq \bigcup \{N(p) \mid [p] \subseteq G_1\}$$
$$G_2^\Delta \triangleq \bigcup \{[p] \mid N(p) \subseteq G_2\}.$$

Remark. Clearly G_1^* and G_2^Δ are open subsets of $\mathbf{F}$ and P respectively.

Theorem 5.10. If G_1 and G_2 are open subsets of P and $\mathbf{F}$ respectively then

1. $G_1 \subseteq G_1^{*\Delta}$.
2. $G_2 \subseteq G_2^{\Delta *}$.

Proof.

1. $a \in G_1 \to [a] \subseteq G_1$
$\to N(a) \subseteq G_1^*$
$\to [a] \subseteq G_1^{*\Delta}$ (since G_1^* is an open subset of $\mathbf{F}$)
$\to a \in G_1^{*\Delta}$.
2. $F \in G_2 \to (\exists a \in F)[N(a) \subseteq G_2]$
$\to (\exists a \in F)[[a] \subseteq G_2^\Delta]$
$\to (\exists a \in F)[N(a) \subseteq G_2^{\Delta *}]$ (since G_2^Δ is an open subset of P)
$\to F \in G_2^{\Delta *}$.

Theorem 5.11. 1. If G_1 and G_2 are open subsets of P then

$$G_1 \subseteq G_2 \to G_1^* \subseteq G_2^*.$$

2. If G_1 and G_2 are open subsets of $\mathbf{F}$ then

$$G_1 \subseteq G_2 \to G_1^\Delta \subseteq G_2^\Delta.$$

Proof. Left to the reader.

Theorem 5.12. If G is an open subset of $\mathbf{F}$ and $[a] \subseteq G^\Delta$, then $N(a) \subseteq G$.

Proof.

$$\begin{aligned}[a] \subseteq G^\Delta &\to a \in G^\Delta \\ &\to (\exists b)[N(b) \subseteq G \wedge a \in [b]] \\ &\to (\exists b \geq a)[N(b) \subseteq G] \\ &\to (\exists b)[N(a) \subseteq N(b) \subseteq G] \\ &\to N(a) \subseteq G.\end{aligned}$$

Theorem 5.13. If G is a regular open subset of P then $G^{*\Delta} = G$.

Proof.

$$\begin{aligned}a \in G^{*\Delta} &\to [a] \subseteq G^{*\Delta} \\ &\to N(a) \subseteq G^* \\ &\to (\forall F)[a \in F \to F \in G^*] \\ &\to (\forall F)[a \in F \to (\exists b \in F)[[b] \subseteq G]].\end{aligned}$$

For each $c \leq a$ there is an ultrafilter F' for **P** such that $c \in F'$. But then, since $c \leq a$ and $c \in F'$ we have $a \in F'$. Consequently if $a \in G^{*\Delta}$ then $\exists b \in F' \cap G$. Since F' is an ultrafilter for **P** and since both c and b are in F'

$$(\exists b' \in F)[b' \leq c \wedge b' \leq b].$$

But $b \in G \wedge b' \leq b$. Therefore $b' \in G$ i.e.,

$$\begin{aligned} a \in G^{*\Delta} &\to (\forall c \leq a)(\exists b \leq c)[b \in G] \\ &\to (\forall c \leq a)[[c] \cap G \neq 0] \\ &\to (\forall c \leq a)[c \in G^{-}] \\ &\to [a] \subseteq G^{-} \\ &\to a \in G^{-0} \\ &\to a \in G. \end{aligned}$$

Then by Theorem 5.10, $G^{*\Delta} = G$.

Theorem 5.14. If G is an open subset of **F** then $G^{\Delta *} = G$.

Proof.

$$\begin{aligned} F \in G^{\Delta *} &\to (\exists a \in F)[[a] \subseteq G^{\Delta}] \\ &\to (\exists a \in F)[N(a) \subseteq G] \\ &\to F \in G. \end{aligned}$$

Therefore by Theorem 5.10, $G^{\Delta *} = G$.

Theorem 5.15. Let G_1 and G_2 be open sets of a topological space $\langle X, T \rangle$. If for each regular open set H

$$G_1 \cap H = 0 \to G_2 \cap H = 0$$

then $G_2 \subseteq G_1{}^{-0}$.

Proof. If $H = (X - G_1)^0$ then H is regular open. If $G_1 \cap H = 0$ then $G_2 \cap H = 0$ and hence

$$G_2 \cap H^{-} = 0.$$

Therefore

$$G_2 \subseteq X - H^{-} = G_1{}^{-0}.$$

Theorem 5.16. 1. If G_1 is an open subset of P then

$$G_1^* = 0 \to G_1 = 0.$$

2. If G_2 is an open subset of **F**

$$G_2^{\Delta} = 0 \to G_2 = 0.$$

Proof. Left to the reader.

Theorem 5.17. 1. If G is a regular open subset of **F** then G^{Δ} is regular open.

2. If G is a regular open subset of P then G^* is regular open.

Proof. 1. Let $G_1 = (G^\Delta)^{-0}$. Then

$$G = G^{\Delta *} \subseteq (G^\Delta)^{-0*} = G_1^*.$$

If G_2 is regular open and $G \cap G_2 = 0$ then

$$(G^\Delta \cap G_2^\Delta)^* \subseteq G^{\Delta *} \cap G_2^{\Delta *} = G \cap G_2 = 0.$$

Therefore $G^\Delta \cap G_2^\Delta = 0$ and hence $G_1 \cap G_2^\Delta = 0$. Furthermore

$$(G_1^* \cap G_2)^\Delta \subseteq G_1^{*\Delta} \cap G_2^\Delta = G_1 \cap G_2^\Delta = 0.$$

Thus $G_1^* \cap G_2 = 0$ and hence, by Theorem 5.15, $G_1^* \subseteq G^{-0} = G$. Consequently

$$G^\Delta = G_1^{*\Delta} = G_1$$

i.e., G^Δ is regular open.

2. Let $G_2 = (G^*)^{-0}$. Then

$$G \subseteq G^{*\Delta} \subseteq G_2^\Delta.$$

If H is regular open and $G \cap H = 0$ then

$$(G^* \cap H^*)^\Delta \subseteq G^{*\Delta} \cap H^{*\Delta} = G \cap H = 0.$$

Therefore $G^* \cap H^* = 0$ and hence $G_2 \cap H^* = 0$. Furthermore

$$(G_2^\Delta \cap H)^* \subseteq G_2^{\Delta *} \cap H^* = G_2 \cap H^* = 0.$$

Consequently $G_2^\Delta \cap H = 0$ and hence $G_2^\Delta \subseteq G^{-0} = G$. Thus, by Theorem 5.14

$$G^* = G_2^{\Delta *} = G_2$$

i.e., G^* is regular open.

Remark. From the foregoing theorems we obtain the following result.

Theorem 5.18. If $\mathbf{P} = \langle P, \leq \rangle$ is a partial order structure, then the Boolean algebra $\mathbf{B}$ of regular open subsets of P is isomorphic to the Boolean algebra of regular open subsets of $\mathbf{F}$.

Proof. The mapping $*$ is a one-to-one, order preserving mapping from the first algebra onto the second.

Remark. As you will see later, it is useful to consider the Boolean algebra of all regular open sets of a product topological space. So we shall show a general theorem about that. If a partial order structure $\mathbf{P} = \langle P, \leq \rangle$ has a greatest element, then we denote it by 1: $(\forall p \in P)[p \leq 1]$. In case P has an element 1, let $P_0 = P - \{1\}$ and $\mathbf{P}_0 = \langle P_0, \leq \rangle$. Then clearly the Boolean algebra of all regular open subsets of $\mathbf{P}$ is isomorphic to that of $\mathbf{P}_0$. Consequently, with regard to Boolean algebras of regular open subsets of partial order structures, we may assume that the partial order structures have a greatest element 1.

Definition 5.19. Let $\mathbf{P}_i = \langle P_i, \leq \rangle$, $i \in I$, (I an index set) be a partial order structure having a greatest element 1_i. Then the product structure $\mathbf{P} \triangleq \prod_{i \in I} \mathbf{P}_i \triangleq \langle P, \leq \rangle$ is the following partial order structure.

1. $P \triangleq \left\{ p \in \prod_{i \in I} P_i \mid p(i) = 1_i \text{ for all but finitely many } i\text{'s} \right\}$.
2. $(\forall p, q \in P)[p \leq q \overset{\Delta}{\leftrightarrow} (\forall i \in I)[p(i) \leq q(i)]]$.
3. $1 \triangleq$ the unique $p \in P$ such that $(\forall i \in I)[p(i) = 1_i]$.

Theorem 5.20. Let $\mathbf{P} = \prod_{i \in I} \mathbf{P}_i$ be given as above and let $\mathbf{F}$ and $\mathbf{F}_i$ be the T_1-spaces corresponding to $\mathbf{P}$, and $\mathbf{P}_i$ respectively in accordance with Definition 5.6. Then $\mathbf{F}$ is homeomorphic to the product space $\prod_{i \in I} \mathbf{F}_i$.

Proof. For a given $i \in I$ and an element $a_i \in P_i$, let $\hat{a}_i$ be the element of P whose ith projection is a_i and whose jth projection is 1_j for $j \neq i$, i.e.,

$$\hat{a}_i(i) = a_i, \ \hat{a}_i(j) = 1_j \quad \text{for} \quad j \neq i.$$

For each $F \in \mathbf{F}$, let $F_i = \{a_i \in P_i \mid \hat{a}_i \in F\}$. Then F_i is a filter for $\mathbf{P}_i$ since $a_i \leq b_i$ implies $\hat{a}_i \leq \hat{b}_i$. Furthermore F_i is maximal: If G_i is a filter for $\mathbf{P}_i$ such that $F_i \subseteq G_i$ and if $G = \{a \in P \mid a(i) \in G_i \wedge (\exists x \in F)(\forall j \neq i)[a(j) \geq x(j)]\}$, then G is a filter and $F \subseteq G$. But F is an ultrafilter. Therefore $G = F$. For each $a \in G_i$ and each $x \in F$ we define a b as follows

$$b(i) = a, \qquad b(j) = x(j) \quad \text{for} \quad j \neq i.$$

Then $b \in G = F$. Hence $\widehat{b(i)} \in F$ and $a \in F_i$, i.e., $G_i \subseteq F_i$. Consequently $F_i = G_i$, that is, F_i is an ultrafilter for $\mathbf{P}_i$.

Thus for each $F \in \mathbf{F}$ and each $i \in I$, F_i is an ultrafilter for $\mathbf{P}_i$, i.e., $F_i \in \mathbf{F}_i$. From this fact we then define a mapping $g: \mathbf{F} \to \prod_{i \in I} \mathbf{F}_i$ by

$$g(F) = \langle F_i \rangle_{i \in I}.$$

The function g is both one-to-one and onto (surjective). To prove this we need only show that each I-sequence, $\langle F_i \rangle_{i \in I}$ uniquely determines an F for which $g(F) = \langle F_i \rangle_{i \in I}$. First we note that if

$$F = \{a \in P \mid (\exists i_1, \cdots i_n \in I)(\exists a_{i_1} \in F_{i_1}) \cdots (\exists a_{i_n} \in F_{i_n})[\hat{a}_{i_1} \cdots \hat{a}_{i_n} \leq a]\}$$

where $\hat{a}_{i_1} \cdots \hat{a}_{i_n}(j) = a_{i_k}$ if $j = i_k$ for some k and $\hat{a}_{i_1} \cdots \hat{a}_{i_n}(j) = 1$ otherwise, then F is an ultrafilter for $\mathbf{P}$ and $g(F) = \langle F_i \rangle_{i \in I}$. Thus g is onto. Second if $\hat{F}_i = \{\hat{a} \mid a \in F_i\}$ and if $g(F) = \langle F_i \rangle_{i \in I}$ then F is the smallest filter for which $\hat{F}_i \subseteq F$, for each $i \in I$. Therefore g is one-to-one.

Now for a given $a \in P$ consider $N(a) = \{F \in \mathbf{F} \mid a \in F\}$. Since $a = \hat{a}_{i_1} \cdots \hat{a}_{i_n}$ for some $a_{i_1} \in P_{i_1}, \ldots, a_{i_n} \in P_{i_n}$, we have

$$\begin{aligned} F \in N(a) &\leftrightarrow a \in F \\ &\leftrightarrow \hat{a}_{i_1} \in F \wedge \cdots \wedge \hat{a}_{i_n} \in F \\ &\leftrightarrow F \in N(\hat{a}_{i_1}) \cap \cdots \cap N(\hat{a}_{i_n}), \end{aligned}$$

i.e., $g``N(a) = N(\hat{a}_{i_n}) \cap \cdots \cap N(\hat{a}_{i_n})$. Therefore g is a topological mapping.

Remark. When we consider regular open sets of a partial order structure **P**, the following notion is very useful.

Definition 5.21. A partial order structure $\mathbf{P} = \langle P, \leq \rangle$ (or a partially ordered set P) is called *fine*, if the following condition is satisfied:

$$(\forall p, q \in P)[q \nleq p \to (\exists r \in P)[r \leq q \wedge \neg \operatorname{Comp}(r, p)]].$$

Lemma 5.22. If **P** is fine, then for each $p \in P$

$$[p]^{-0} = [p].$$

Proof. We have only to show $[p]^{-0} \subseteq [p]$. Let $q \in P$ such that $q \notin [p]$, i.e., $q \nleq p$. Then, by Definition 5.21, $(\exists r \in P)[r \leq q \wedge \neg \operatorname{Comp}(r, p)]$. Therefore, $[r] \cap [p] = 0$ and hence $r \notin [p]^-$. This implies $[q] \nsubseteq [p]^-$. Consequently $q \notin [p]^{-0}$, i.e., if $q \in [p]^{-0}$ then $q \in [p]$.

Remark. Many **P**'s used in later sections are fine.

6. Boolean-Valued Structures

The notion of a Boolean-valued structure is obtained from the definition of an ordinary 2-valued structure by replacing the Boolean algebra **2** of two truth values "truth" and "falsehood" by any complete Boolean algebra **B**. While some of the basic definitions and theorems can be generalized to the **B**-valued case almost mechanically the intuitive ideas behind these general notions are more difficult to perceive.

Throughout this section $\mathbf{B} = \langle B, +, \cdot, ^{-}, \mathbf{0}, \mathbf{1} \rangle$ denotes a complete Boolean algebra.

Definition 6.1. If $\mathscr{L}$ is a first order language with individual constants

$$c_0, c_1, \ldots, c_i, \ldots \qquad i < \alpha$$

and predicate constants

$$R_0, R_1, \ldots, R_j, \ldots \qquad j < \beta.$$

Then a **B**-valued interpretation of $\mathscr{L}$ is a pair $\langle A, \phi \rangle$ where A is a nonempty set and ϕ is a mapping defined on the set of constants of the language $\mathscr{L}$ satisfying the following,

1. $\phi(c_i) \in A$, $i < \alpha$.
2. $\phi(R_j)\colon A^{n_j} \to B$, for $j < \beta$ where n_j is the number of arguments of R_j.

Remark. In order to define a truth value for closed formulas of $\mathscr{L}$ under a given **B**-valued interpretation we first extend $\mathscr{L}$ to a new language $\mathscr{L}^* \triangleq \mathscr{L}(C(A))$ by introducing new individual constants c_a for each $a \in A$.

Definition 6.2. If φ is a closed formula of $\mathscr{L}^*$ then $[\![\varphi]\!]$ is an element of B defined recursively in the following way

1. $[\![R_j(c_1, \ldots, c_{n_j})]\!] \triangleq \phi(R_j)(\phi(c_1), \ldots, \phi(c_{n_j}))$ for every finite sequence of constants $c_1, \ldots, c_{n_j}$ of $\mathscr{L}^*$ and $\phi(c_a) = a$ for the new constants c_a, $a \in A$.

2. $[\![\neg\psi]\!] \triangleq {}^{-}[\![\psi]\!]$.
3. $[\![\psi \wedge \eta]\!] \triangleq [\![\psi]\!]\,[\![\eta]\!]$.
4. $[\![\psi \vee \eta]\!] \triangleq [\![\psi]\!] + [\![\eta]\!]$.
5. $[\![(\forall x)\psi(x)]\!] \triangleq \prod_{a \in A} [\![\psi(c_a)]\!]$.
6. $[\![(\exists x)\psi(x)]\!] \triangleq \sum_{a \in A} [\![\psi(c_a)]\!]$.

Remark. For the special case $\mathbf{B} = \mathbf{2}$, Definition 6.2 is the usual definition of satisfaction in ordinary 2-valued logic.

We will write $[\![\psi(a_1, \ldots, a_n)]\!]$ for $[\![\psi(c_{a_1}, \ldots, c_{a_n})]\!]$, $\bar{R}_j$ for $\phi(R_j)$ and $\bar{c}$ for $\phi(c)$. If $\langle A, \phi \rangle$ is a $\mathbf{B}$-valued interpretation then

$$\mathbf{A} \overset{\Delta}{=} \langle A, \bar{R}_0, \bar{R}_1, \ldots, \bar{c}_0, \bar{c}_1, \ldots \rangle$$

is a $\mathbf{B}$-valued structure.

We will occasionally consider several interpretations in the same context. We will write $[\![\varphi]\!]_{\mathbf{A}}$ to indicate that element of B determined by φ and the interpretation $\langle A, \phi \rangle$ of the $\mathbf{B}$-valued structure $\mathbf{A}$.

Definition 6.3. $\mathbf{A} \models \varphi \overset{\Delta}{\leftrightarrow} [\![\varphi]\!] = \mathbf{1}$.

Remark. $\mathbf{A} \models \varphi$ is read "$\mathbf{A}$ satisfies φ" or "φ is true in $\mathbf{A}$."

The usual axioms of the predicate calculus are also valid in every $\mathbf{B}$-valued structure as we now show.

Theorem 6.4. If φ is a closed formula of the language $\mathscr{L}$ then φ is satisfied in every $\mathbf{B}$-valued structure iff φ is logically valid.

Proof. φ is logically valid iff φ is satisfied in every $\mathbf{2}$-valued structure. Since $\mathbf{2}$ is a complete subalgebra of $\mathbf{B}$ every $\mathbf{2}$-valued structure is a $\mathbf{B}$-valued structure. Conversely if φ is not satisfied by some $\mathbf{B}$-valued structure $\mathbf{A}$ i.e.,

$$[\![\varphi]\!]_{\mathbf{A}} = b \neq \mathbf{1}$$

then the computation of $[\![\varphi]\!]_{\mathbf{A}}$ requires only a finite number of applications of Definition 6.2 (5), (6), say

$$b_1 = \prod_{\alpha \in A_1} b_{1\alpha}, \ldots, b_n = \prod_{\alpha \in A_n} b_{n\alpha}$$

$$b_1' = \sum_{\alpha \in B_1} b_{1\alpha}', \ldots, b_m' = \sum_{\alpha \in B_n} b_{n\alpha}'.$$

By the Rasiowa-Sikorski Theorem (Theorem 2.14) there is a homomorphism $h: |\mathbf{B}| \to |\mathbf{2}|$ such that $h(b) = \mathbf{0}$ and h preserves sums and hence products. If $\langle A, \phi \rangle$ is the $\mathbf{B}$-valued interpretation that determines $\mathbf{A}$ and $\phi' \overset{\Delta}{=} h \circ \phi$ then $\langle A, \phi' \rangle$ determines a $\mathbf{2}$-valued structure $\mathbf{A}'$. Since h is a homomorphism that preserves sup's and inf's, and $h(b) = \mathbf{0}$, it follows that φ does not hold in $\mathbf{A}'$.

Exercises.

1. Let $\mathbf{B}_1, \mathbf{B}_2$ be complete Boolean algebras and h a complete homomorphism of $\mathbf{B}_1$ into $\mathbf{B}_2$. If $\langle A, \phi_1 \rangle$ is a $\mathbf{B}_1$-interpretation, if $\phi_2 = h \circ \phi_1$ and $\phi_2 = \phi_1$ on the constants of $\mathscr{L}$ then $\langle A, \phi_2 \rangle$ is a $\mathbf{B}_2$-interpretation and for each closed formula φ of the language $\mathscr{L}$

$$[\![\varphi]\!]_{\mathbf{A}_2} = h([\![\varphi]\!]_{\mathbf{A}_1})$$

where $\mathbf{A}_1$ is the structure determined by $\langle A, \phi_1 \rangle$.

2. Let $\mathbf{P}$ be a partial order structure, and $\mathbf{B}$ the complete Boolean algebra of all regular open sets in $\mathbf{P}$. If π is an automorphism of $\mathbf{P}$ then π induces an

automorphism $\tilde{\pi}$ of **B**. ($\pi: P \to P$ is an automorphism iff π is one-to-one and onto.) If in addition

$$(\forall p_1, p_2 \in P)(\exists \pi)[\pi \text{ is an automorphism of } \mathbf{P} \wedge \operatorname{Comp}(p_1, \pi(p_2))]$$

then

$$(\forall b_1, b_2 \in B - \{\mathbf{0}\})\ (\exists \pi)[\pi \text{ is an automorphism of } \mathbf{B} \wedge b_1\pi(b_2) \neq \mathbf{0}].$$

Remark. We turn next to languages with equality. As in the 2-valued case we have special axioms for equality that are easily generalized for the **B**-valued case.

Definition 6.5. If $\mathscr{L}$ is a first order language with equality then by a **B**-valued interpretation of $\mathscr{L}$ we mean a **B**-valued interpretation in the sense of Definition 6.1 that in addition satisfies the following Axioms of Equality

1. $[\![c = c]\!] = \mathbf{1}$.
2. $[\![c_1 = c_2]\!] = [\![c_2 = c_1]\!]$.
3. $[\![c_1 = c_2]\!][\![c_2 = c_3]\!] \leq [\![c_1 = c_3]\!]$.
4. For each n-ary predicate constant R of the language

$$[\![c_1 = c_1']\!] \cdots [\![c_n = c_n']\!][\![R(c_1, \ldots, c_n)]\!] \leq [\![R(c_1', \ldots, c_n')]\!].$$

Theorem 6.6. Every logically valid sentence of a first order language $\mathscr{L}$ with equality is satisfied by a **B**-valued structure of $\mathscr{L}$.

In particular

$$[\![c_1 = c_1']\!] \cdots [\![c_n = c_n']\!][\![\varphi(c_1, \ldots, c_n)]\!] \leq [\![\varphi(c_1', \ldots, c_n')]\!].$$

Remark. Note that $[\![c_1 = c_2]\!]$ may be different from $\mathbf{0}$ and from $\mathbf{1}$. Also, we may have $[\![c_1 = c_2]\!] = \mathbf{1}$ but $c_1 \neq c_2$. To exclude this last possibility we introduce the separated **B**-valued structures.

Definition 6.7. A **B**-valued structure $\mathbf{A} = \langle A, \bar{=}, \bar{R}_0, \ldots, \bar{c}_0, \ldots \rangle$ is *separated* iff

$$(\forall a_1, a_2 \in A)[[\![a_1 = a_2]\!] = \mathbf{1} \to a_1 = a_2].$$

Remark. Every **B**-valued structure **A** is equivalent to a separated **B**-valued structure $\langle \hat{A}, \hat{=}, \hat{R}_0, \ldots, \hat{c}_0, \ldots \rangle$ obtained from **A** by considering the equivalence classes of the relation $\{\langle a, b\rangle \in A^2 \mid [\![a = b]\!] = \mathbf{1}\}$. If $\hat{A}$ is the set of these equivalence classes there are **B**-valued relations , $\hat{=}$ $\hat{R}_0, \ldots,$ on $\hat{A}$ and members $\hat{c}_0, \ldots$ of $\hat{A}$ (which are uniquely determined) such that for every formula φ of $\mathscr{L}$ and any $a_1, \ldots, a_n \in A$

$$[\![\varphi(a_1, \ldots, a_n)]\!]_{\mathbf{A}} = [\![\varphi(\hat{a}_1, \ldots, \hat{a}_n)]\!]_{\hat{\mathbf{A}}}$$

where $\hat{a}_i$ is the equivalence class containing a_i.

Definition 6.8. A *partition of unity* is an indexed family $\langle b_i \mid i \in I\rangle$ of elements of B such that

$$\sum_{i \in I} b_i = \mathbf{1} \wedge (\forall i, j \in I)[i \neq j \to b_i b_j = \mathbf{0}].$$

A **B**-valued structure $\mathbf{A} = \langle A, \equiv, \bar{R}_0, \ldots, \bar{c}_0, \ldots \rangle$ is *complete* iff whenever $\langle b_i \mid i \in I \rangle$ is a partition of unity and $\langle a_i \mid i \in I \rangle$ is any family of elements of A then

$$(\exists a \in A)(\forall i \in I)[b_i \leq [\![a = a_i]\!]].$$

Remark. The set a, of Definition 6.8, is unique in the following sense.

Theorem 6.9. If $\langle b_i \mid i \in I \rangle$ is a partition of unity, if

$$(\forall i \in I)[b_i \leq [\![a = a_i]\!]]$$

and

$$(\forall i \in I)[b_i \leq [\![a' = a_i]\!]]$$

then $[\![a = a']\!] = \mathbf{1}$.

Proof. $b_i \leq [\![a = a_i]\!][\![a' = a_i]\!] \leq [\![a = a']\!]$, $i \in I$

$$\mathbf{1} = \sum_{i \in I} b_i \leq [\![a = a']\!] \leq \mathbf{1}.$$

Hence $[\![a = a']\!] = \mathbf{1}$.

Remark. This unique a is sometimes denoted by $\sum_{i \in I} b_i a_i$. We next provide an example of a **B**-valued structure that is separated and complete.

Example. Define $\mathbf{A} = \langle B, \equiv \rangle$ by

$$[\![b_1 = b_2]\!] = b_1 b_2 + (^-b_1)(^-b_2)$$

i.e., $[\![b_1 = b_2]\!]$ is the Boolean complement of the symmetric difference of b_1 and b_2.

It is easily proved that $\mathbf{A}$ is a **B**-valued structure that satisfies 1–3 of Definition 6.5. Since

$$\begin{aligned}
[\![b_1 = b_2]\!] = \mathbf{1} &\rightarrow b_1 b_2 + (^-b_1)(^-b_2) = \mathbf{1} \\
&\rightarrow b_1 b_2 = {}^-(^-b_1)(^-b_2) = b_1 + b_2 \\
&\rightarrow b_1 b_2 = b_1 \wedge b_2 b_1 = b_2 \\
&\rightarrow b_2 \leq b_1 \wedge b_1 \leq b_2 \\
&\rightarrow b_1 = b_2,
\end{aligned}$$

$\mathbf{A}$ is separated.

If $\langle b_i \mid i \in I \rangle$ is a partition of unity and $\langle a_i \mid i \in I \rangle$ is a family of sets of B then since **B** is complete

$$a \triangleq \sum_{i \in I} b_i a_i \in B.$$

Then

$$\begin{aligned}
b_i [\![a = a_i]\!] &= b_i a_i \sum_{j \in I} a_j b_j + b_i(^-a_i)\left(^-\sum_{j \in I} a_j b_j\right) \\
&= (a_i b_i + (^-a_i) b_i)\left(\sum_{j \in I} a_j b_j + {}^-\sum_{j \in I} a_j b_j\right).
\end{aligned}$$

But the b_i's are pairwise disjoint. Then

$$({}^{-}a_i)b_i \sum_{j \in I} a_j b_j = ({}^{-}a_i)a_i b_i = 0$$

$$a_i b_i \left({}^{-}\sum_{j \in I} a_j b_j \right) = 0.$$

Then

$$b_i[\![a = a_i]\!] = a_i b_i + ({}^{-}a_i)b_i = b_i.$$

Therefore $(\forall i \in I)[b_i \le [\![a = a_i]\!]]$ i.e., **A** is complete.

7. Relative Constructibility

Godel's constructibility was generalized, in a natural way, by Levy and Shoenfield to a relative constructibility which assures us of the existence of a standard transitive model $L[a]$ of ZF for each set a. Levy-Shoenfield's relative constructibility is rather narrow but quite easily generalized. In this section we will study a general theory of relative constructibility and deal with several basic relative constructibilities as special cases. Later we will extend our relative constructibility to Boolean valued relative constructibility from which we will in turn define forcing.

There is a modern tendency to avoid the rather cumbersome theory of relative constructibility. We believe this to be a mistake. Although we do not pursue the subject, it is clear that one can consider wider and wider types of relative constructibility. Accordingly, we have many types of Boolean valued relative constructibility. We feel that these sometimes wild Boolean valued relative constructibilities might be very important for future work. Indeed, it is not at all clear whether the structures they produce can be constructed by the usual method of Scott-Solovay's Boolean valued models without using relative constructibility.

If a and b are sets there are two different definitions of the notion "b is constructible from a" namely $b \in L_a$ or $b \in L[a]$ where

L_a is the smallest class M satisfying

1. M is a standard transitive model of ZF.
2. $On \subseteq M$.
3. $(\forall x \in M)[x \cap a \in M]$.

$L[a]$ is the smallest class M satisfying

1. M is a standard transitive model of ZF.
2. $On \subseteq M$.
3. $a \in M$.

Obviously, $L_a \subseteq L[a]$.

In this section we will show, by a modification of Gödel's methods used to define the class L of constructible sets, that the classes L_a and $L[a]$ exist. It should be noted that neither the characterization of L_a nor of $L[a]$ can be formalized in ZF.

The main difference between L_a and $L[a]$ as we will see is that L_a satisfies the AC while $L[a]$ need not. Since we will eventually wish to prove the independence of the AC from the axioms of ZF using results of this and later sections we must exercise care to avoid the use of the AC in proving the following results.

It is of interest to consider a slightly more general situation allowing a to be a proper class A. L_A can be characterized exactly as L_a was. For $L[A]$ there is however a problem in that we cannot have $A \in M$. Instead we define:

$$L[A] = \bigcup_{\alpha \in 0_n} L[A \cap R'\alpha].$$

We first develop a general theory that allows us to treat L_A and $L[A]$ simultaneously. Let $\mathscr{L}$ be a language with predicate constants

$$R_0, \ldots, R_n$$

and individual constants

$$c_0, \ldots, c_m.$$

Some results of this section remain true if we allow $\mathscr{L}$ to have an arbitrary well-ordered set (possible even uncountably many) of constants.

Definition 7.1. If $\mathbf{A} = \langle A, R_0^{\mathbf{A}}, \ldots, R_n^{\mathbf{A}}, c_0^{\mathbf{A}}, \ldots, c_m^{\mathbf{A}} \rangle$ and

$$\mathbf{B} = \langle B, R_0^{\mathbf{B}}, \ldots, R_n^{\mathbf{B}}, c_0^{\mathbf{B}}, \ldots, c_m^{\mathbf{B}} \rangle$$

are two structures for the language $\mathscr{L}$, then $\mathbf{A}$ is a substructure of $\mathbf{B}$, ($\mathbf{A} \subseteq \mathbf{B}$) iff

1. $A \subseteq B$.
2. For each R_i, $i = 0, \ldots, n$ if R_i is p-ary then $\forall a_1, \ldots, a_p \in A$

$$R_i^{\mathbf{A}}(a_1, \ldots, a_p) \leftrightarrow R_i^{\mathbf{B}}(a, \ldots, a_p).$$

3. $c_j^{\mathbf{A}} = c_j^{\mathbf{B}} \qquad j = 0, \ldots, m.$

Exercise. If $\mathbf{B} = \langle B, R_0^{\mathbf{B}}, \ldots, R_n^{\mathbf{B}}, c_0^{\mathbf{B}}, \ldots, c_m^{\mathbf{B}} \rangle$ is a structure for $\mathscr{L}$ if $A \subseteq B$ and $c_j \in A, j \leq m$, then there is a unique substructure $\mathbf{A} \subseteq \mathbf{B}$ such that $|\mathbf{A}| = A$ ($|\mathbf{A}|$ denotes the universe of $\mathbf{A}$). This structure we denote by $\mathbf{B} \ulcorner A$.

Definition 7.2. $C(A) \triangleq \{c_a \mid a \in A\}$. $\mathscr{L}(C(A))$ is the language obtained from $\mathscr{L}$ by adding c_a for each $a \in A$ as new individual constants. $\mathscr{L}_0$ is always understood to be the first order language whose only constant is $\in$.

Remark. Hereafter we assume that $R_0 = \in$, i.e., $\mathscr{L}$ is an extension of $\mathscr{L}_0$. We will be mostly interested in structures $\langle A, \in, \ldots \rangle$ where A is transitive. In this case we do not list $\in$ explicitly. In particular we call a structure $\langle M, \bar{\in} \rangle$ for $\mathscr{L}_0$ transitive iff M is transitive and $\bar{\in} = \in$. In this case we write M for $\langle M, \in \rangle$.

For the following we assume a suitable Gödelization of the formulas of $\mathscr{L}(C(A))$ in ZF and a formalization of several syntactical and semantical notions meeting certain requirements on definability and absoluteness with respect to transitive models of ZF. In particular there is a formula F of $\mathscr{L}_0$ involving A as a constant such that $F(x)$ formalizes the notion

"x is the Gödel number of a formula of $\mathscr{L}(C(A))$"

and such that F is absolute with respect to any transitive model M of ZF for which $A \in M$. We then define

$Fml(\mathbf{A}) \triangleq \{x \mid F(x)\}$.

$C(x)$: "x is the Gödel number of a closed *wff* of $\mathscr{L}(C(A))$."

$\bar{F}(x)$: "x is the Gödel number of a *wff* of $\mathscr{L}(C(A))$ having at most one free variable."

$Fml^0(\mathbf{A}) \triangleq \{x \mid C(x)\}$.

$Fml^1(\mathbf{A}) \triangleq \{x \mid \bar{F}(x)\}$.

Definition 7.3. A class A is definable in $\mathscr{L}_0$ iff there is a formula $\varphi(x)$ of $\mathscr{L}_0$ containing no free variable other than x such that $A = \{x \mid \varphi(x)\}$. In this case φ is called a defining formula for A. Moreover, if φ contains parameters $y_1, \ldots, y_k$ from a given set c then we say A is definable in $\mathscr{L}_0$ from c.

Definition 7.4. Let M be a standard transitive model of ZF and let $\varphi(x_1, \ldots, x_n)$ be a formula of $\mathscr{L}_0$ containing no free variable other than $x_1, \ldots, x_n$. Then φ is absolute with respect to M iff

$$(\forall x_1, \ldots, x_n \in M)[\varphi(x_1, \ldots, x_n) \leftrightarrow \varphi^M(x_1, \ldots, x_m)],$$

where φ^M is the formula obtained from φ by replacing $\exists y$ and $\forall y$ by $\exists y \in M$ and $\forall y \in M$ respectively. Moreover if φ contains a set $c \in M$, $\varphi(c, x_1, \ldots, x_n)$, then we say that φ is absolute with respect to M regarding c as a constant.

A class A definable in $\mathscr{L}_0$, is absolute with respect to M iff its defining formula is absolute with respect to M.

Theorem 7.5. If $\mathbf{A} = \langle A, \bar{R}_0, \ldots, \bar{R}_n, \bar{c}_0, \ldots, \bar{c}_m\rangle$ is a transitive structure for $\mathscr{L}$, where A is a set, then there is a wff ψ of $\mathscr{L}_0$ such that for every closed wff φ of $\mathscr{L}(C(A))$

$$\mathbf{A} \models \varphi \leftrightarrow \psi(\mathbf{A}, \ulcorner\varphi\urcorner).$$

Furthermore if M is a standard transitive model of ZF and $\mathbf{A} \in M$ then ψ is absolute with respect to M (regarding $\mathbf{A}$ as a constant).

Remark. Since we did not formalize explicitly all of the necessary syntactical notions we can only give an outline of a proof.

We define a formula $\psi_0(f, \mathbf{A})$ in the language $\mathscr{L}_0$ that formalizes the notion "f is the characteristic function of the closed wffs of $\mathscr{L}(C(A))$ that are true in $\mathbf{A}$"

i.e., $\psi_0(f, \mathbf{A})$ is the conjunction of the following formulas:

1. $f\colon Fml^0(\mathbf{A}) \to 2$.
2. $\forall \ulcorner\varphi\urcorner[f(\ulcorner\neg\varphi\urcorner) = 1 \leftrightarrow f(\ulcorner\varphi\urcorner) = 0]$.
3. $\forall \ulcorner\varphi_1\urcorner \forall \ulcorner\varphi_2\urcorner[[f(\ulcorner\varphi_1 \wedge \varphi_2\urcorner) = 1 \leftrightarrow f(\ulcorner\varphi_1\urcorner) = 1 \wedge f(\ulcorner\varphi_2\urcorner) = 1]$
$\wedge\ [f(\ulcorner\varphi_1 \vee \varphi_2\urcorner) = 1 \leftrightarrow f(\ulcorner\varphi_1\urcorner) = 1 \vee f(\ulcorner\varphi_2\urcorner) = 1]]$.

4. $\forall^{\ulcorner}(\forall x)\varphi(x)^{\urcorner}[f(^{\ulcorner}(\forall x)\varphi(x)^{\urcorner}) = 1 \leftrightarrow (\forall a \in A)[f(^{\ulcorner}\varphi(c_a)^{\urcorner}) = 1]]$.
5. $\forall^{\ulcorner}(\exists x)\varphi(x)^{\urcorner}[f(^{\ulcorner}(\exists x)\varphi(x)^{\urcorner}) = 1 \leftrightarrow (\exists a \in A)[f(^{\ulcorner}\varphi(c_a)^{\urcorner}) = 1]]$.
6. $(\forall a_1 \in A)(\forall a_2 \in A)[f(^{\ulcorner}c_{a_1} \in c_{a_2}{}^{\urcorner}) = 1 \leftrightarrow a_1 \in a_2]$.
7. $(\forall a_1 \in A) \cdots (\forall a_{n_i} \in A)[f(^{\ulcorner}R_j(c_{a_1}, \ldots, c_{a_{n_i}})^{\urcorner}) = 1 \leftrightarrow \bar{R}_j(a_1, \ldots, a_{n_i})$.

for each predicate symbol R_j (except $\in$) where n_j is the number of argument places of R_j. (For simplicity we assume that $\mathscr{L}$ has no individual constants.) Here $^{\ulcorner}\varphi^{\urcorner}$ is to range over $Fml^0(\mathbf{A})$ which is a set, since A is a set. $^{\ulcorner}(\forall x)\varphi(x)^{\urcorner}$ in 4 ranges over all (Gödel numbers of) closed formulas of $\mathscr{L}(C(A))$, that are of the form $(\forall x)\varphi(x)$. In 4 we also assume a suitable formalization of substitution. Since $Fml^0(\mathbf{A})$ is a set we can prove the following in ZF.

Theorem 7.6.

$$\begin{aligned} \mathbf{A} \models \varphi &\leftrightarrow (\forall f)[\psi_0(f, \mathbf{A}) \rightarrow f(^{\ulcorner}\varphi^{\urcorner}) = 1] \\ &\leftrightarrow (\exists f)[\psi_0(f, \mathbf{A}) \wedge f(^{\ulcorner}\varphi^{\urcorner}) = 1] \end{aligned}$$

Remark. Thus if

$$\psi(A, {}^{\ulcorner}\varphi^{\urcorner}) \overset{\Delta}{\leftrightarrow} (\forall f)[\psi_0(f, A) \rightarrow f(^{\ulcorner}\varphi^{\urcorner}) = 1]$$

then

$$A \models \varphi \leftrightarrow \psi(A, {}^{\ulcorner}\varphi^{\urcorner})$$

Futhermore, let M be a standard transitive model of ZF with $\mathbf{A} \in M$. Then $\psi_0(f, \mathbf{A})$ is absolute with respect to M since $Fml^0(\mathbf{A}) \in M$ and all quantifiers in 1–7 of the definition of $\psi_0(f, \mathbf{A})$ can be restricted to M. It then follows that $\psi(\mathbf{A}, x)$ is absolute with respect to M. (See Theorem 13.8, GTM Vol. 1.)

If we allow A to be a class then Theorem 7.5 no longer holds for otherwise we would obtain a truth definition for V definable in the language of ZF. $\psi_0(f, \mathbf{A})$ can still be defined as above even if A is a proper class, however, in this case f is a class variable. Consequently we would have a bound second order variable in Theorem 7.6.

In the language of Gödel-Bernays (GB) set theory $\mathbf{A} \models \varphi$ can be defined by Theorem 7.6 however we cannot prove in GB that it has the desired properties unless we assume some further axioms, e.g., mathematical induction or the comprehension axiom for formulas involving bound class variables.

Later we will encounter a similar situation when considering the definability of forcing for unlimited formulas. On the other hand we can prove the following theorem in ZF.

Theorem 7.7. If A is a class then for *each* formula φ of $\mathscr{L}$ with free variables $a_1, \ldots, a_k$ there is a formula ψ of $\mathscr{L}_0$ for which

$$(\forall a_1, \ldots, a_k \in A)[\mathbf{A} \models \varphi(a_1, \ldots, a_k) \leftrightarrow \psi(A, a_1, \ldots, a_k)].$$

Proof. For each $a_1, \ldots, a_k \in A$

$$\mathbf{A} \models \varphi(a_1, \ldots, a_k) \leftrightarrow \bar{\varphi}(a_1, \ldots, a_k)$$

where $\bar{\varphi}(a_1, \ldots, a_k)$ is $\varphi^A(a_1, \ldots, a_k)$ with each occurrence of R_i replaced by $\bar{R}_i$ and each occurrence of c_j replaced by $\bar{c}_j$.

Remark. As mentioned above, in the general case where A may be a proper class we do not have a single formula ψ of $\mathscr{L}_0$ such that Theorem 7.7 holds simultaneously for all formulas of $\mathscr{L}$, (ψ having a further argument $\ulcorner\varphi\urcorner$). This result can however be obtained if we restrict ourselves to formulas φ with less than n quantifiers, where n is a fixed natural number.

Definition 7.8. If A is a set then

$$Df(\mathbf{A}) \triangleq \{a \mid \exists \ulcorner\varphi(x)\urcorner \in Fml^1(\mathbf{A})[a = \{x \in A \mid \mathbf{A} \models \varphi(x)\}]\}.$$

Remark. $Df(\mathbf{A})$ is the set of sets definable from $\mathbf{A}$ (using elements from A as parameters). Here we need Theorem 7.6 to show that $\{x \in A \mid \mathbf{A} \models \varphi(x)\}$ is a set in ZF.

Taking $\varphi(x)$ as $x = x$ we obtain the following.

Theorem 7.9. $A \in Df(\mathbf{A})$ provided A is a set.

Theorem 7.10. If A is a set then $Df(\mathbf{A})$ is a set definable in $\mathscr{L}_0$ from $\mathbf{A}$. If, in addition, M is a standard transitive model of ZF and $\mathbf{A} \in M$ then $Df(\mathbf{A})$ is absolute with respect to M.

Remark. For our general theory let $\langle \mathbf{M}_\alpha \mid \alpha \in On\rangle$ be a sequence of transitive structures of $\mathscr{L}$ such that $M_\alpha = |\mathbf{M}_\alpha|$ is a set and the following three conditions are satisfied:

1. $\alpha < \beta \rightarrow \mathbf{M}_\alpha$ is a substructure of $\mathbf{M}_\beta$.
2. $M_\alpha = \bigcup_{\beta<\alpha} M_\beta$, $\alpha \in K_{\mathrm{II}}$
3. $Df(\mathbf{M}_\alpha) \subseteq M_{\alpha+1}$.

$$M = \bigcup_{\alpha\in On} M_\alpha, \qquad \mathbf{M} = \langle M, R_0^{\mathbf{M}}, \ldots\rangle.$$

Remark. Since each M_α, $\alpha \in On$, is transitive, so is M. Moreover R_i^M is defined by

$$R_i^{\mathbf{M}}(a_1, \ldots, a_{n_i}) \leftrightarrow R_i^{\mathbf{M}_\alpha}(a_1, \ldots, a_{n_i}), a_1, \ldots, a_{n_i} \in M_\alpha.$$

In view of 1 this definition is unambiguous. Furthermore $\mathbf{M}_\alpha \subseteq \mathbf{M}$ for each $\alpha \in On$.

We now wish to prove that $\mathbf{M}$ is a standard transitive model of ZF.

Theorem 7.11.

1. $M_\alpha \in M_{\alpha+1}$.
2. $M_\alpha \in M$.

Proof. Theorems 7.9 and 3 above.

Remark. Since M is transitive, $\mathbf{M}$ satisfies the axioms of extensionality and regularity. It is easy to check that the axioms of Pairing and Union hold in $\mathbf{M}$. Since $\alpha \in M_{\alpha+1}$, $On \subseteq M$ and $\omega \in M$. Therefore $\mathbf{M}$ satisfies the Axiom of Infinity.

The main idea used for the proof of the remaining axioms is contained in the proof of the following proposition.

Theorem 7.12. $a \subseteq M \rightarrow (\exists \alpha)[a \subseteq M_\alpha]$.

Proof. From the Axiom Schema of Replacement it follows that

$$(\exists \alpha)\left[\alpha = \bigcup_{x \in a} \mu_\beta(x \in M_\beta)\right]$$

then $a \subseteq M_\alpha$.

Remark. In order to prove the Axiom of Separation in **M** we first prove a kind of reflection principle in **M**. This proof requires several preliminary results.

Definition 7.13. A function F: $On \rightarrow On$ is semi-normal iff

1. $(\forall \alpha)[a \leq F(\alpha)]$.
2. $(\forall \alpha, \beta)[\alpha < \beta \rightarrow F(\alpha) \leq F(\beta)]$.
3. $(\forall \alpha \in K_{\text{II}})\left[F(\alpha) = \bigcup_{\beta < \alpha} F(\beta)\right]$.

Definition 7.14. A function F: $On \rightarrow On$ is a normal function iff

1. $(\forall \alpha, \beta)[\alpha < \beta \rightarrow F(\alpha) < F(\beta)]$.
2. $(\forall \alpha \in K_{\text{II}})\left[F(\alpha) = \bigcup_{\beta < \alpha} F(\beta)\right]$.

Remark. Every normal function is a strictly monotonic ordinal function. Since we have $\alpha \leq F(\alpha)$ for every strictly monotonic ordinal function it follows that every normal function is also semi-normal.

Theorem 7.15. If $F_1, \ldots, F_n$ are semi-normal functions then

$$(\forall \alpha)(\exists \beta > \alpha)[\beta = F_1(\beta) = \cdots = F_n(\beta)].$$

Proof. We define an ω-sequence $\langle \alpha_m \mid m \in \omega \rangle$ by recursion:

$$\alpha_1 = \alpha + 1,\ \alpha_2 = F_1(\alpha_1), \ldots, \alpha_{n+1} = F_n(\alpha_n)$$

$$\alpha_{k+i} = F_i(\alpha_k),\ i \equiv k(\text{mod } n),\ i = 1, \ldots, n.$$

If $\beta = \bigcup_{m \in \omega} \alpha_m$ then $\alpha < \beta$ and the sequence $\langle \alpha_m \mid m \in \omega \rangle$ is nondecreasing. If $\beta \in K_{\text{I}}$ then $m \in \omega$, $\beta = \alpha_m = \alpha_{m+1} = \cdots$, and hence

$$\beta = F_1(\beta) = \cdots = F_n(\beta).$$

If $\beta \in K_{\text{II}}$ then

$$\begin{aligned} F_i(\beta) &= \bigcup_{k \in \omega} F_i(\alpha_k) \\ &= \bigcup_{k \in \omega} \{\alpha_{k+i} \mid i \equiv k(\text{mod } n)\}\left(= \bigcup_{k \in \omega} \alpha_{kn+i+1}\right) \\ &= \beta. \end{aligned}$$

Theorem 7.16. If $\varphi(a_0, \ldots, a_n)$ is a formula of $\mathscr{L}$ then

$$(\forall\alpha)(\exists\beta \geq \alpha)(\forall a_1, \ldots, a_n \in M_\alpha)[\mathbf{M} \models (\exists x)\varphi(x, a_1, \ldots, a_n) \leftrightarrow (\exists a \in M_\beta)[M \models \varphi(a, a_1, \ldots, a_n)].$$

Proof. By Theorem 7.7.

$$\mathbf{M} \models \varphi(a, a_1, \ldots, a_n)$$

is expressible by a formula of $\mathscr{L}_0$. Using the fact that M_α is a set we can then define

$$\beta = \sup_{a_1, \ldots, a_n \in M_\alpha} \mu_{\beta'}(\beta' \geq \alpha \wedge (\exists a \in M_{\beta'})[\mathbf{M} \models \varphi(a, a_1, \ldots, a_n)]).$$

This β has the desired properties.

Theorem 7.17. For each formula $\varphi(a_0, \ldots, a_n)$ of $\mathscr{L}$ there exists a semi-normal function F such that

$$(\forall\alpha)(\forall\beta)[\beta = F(\alpha) \rightarrow (\forall a_1, \ldots, a_n \in M_\alpha)[\mathbf{M} \models (\exists x)\varphi(x, a_1, \ldots, a_n) \leftrightarrow (\exists a \in M_\beta)[\mathbf{M} \models \varphi(a, a_1, \ldots, a_n)]]].$$

Proof. From Theorem 7.16

$$(\exists\beta \geq \alpha)(\forall a_1, \ldots, a_n \in M_\alpha)[\mathbf{M} \models (\exists x)\varphi(x, a_1, \ldots, a_n) \leftrightarrow (\exists a \in M_\beta)[\mathbf{M} \models \varphi(a, a_1, \ldots, a_n)]]$$

therefore if

$$F(\alpha) = \mu_\beta(\beta \geq \alpha \wedge (\forall a_1, \ldots, a_n \in M_\alpha)[\mathbf{M} \models (\exists x)\varphi(x, a_1, \ldots, a_n) \leftrightarrow (\exists a \in M_\beta)[\mathbf{M} \models \varphi(a, a_1, \ldots, a_n)]])$$

then $\alpha \leq F(\alpha)$. Furthermore F is nondecreasing and continuous hence semi-normal.

Corollary 7.18. For each formula $\varphi(a_0, \ldots, a_n)$ of $\mathscr{L}$ there exists a semi-normal function F such that

$$(\forall\beta)[\beta = F(\beta) \rightarrow (\forall a_1, \ldots, a_n \in M_\beta)[\mathbf{M} \models (\exists x)\varphi(x, a_1, \ldots, a_n)] \leftrightarrow (\exists a \in M_\beta)[\mathbf{M} \models \varphi(a, a_1, \ldots, a_n)]].$$

Theorem 7.19. For each formula $\varphi(a_1, \ldots, a_n)$ of $\mathscr{L}$ there are finitely many semi-normal functions $F_1, \ldots, F_n$ such that

$$(\forall\beta)[\beta = F_1(\beta) = \cdots = F_m(\beta) \rightarrow (\forall a_1, \ldots, a_n \in M_\beta)[\mathbf{M} \models \varphi(a_1, \ldots, a_n) \leftrightarrow \mathbf{M}_\beta \models \varphi(a_1, \ldots, a_n)]].$$

Proof. (By induction on the number of logical symbols in φ.) If φ is atomic the theorem follows from the fact that $(\forall\alpha)[\mathbf{M}_\alpha \subseteq \mathbf{M}]$. If φ is of the form $\neg\psi$ or $\psi \wedge \eta$ the conclusion is obvious. If $\varphi(a_1, \ldots, a_n)$ is of the form $(\exists x)\psi(x, a_1, \ldots, a_n)$, then from the induction hypothesis there are semi-normal functions $F_1, \ldots, F_n$ such that

$$\beta = F_1(\beta) = \cdots = F_m(\beta) \rightarrow (\forall a_0, \ldots, a_n \in M_\beta)[\mathbf{M} \models \psi(a_0, \ldots, a_n) \leftrightarrow \mathbf{M}_\beta \models \psi(a_0, \ldots, a_n)].$$

By Corollary 7.18 there exists a semi-normal function F_0 such that

$$\beta = F_0(\beta) \to (\forall a_1, \ldots, a_n \in M_\beta)[\mathbf{M} \models (\exists x)\psi(x, a_1, \ldots, a_n) \leftrightarrow (\exists a \in M_\beta)[\mathbf{M} \models \psi(a, a_1, \ldots, a_n)]].$$

Therefore if $\beta = F_0(\beta) = \cdots = F_m(\beta)$, and $a_1, \ldots, a_n \in M_\beta$ then

$$\begin{aligned} \mathbf{M} \models \varphi(a_1, \ldots, a_n) &\leftrightarrow (\exists a \in M_\beta)[\mathbf{M} \models \psi(a, a_1, \ldots, a_n)] \\ &\leftrightarrow (\exists a \in M_\beta)[\mathbf{M}_\beta \models \psi(a, a_1, \ldots, a_n)] \\ &\leftrightarrow \mathbf{M}_\beta \models (\exists x)\psi(x, a_1, \ldots, a_n) \\ &\leftrightarrow \mathbf{M}_\beta \models \varphi(a_1, \ldots, a_n). \end{aligned}$$

Corollary 7.20. For each formula $\varphi(a_1, \ldots, a_n)$ of $\mathscr{L}$

$$(\forall \alpha)(\exists \beta)[\beta \geq \alpha \wedge (\forall a_1, \ldots, a_n \in M_\beta)[\mathbf{M} \models \varphi(a_1, \ldots, a_n) \leftrightarrow \mathbf{M}_\beta \models \varphi(a_1, \ldots, a_n)]].$$

Remark. From Corollary 7.20 we can easily prove that $\mathbf{M}$ satisfies the remaining axioms of ZF.

Theorem 7.21. $\mathbf{M}$ satisfies the Axiom of Separation.

Proof. If $\varphi(x, y_1, \ldots, y_n)$ is a formula of $\mathscr{L}$, if $a, a_1, \ldots, a_n \in M$, and if

$$A \triangleq a \bigcup \{a, a_1, \ldots, a_n\}$$

then A is a subset of M. Thus $(\exists \alpha)[A \subseteq M_\alpha]$. By Corollary 7.20

$$(\exists \beta \geq \alpha)(\forall b \in M_\beta)[\mathbf{M} \models b \in a \wedge \varphi(b, a_1, \ldots, a_n) \leftrightarrow \mathbf{M}_\beta \models b \in a \wedge \varphi(b, a_1, \ldots, a_n)].$$

Then

$$\{x \mid \mathbf{M} \models x \in a \wedge \varphi(x, a_1, \ldots, a_n)\} = \{x \in M_\beta \mid \mathbf{M}_\beta \models x \in a \wedge \varphi(x, a_1, \ldots, a_n)\}.$$

Consequently

$$\{x \mid \mathbf{M} \models x \in a \wedge \varphi(x, a_1, \ldots, a_n)\} \in Df(\mathbf{M}_\beta) \subseteq M_{\beta+1} \subseteq M.$$

Theorem 7.22. $\mathbf{M}$ satisfies the Power Set Axiom.

Proof. If $a \in M$ then $\mathscr{P}(a) \cap M$ is a subset of M. Hence

$$(\exists \alpha)[\mathscr{P}(a) \cap M \subseteq M_\alpha \in M].$$

Since, by Theorem 7.21, $\mathbf{M}$ satisfies the Axiom of Separation and $M_\alpha \in M$ it follows that $\mathscr{P}(a) \cap M$ is an element of M.

Theorem 7.23. $\mathbf{M}$ satisfies the Axiom Schema of Replacement.

Proof. If $\varphi(a_0, \ldots, a_n)$ is a formula of $\mathscr{L}$ such that

$$(\forall a_2, \ldots, a_n \in M)(\forall x \in M)(\exists !\, y \in M)[\mathbf{M} \models \varphi(x, y, a_2, \ldots, a_n)]$$

if $a \in M$, and

$$C \triangleq \{y \in M \mid (\exists x \in a)[\mathbf{M} \models \varphi(x, y, a_2, \ldots, a_n)]\}$$

then C is a subset of M. Thus

$$(\exists\alpha)[C \subseteq M_\alpha \in M].$$

Then by Theorem 7.21

$$C = C \cap M_\alpha \in M.$$

Remark. We have established that **M** is a standard transitive model of ZF that contains all of the ordinals. The method by which **M** was constructed is of particular interest. **M** was defined from a sequence $\langle \mathbf{M}_\alpha \mid \alpha \in On\rangle$ of transitive structures satisfying the conditions 1–3 of page 68. We next consider two applications of this same general method.

Definition 7.24. If K is a class, if $\mathscr{L}(\{K(\ )\})$ is the language obtained from $\mathscr{L}_0$ by adding a unary predicate symbol $K(\ )$ and if $\langle \mathbf{A}_\alpha \mid \alpha \in On\rangle$ is a sequence of structure $\mathbf{A}_\alpha = \langle A_\alpha, \overline{K}_\alpha\rangle$ defined by transfinite induction on α by

$$\begin{aligned} A_0 &\triangleq 0 \\ \overline{K}_\alpha &\triangleq A_\alpha \cap K \\ A_{\alpha+1} &\triangleq Df(\mathbf{A}_\alpha) \\ A_\alpha &\triangleq \bigcup_{\beta<\alpha} A_\beta,\ \alpha \in K_{\mathrm{II}}, \end{aligned}$$

then

$$L_K \triangleq \bigcup_{\alpha\in On} A_\alpha.$$

Remark. It can then be easily shown that $\mathbf{A}_\alpha$ satisfies the conditions 1–3 of page 68. From this, as before, we can prove the following theorem.

Theorem 7.25. L_K is a transitive model of ZF and $On \subseteq L_K$.

Remark. Definability in $\mathscr{L}_0(\{K(\ )\})$ for classes is defined in the same way as in Definition 7.3.

Theorem 7.26.

1. L_K is definable in $\mathscr{L}_0(\{K(\ )\})$.
2. $a \in L_K \rightarrow a \cap K \in L_K$.
3. $\mathscr{M}(K)^* \rightarrow K \cap L_K \in L_K$.

Proof. 1. Obvious from Theorem 7.10 and the definition of L_K.

2. If $a \in L_K$ then $(\exists\alpha)(a \in A_\alpha)$. Therefore $a \subseteq A_\alpha$ and $a \cap K \subseteq \overline{K}_\alpha \subseteq A_\alpha$. Then $a \cap K = \{x \in A_\alpha \mid \langle A_\alpha, \overline{K}_\alpha\rangle \vDash x \in c_a \wedge K(x)\}$ where $a \in A_\alpha$ i.e.,

$$a \cap K \in Df(\mathbf{A}_\alpha) = A_{\alpha+1} \subseteq L_K.$$

3. If K is a set and $k_0 \triangleq K \cap L_K$ then k_0 is a subset of A_α for some α. Then

$$k_0 = \{x \in A_\alpha \mid \langle A_\alpha, \overline{K}_\alpha\rangle \vDash K(x)\} \in Df(\mathbf{A}_\alpha) = A_{\alpha+1}$$

and hence $k_0 \in L_K$.

* $\mathscr{M}(K)$ means "K is a set."

Remark. We say M is a standard transitive model of ZF in the language $\mathscr{L}_0(\{K(\ )\})$ if the following conditions are satisfied:

1. M is transitive.
2. There is a class $K \subseteq M$ such that $(\forall x \in M)[x \cap K \in M]$.
3. M satisfies the axioms of ZF described in the language $\mathscr{L}_0(\{K(\ )\})$ by interpreting $K(\ )$ by K.

Let M be such a model and let $\varphi(x_1, \ldots, x_n)$ be a formula of $\mathscr{L}_0(\{K(\ )\})$ containing no free variables other than $x_1, \ldots, x_n$. Then φ is absolute with respect to M iff

$$(\forall x_1, \ldots, x_n \in M)[\varphi(x_1, \ldots, x_n) \leftrightarrow \varphi^M(x_1, \ldots, x_n)].$$

Note that in making φ^M from φ the symbol K, if it occurs in φ, is left invariant.

Theorem 7.27. If M is a standard transitive model of ZF in the language $\mathscr{L}_0(\{K(\ )\})$ and if $On \subseteq M$, then $L_K \subseteq M$.

Proof. We prove by induction that $A_\alpha \in M$. Clearly $A_0 = 0 \in M$. If $A_\alpha \in M$ then $\overline{K}_\alpha = A_\alpha \cap K \in M$, hence $\mathbf{A}_\alpha = \langle A_\alpha, \overline{K}_\alpha \rangle \in M$. By Theorem 7.10, $Df(\mathbf{A}_\alpha)$ is absolute with respect to M. Therefore $A_{\alpha+1} \subseteq M$. But A a set implies that $Df(\mathbf{A})$ is a set. Therefore $Df(\mathbf{A}_\alpha) \in M$, i.e.,

$$A_{\alpha+1} \in M.$$

If $\alpha \in K_{\mathrm{II}}$ and $\forall \beta < \alpha$, $A_\beta \in M$ then since the sequence $\langle A_\beta \mid \beta < \alpha \rangle$ is definable in M*,

$$A_\alpha = \bigcup_{\beta \in \alpha} A_\beta \in M.$$

Then $L_K \subseteq M$, since M is transitive.

Theorem 7.28. If $K_0 = K \cap L_K$

1. $L_K = L_{K_0} \wedge K_0 \subseteq L_{K_0}$ (Therefore $K_0 \in L_{K_0}$ if K_0 is a set).
2. $L_{K_0} \models V = L_{K_0}$.
3. $L_{K_0} \models AC$.

Proof. 1. $A_\alpha \cap K_0 = A_\alpha \cap K \cap L_K = A_\alpha \cap K$, since $A_\alpha \subseteq L_K$. Therefore $L_K = L_{K_0}$ and $K_0 \subseteq L_K = L_{K_0}$.

2. If $\mathbf{A}'_\alpha = \langle A'_\alpha, K'_\alpha \rangle$ is $\mathbf{A}_\alpha = \langle A_\alpha, \overline{K}_\alpha \rangle$ relativized to L_{K_0} then we prove by induction on α that $\mathbf{A}'_\alpha = \mathbf{A}_\alpha$. Obviously $\mathbf{A}'_0 = \mathbf{A}_0$. If $\mathbf{A}'_\alpha = \mathbf{A}_\alpha$ then $\mathbf{A}_\alpha \in L_{K_0}$ and L_{K_0} is a transitive model of ZF. Therefore

$$A'_{\alpha+1} = Df(\mathbf{A}_\alpha) = A_{\alpha+1}$$

since $\overline{K}'_\alpha = \overline{K}_\alpha$ we have $\overline{K}'_{\alpha+1} = \overline{K}_{\alpha+1}$. The case $\alpha \in K_{\mathrm{II}}$ is obvious.

3. We first prove in ZF that if a is well ordered then $a' = Df(\langle a, k \rangle)$ is well ordered.

* A class $A(\subseteq M)$ is definable in M iff there is a formula $\varphi(x)$ of $\mathscr{L}(\{K(\ )\})$ containing no free variables other than x such that $A = \{x \in M \mid \varphi^M(x)\}$.

If $b \in a'$ then there is a formula φ of $\mathscr{L}(\{K(\ )\})$ and a finite set of constants $\{c_1, \ldots, c_n\} \subseteq a$ for which

$$b = \{x \in a \mid \langle a, k \rangle \models \varphi(x, c_1, \ldots, c_n)\}$$

Thus b is determined by $\langle \ulcorner\varphi\urcorner^*, \{c_1, \ldots, c_n\}\rangle$. The set of formulas of $\mathscr{L}(\{K(\ )\})$ is countable and the finite subsets of constants from a can be well ordered since a is well ordered. This gives a well ordering of a'.

Since L_{K_0} is a model of ZF it then follows from the foregoing argument that $A_{\alpha+1}$ is well ordered in L_{K_0} if A_α is well ordered in L_{K_0}. Thus by induction on α there are relations $<_\alpha$ in L_{K_0} such that $<_\alpha$ well orders A_α($<_\alpha$ is definable uniformly for all α in L_{K_0}).

If $Od(a) \triangleq \mu_\alpha(a \in A_\alpha)$, $a \in L_K$ and if

$$a < b \leftrightarrow a \in L_{K_0} \wedge b \in L_{K_0} \wedge [Od(a) < Od(b) \vee [Od(a) = Od(b) \wedge a <_{Od(a)} b]]$$

then $<$ is a well ordering of L_{K_0} that is definable in L_{K_0}. In particular each $a \in L_{K_0}$ is well ordered by $\{\langle x, y\rangle \mid x < y \wedge x, y \in a\} \in L_{K_0}$.

Corollary 7.29. If there exists a standard transitive model of ZF then there exists a standard transitive model of $ZF + AC + V = L_0$.

Exercise. Show that L_0 is Gödel's class of constructible sets.

Remark. In Introduction to Axiomatic Set Theory we proved that $V = L_0$ implies the GCH. We now wish to prove a corresponding result namely $V = L_a \rightarrow GCH$. For this proof we require the following.

Definition 7.30. 1. If $\mathbf{A}$ is a structure and $\varphi(a_0, a_1, \ldots, a_n)$ is a formula in the language of $\mathbf{A}$, then a function $f \colon A^n \rightarrow A$ is a Skolem function for $(\exists x)\varphi(x, a_1, \ldots, a_n)$ with respect to $\mathbf{A}$ iff

$$(\forall x_1, \ldots, x_n \in A)[\mathbf{A} \models (\exists x)\varphi(x, x_1, \ldots, x_n) \leftrightarrow \mathbf{A} \models \varphi(f(x_1, \ldots, x_n), x_1, \ldots, x_n)].$$

2. $\mathbf{B}$ is an elementary substructure of $\mathbf{A}$ (written $\mathbf{B} \prec \mathbf{A}$) iff $\mathbf{B}$ is a substructure of $\mathbf{A}$ and for every formula φ of the language of $\mathbf{A}$ i.e., $\mathscr{L}(C(A))$, and $\forall a_1, \ldots, a_n \in B$

$$\mathbf{B} \models \varphi(a_1, \ldots, a_n) \leftrightarrow \mathbf{A} \models \varphi(a_1, \ldots, a_n).$$

Remark. We next show how to obtain an elementary substructure of $\mathbf{A}$ that contains a given subset of A, provided that we have a family of Skolem functions for all formulas of the language of $\mathbf{A}$.

Theorem 7.31. If $\mathbf{A}$ is a structure and F a set of Skolem functions such that for every formula $(\exists x)\varphi(x, a_1, \ldots, a_n)$ of the language of $\mathbf{A}$ there exists in F a Skolem function for that formula with respect to $\mathbf{A}$, if $B \subseteq A$ and if B is closed under the functions of F then $\mathbf{B} \triangleq \mathbf{A} \upharpoonright B$ is an elementary substructure of $\mathbf{A}$.

* $\ulcorner\varphi\urcorner$ is the Gödel number of φ.

Proof. By induction on the number of logical symbols in φ. If φ is atomic or of the form $\neg\psi$ or $\psi \wedge \eta$ the conclusion is obvious. If $\varphi(a_1, \ldots, a_n)$ is $(\exists x)\psi(x, a_1, \ldots, a_n)$ and if $b_1, \ldots, b_n \in B$ then

$$\begin{aligned}\mathbf{B} \models (\exists x)\psi(x, b_1, \ldots, b_n) &\to (\exists b \in B)[\mathbf{B} \models \psi(b, b_1, \ldots, b_n)] \\ &\to (\exists b \in B)[\mathbf{A} \models \psi(b, b_1, \ldots, b_n)] \\ &\to \mathbf{A} \models (\exists x)\psi(x, b_1, \ldots, b_n) \\ &\to (\exists f \in F)[\mathbf{A} \models \psi(f(b_1, \ldots, b_n), b_1, \ldots, b_n)] \\ &\to (\exists x \in B)[\mathbf{A} \models \psi(x, b_1, \ldots, b_n)] \\ &\to \mathbf{B} \models (\exists x)\psi(x, b_1, \ldots, b_n).\end{aligned}$$

Lemma. If A is a set, and if

$$(\forall x, y \in A)[x \neq y \to (\exists z \in A) \neg [z \in x \leftrightarrow z \in y]]$$

then there exists a transitive set a and a function f such that

$$f\colon A \xrightarrow[\text{onto}]{1-1} a$$

and $(\forall x, y \in A)[x \in y \leftrightarrow f(x) \in f(y)]$. Moreover if b is a transitive subset of A then $f \restriction b = I \restriction b$.*

Proof. We define f recursively by

$$f(y) = \{f(x) \mid x \in A \cap y\}.$$

The conclusion is then immediate from the definition of f, that is,

$$(\forall x, y \in A)[x \in y \leftrightarrow f(x) \in f(y)].$$

Also, by $\in$-induction, it follows that if b is a transitive subset of A then $f \restriction b = I \restriction b$. (For details see Takeuti and Zaring: *Introduction to Axiomatic Set Theory*, Springer-Verlag, 1971, p. 19.)

Remark. In the foregoing Lemma both f and a are unique.

Theorem 7.32. If A is a transitive set, if $k \in A$ and if $\langle A, \in, k\rangle$ is a model of $ZF + V = L_k$ then $(\exists\alpha)[A = A_\alpha]$ where A_α is as in Definition 7.24.

Proof. Since $L_k = \bigcup_{\alpha \in On} A_\alpha$ and A_α is absolute with respect to A for each $\alpha \in A \cap On$, $A = \bigcup_{\alpha \in A \cap On} A_\alpha$. Furthermore, because A is transitive, $A \cap On = \bigcup_{\alpha \in A} \alpha \triangleq \beta$. Therefore since A is a model of ZF, $\beta \in K_{\text{II}}$ and hence

$$A = \bigcup_{\alpha \in \beta} A_\alpha = A_\beta.$$

Theorem 7.33. If k_0 is the transitive closure of k then

$$V = L_k \to (\forall\alpha)[\bar{\bar{k}}_0 \leq \aleph_\alpha \to 2^{\aleph_\alpha} = \aleph_{\alpha+1}].$$

Proof. If $V = L_k$ then $k \in L_k$. Let F be a countable family of Skolem functions, with respect to L_k, for all formulas of the language $\mathscr{L}_0(\{k(\ )\})$. If $a \subseteq \aleph_\alpha$, $\bar{\bar{k}}_0 \leq \aleph_\alpha$ and

$$b \triangleq \{a\} \cup \aleph_\alpha \cup k_0 \cup \{k\}$$

* $I \triangleq \{\langle x, x\rangle \mid x \in V\}$, $f \restriction b \triangleq \{\langle x, y\rangle \in f \mid x \in b\}$.

then b is transitive and $\bar{\bar{b}} = \aleph_\alpha$. Let A be the closure of b under all of the functions in F. Then $\bar{\bar{A}} = \bar{\bar{b}} = \aleph_\alpha$ and, by Theorem 7.31

$$\langle A, \in, k\rangle \models ZF + V = L_k.$$

From the Lemma there exists a transitive set a_0 and a function f from A one-to-one onto a_0 such that f preserves the $\in$-relation. Since b is a transitive subset of A, f is the identity function on b, in particular $f(k) = k$. Therefore $\langle a_0, \in, k\rangle \models ZF + V = L_k$. By Theorem 7.32

$$(\exists\beta)[a_0 = A_\beta].$$

But $\bar{\bar{a}}_0 = \bar{\bar{A}} = \aleph_\alpha$. Hence $\bar{\bar{\beta}} \leq \aleph_\alpha$ i.e., $\beta < \aleph_{\alpha+1}$. Since $a = f(a) \in f``A = a_0$, this proves that $(\forall a \subseteq \aleph_\alpha)(\exists\beta < \aleph_{\alpha+1})[a \in A_\beta]$. Therefore

$$\mathscr{P}(\aleph_\alpha) \subseteq A_{\aleph_{\alpha+1}}.$$

But $\bar{\bar{A}}_{\aleph_{\alpha+1}} = \aleph_{\alpha+1}$. Hence $\overline{\overline{\mathscr{P}(\aleph_\alpha)}} = \aleph_{\alpha+1}$.

Remark. Note that $V = L_k$ can be expressed as a simple sentence $V = \bigcup_{\alpha\in On} A_\alpha$ in the language $\mathscr{L}(\{k(\ )\})$. To prove the preceding theorem assuming the axioms of ZF and $V = L_k$ we note that in fact we used only finitely many axioms $\varphi_0, \ldots, \varphi_n$. Let F_0 be the family of Skolem functions for the finitely many subformulas of $\varphi_0, \ldots, \varphi_n$. Then F_0 can be defined in the language of ZF. The proof can then be carried out with F replaced by F_0.

As a corollary we have $V = L \rightarrow GCH$ and hence the following theorem.

Theorem 7.34. If there exists a standard transitive model of ZF then there exists a standard transitive model of $ZF + AC + GCH$.

Remark. For our second application of our general theory we define $L[A]$.

Definition 7.35. If K is a transitive class, if $F \subseteq K$, if $\mathscr{L} = \mathscr{L}(\{K(\ ), F(\ )\})$ and if $\mathbf{B}_\alpha = \langle B_\alpha, \bar{K}_\alpha, \bar{F}_\alpha\rangle$ are structures for $\mathscr{L}$ defined recursively by

1. $B_0 \triangleq 0$.
2. $\bar{K}_\alpha \triangleq R(\alpha) \cap K \wedge \bar{F}_\alpha = R(\alpha) \cap F$.
3. $B_{\alpha+1} \triangleq Df(\mathbf{B}_\alpha) \cup \bar{K}_{\alpha+1}$.
4. $B_\alpha \triangleq \bigcup_{\beta\in\alpha} B_\beta,\ \alpha \in K_{II}$

then

$$L[K; F] \triangleq \bigcup_{\alpha\in On} B_\alpha.$$

Remark. Since K is transitive, B_α is transitive for each α. Then $\langle \mathbf{B}_\alpha \mid \alpha \in On\rangle$ satisfies the conditions 1–3 of page 68. Consequently we can prove the following

Theorem 7.36. $L[K; F]$ is a standard transitive model of ZF and $On \subseteq L[K; F]$.

Definition 7.37. 1. $L[F] \triangleq L[K_0; F]$ where K_0 is the transitive closure of F.

2. If M is a standard transitive model of ZF, $On \subseteq M$, and $F \subseteq M$ then

$$M[F] \triangleq L[M; F].$$

Theorem 7.38. 1. $L[K; F]$ and $L[F]$ are definable in $\mathscr{L}_0(\{K(\ ), F(\ )\})$ and $\mathscr{L}_0(\{F(\ )\})$ respectively.

2. $a \in L[a]$.
3. $L[K] = \bigcup_{\alpha \in On} L[K \cup R(\alpha)]$.

Proof. 1. Similar to Theorem 7.10 for a language with constants $K(\ )$ and $F(\ )$.

2. If $\alpha = \text{rank}(a)$ then $a \subseteq R(\alpha)$. Therefore $\bar{F}_\alpha = a$ and hence

$$a \in Df(\mathbf{B}_\alpha) = B_{\alpha+1}.$$

3. For each γ it is easily shown by transfinite induction on α that $B_\alpha(\gamma) \subseteq L[K]$, where $L[K \cap R(\gamma)] = \bigcup_{\alpha \in On} B_\alpha(\gamma)$. Conversely, if $a \in L[K]$ $(\exists \alpha)[a \in B_\alpha]$. But for each α there is a $\gamma(\geq \alpha)$ such that $B_\alpha \subseteq L[K \cap R(\gamma)]$. Therefore

$$a \in \bigcup_{\alpha \in On} L[K \cap R(\alpha)].$$

Theorem 7.39. If M is a standard transitive model of ZF in the language $\mathscr{L}_0(\{a(\ )\})$ such that

1. $On \subseteq M$,
2. $a \in M$,

then $L[a] \subseteq M$.

Proof. If a_0 is the transitive closure of a then since $a \in M$ and M is a model of ZF, $a_0 \in M$ and a_0 is the transitive closure of a in M. Since M is transitive $a \subseteq M$ and $a_0 \subseteq M$. Also since the rank function is absolute with respect to standard transitive models of ZF, $[R(\alpha) \cap a]^M = R(\alpha) \cap M \cap a = R(\alpha) \cap a$.

Then

$$(\forall \alpha \in On)[\bar{K}_\alpha, \bar{F}_\alpha \in M].$$

Clearly $B_0 \in M$. If $B_\alpha \in M$ then $Df(\mathbf{B}_\alpha) \in M$ and hence $B_{\alpha+1} \in M$. Thus by transfinite induction

$$(\forall \alpha \in On)[B_\alpha \in M].$$

Therefore $L[a] \subseteq M$.

Theorem 7.40. If a has a well ordering in $L[a]$ then $L[a]$ satisfies the AC.

Proof. The proof is similar to that of Theorem 7.28 and is left to the reader.

Theorem 7.41.

1. $a \subseteq L \to L_a = L[a]$.
2. $L[a] \models V = L[a]$.

Proof. 1. If $a \subseteq L$ then $a \subseteq L_a$ since $L \subseteq L_a$. Therefore $a \in L_a$. Since $L[a]$ is the least standard transitive model of ZF that contains a and all the ordinals as elements we have $L[a] \subseteq L_a$.

But, since $a \in L[a]$.

$$(\forall x \in L[a])[x \cap a \in L[a]].$$

Therefore $L_a \subseteq L[a]$.

2. The proof is left to the reader.

Exercises. 1. If M is a standard transitive model of ZF, $On \subseteq M$, and $a \subseteq M$ then $M[a]$ is the smallest standard transitive model N of ZF in the language $\mathscr{L}_0(\{M(\ ), C_a\})$ such that (i) $M \subseteq N$ and (ii) $a \in N$.

2. If M is a standard transitive model of ZF, $On \subseteq M$, $K \subseteq M$ and $\mathscr{L} = \mathscr{L}_0(\{M(\ ), K(\ )\})$ we define $\mathbf{C} = \langle C_\alpha, \overline{M}_\alpha, \overline{K}_\alpha \rangle$ and M_K by recursion:

i) $C_0 = 0$.
ii) $\overline{M}_\alpha = M \cap R(\alpha)$, $\overline{K}_\alpha = C_\alpha \cap K$.
iii) $C_{\alpha+1} = Df(\mathbf{C}_\alpha) \cup \overline{M}_{\alpha+1} \cup \overline{K}_\alpha$.
iv) $C_\alpha = \bigcup_{\beta<\alpha} C_\beta$, $\alpha \in K_{\text{II}}$.
v) $M_K = \bigcup_{\alpha \in On} C_\alpha$.

Then $M_K = M[K]$.

8. Relative Constructibility and Ramified Languages

Using a ramified language we shall give another definition of $L[K; F]$ a definition that has many applications since it only uses the concepts of ordinal number and transfinite induction. On the other hand, to carry out the actual induction steps may become rather complicated in particular cases where definitions by simultaneous recursion are involved.

The symbols of the ramified language $R(K, F)$ are the following.

Variables: $x_0, x_1, \dots, x_n, \dots \quad n \in \omega \quad$ (unranked).
$x_0{}^\alpha, x_1{}^\alpha, \dots, x_n{}^\alpha, \dots \quad n \in \omega, \alpha \in On \quad$ (ranked).

Predicate constants: $\in$, $K(\ )$, $F(\ )$.
Individual constants: $\underline{k}$ for each $k \in K$, where K is a given class.
Logical symbols: $\neg$, $\wedge$, $\forall$.
Abstraction operator: $\hat{x}_n{}^\alpha$.
Parentheses: (,).

Definition 8.1. Limited formulas and abstraction terms are defined simultaneously by the following recursion.

1. If each of t_1, t_2 is either an individual constant, a ranked variable, or an abstraction term then

$$F(t_1), \qquad K(t_1), \qquad t_1 \in t_2$$

are limited formulas.

2. If φ and ψ are limited formulas then $\neg\varphi$, and $\varphi \wedge \psi$ are limited formulas.

3. If $\varphi(x^\alpha)$ is a limited formula that does not contain x^α as a bound variable then $(\forall x^\alpha)\varphi(x^\alpha)$ is a limited formula.

4. If $\varphi(x^\alpha)$ is a limited formula satisfying the following,

a. $\varphi(x^\alpha)$ contains no free variables other than x^α,
b. if $\underline{k}$ is an individual constant occurring in $\varphi(x^\alpha)$ then rank $(k) < \alpha$,
c. if an abstraction term $\hat{x}^\beta\psi(x^\beta)$ occurs in $\varphi(x^\alpha)$ then $\beta < \alpha$,
d. if a quantifier $\forall y^\beta$ occurs in $\varphi(x^\alpha)$ then $\beta \leq \alpha$,

then $\hat{x}^\alpha\varphi(x^\alpha)$ is an abstraction term.

5. A formula is a limited formula iff its being so is deducible from 1–4. An expression is an abstraction term, iff its being so is deducible from 1–4.

Remark. The requirements for φ in 4 are chosen to assure that sets are built up in a predicative way (disregarding the constants $k \in K$) i.e., if a set b is determined by $\varphi(x^\alpha)$, (1) φ should not contain any free variables other than x^α, (2) any individual constant occurring in φ should be of rank less than α, (3) any set occurring in φ as a constant should already be defined at a previous level, and finally (4) φ should contain no quantification over levels to which it itself belongs. One might also think of x^α as ranging over B_α (Definition 7.35) therefore $\hat{x}^\alpha\varphi(x^\alpha)$ should be an element of $B_{\alpha+1} = Df(\mathbf{B}_\alpha) \cup \bar{K}_{\alpha+1}$ which provides another motivation for the conditions in 4.

Definition 8.2. A constant term is either an individual constant or an abstraction term. We define the rank ρ of a constant term:

$$\rho(\underline{k}) \triangleq \text{rank}\,(k) \qquad k \in K.$$
$$\rho(\hat{x}^\alpha\varphi(x^\alpha)) \triangleq \alpha.$$
$$T_\alpha \triangleq \{t \mid t \text{ is a constant term and } \rho(t) < \alpha\}.$$
$$T \triangleq \bigcup_{\alpha \in On} T_\alpha.$$

Definition 8.3. Unlimited formulas of $R(K, F)$ (or simply formulas of $R(K, F)$) are defined as follows.

1. If each of t_1 and t_2 is a constant term or a variable, then $K(t_1)$, $F(t_1)$, and $t_1 \in t_2$ are unlimited formulas.
2. If φ and ψ are unlimited formulas, then $\neg\varphi$ and $\varphi \wedge \psi$ are unlimited formulas.
3. If $\varphi(x)$ is an unlimited formula in which x is a variable, ranked or unranked, that does not occur as a bound variable in φ, then $(\forall x)\varphi(x)$ is an unlimited formula.
4. A formula is unlimited iff its being so is deducible from 1–3.

Remark. For induction on limited formulas we need the following notions:

Definition 8.4. Let φ be a limited sentence and t_1, t_2 be constant terms.

1. The grade g of a constant term t or of a quantified ranked variable $\forall x^\alpha$ is defined by

$$g(t) = 2\rho(t) + 2$$
$$g(\forall x^\alpha) = 2\alpha + 1.$$

2. $\text{Ord}^1(\varphi)$ is the maximum of $g(t)$ and $g(\forall x^\alpha)$ for all t and $\forall x^\alpha$ that occur in φ.
3. $\text{Ord}^2(\varphi) = 0$ if φ has no subformulas of the form $t_1 \in t_2$ where $g(t) = \text{Ord}^1(\varphi)$ and no subformula $K(t)$ nor subformula $F(t)$ where t is a constant term and $g(t) = \text{Ord}^1(\varphi)$.

   $\text{Ord}^2(\varphi) = 1$ otherwise.

4. $\mathrm{Ord}^3(\varphi)$ is the length of φ i.e., the number of logical symbols in φ where atomic formulas $t_1 \in t_2$, $K(t)$, $F(t)$ are assigned length 1.

5. $\mathrm{Ord}(\varphi) = \omega^2 \cdot \mathrm{Ord}^1(\varphi) + \omega \cdot \mathrm{Ord}^2(\varphi) + \mathrm{Ord}^3(\varphi)$.

Remark. Note that $\mathrm{Ord}^2(\varphi)$ and $\mathrm{Ord}^3(\varphi)$ are natural numbers. Proof by induction on $\mathrm{Ord}(\varphi)$ is illustrated by the following in which $P(\ )$ stands for $K(\ )$ or $F(\ )$.

Theorem 8.5. If t is a term in φ such that $\mathrm{Ord}^1(\varphi) = g(t)$ then t does not occur in any other abstraction term of φ.

Proof. If $t_1 = \hat{x}^\alpha \psi(x^\alpha)$ is an abstraction term such that t occurs in t_1 and t_1 occurs in φ, then t occurs in $\psi(x^\alpha)$. Hence by Definition 8.4.

$$g(t_1) = 2\alpha + 2 \le \mathrm{Ord}^1(\varphi).$$

If $t = \underset{\sim}{k}$ for some $k \in K$ then rank $(k) = \rho(k) < \alpha$.

If $t = \hat{y}^\beta \psi(y^\beta)$ with $\beta < \alpha$ then $g(t) < 2\alpha + 1 < \mathrm{Ord}^1(\varphi)$.

So if t occurs in an abstraction term of φ then $\mathrm{Ord}^1(\varphi) > g(t)$.

Remark. Thus any term of maximal grade occurring in φ cannot occur within another term of φ. We shall use this result frequently in proofs to follow.

Theorem 8.6. $t \in T_\alpha \to \mathrm{Ord}(\varphi(t)) < \mathrm{Ord}((\forall x^\alpha)\varphi(x^\alpha))$.

Proof. If $t \in T_\alpha$ then $\rho(t) < \alpha$. Hence

$$g(t) < 2\alpha + 1 = g(\forall x^\alpha).$$

Therefore

$$\mathrm{Ord}^1(\varphi(t)) \le \mathrm{Ord}^1((\forall x^\alpha)\varphi(x^\alpha)).$$

We then need only consider the case

$$\alpha_0 \triangleq \mathrm{Ord}^1(\varphi(t)) = \mathrm{Ord}^1((\forall x^\alpha)\varphi(x^\alpha)).$$

Clearly,

$$\mathrm{Ord}^3(\varphi(t)) < \mathrm{Ord}^3((\forall x^\alpha)\varphi(x^\alpha)).$$

If $\mathrm{Ord}^2(\varphi(t)) = 0$ then $\mathrm{Ord}^2(\varphi(t)) \le \mathrm{Ord}^2((\forall x^\alpha)\varphi(x^\alpha))$ and hence $\mathrm{Ord}(\varphi) < \mathrm{Ord}((\forall x^\alpha)\varphi(x^\alpha))$.

If $\mathrm{Ord}^2(\varphi(t)) = 1$ then for some t_1, t_2

$$t_1 \in t_2 \text{ or } P(t_1) \text{ occurs in } \varphi(t) \text{ and } g(t_1) = \alpha_0.$$

Since $g(t) < g(\forall x^\alpha)$ t does not have maximal grade. Therefore t_1 is not t and t_1 does not occur in t. Therefore $t_1 \in t_2$ or $t_1 \in x^\alpha$ or $P(t_1)$ occurs in $(\forall x^\alpha)\varphi(x^\alpha)$. Since $\mathrm{Ord}^1((\forall x^\alpha)\varphi(x^\alpha)) = \mathrm{Ord}^1(\varphi(t))$ we have

$$\mathrm{Ord}^2((\forall x^\alpha)\varphi(x^\alpha)) = 1$$

hence

$$\mathrm{Ord}(\varphi(t)) < \mathrm{Ord}((\forall x^\alpha)\varphi(x^\alpha)).$$

Definition 8.7.

$$t_1 \simeq t_2 \overset{\Delta}{\leftrightarrow} (\forall x^\beta)[x^\beta \in t_1 \overset{\Delta}{\leftrightarrow} x^\beta \in t_2],\ \beta = \max(\rho(t_1), \rho(t_2)).$$

Remark. $t_1 \simeq t_2$ is defined by a limited formula whereas $t_1 = t_2$ is an unlimited formula.

Theorem 8.8. $t_1 \in T_\alpha \to \mathrm{Ord}\,(t \simeq t_1) < \mathrm{Ord}\,(t \in \hat{x}^\alpha\varphi(x^\alpha))$.

Proof. If $t_1 \in T_\alpha$ we obtain, as in the proof of Theorem 8.6

$$\mathrm{Ord}^1\,(t \simeq t_1) \leq \mathrm{Ord}^1\,(t \in \hat{x}^\alpha\varphi(x^\alpha)).$$

Again we need only consider the case

$$\alpha_0 \overset{\Delta}{=} \mathrm{Ord}^1\,(t \simeq t_1) = \mathrm{Ord}^1\,(t \in \hat{x}^\alpha\varphi(x^\alpha)).$$

Then $\alpha_0 = \max(g(t), g(\hat{x}^\alpha\varphi(x^\alpha)) = \max(g(t), g(t_1)) = g(t)$, since $t_1 \in T_\alpha \to \rho(t_1) < g(\hat{x}^\alpha\varphi(x^\alpha))$. Therefore $\mathrm{Ord}^1\,(t \in \hat{x}^\alpha\varphi(x^\alpha)) = g(t)$ and hence

$$\mathrm{Ord}^2\,(t \in \hat{x}^\alpha\varphi(x^\alpha)) = 1.$$

On the other hand since $\alpha_0 = \mathrm{Ord}^1\,(t \simeq t_1) = g(t)$ and $g(t) \neq g(\forall x^\beta) = 2\beta + 1$ where $\beta = \max(\rho(t), \rho(t_1))$,

$$\mathrm{Ord}^2\,(t \simeq t_1) = 0.$$

Therefore

$$\mathrm{Ord}\,(t \simeq t_1) < \mathrm{Ord}\,(t \in \hat{x}^\alpha\varphi(x^\alpha)).$$

Theorem 8.9. $t \in T_\alpha \to \mathrm{Ord}\,(\varphi(t)) < \mathrm{Ord}\,(t \in \hat{x}^\alpha\varphi(x^\alpha))$.

Proof. If $t \in T_\alpha$ then $\rho(t) < \alpha$. Therefore since $\hat{x}^\alpha\varphi(x^\alpha)$ is an abstraction term and hence φ must satisfy 4 of Definition 8.1,

$$\mathrm{Ord}^1\,(\varphi(t)) < g(\hat{x}^\alpha\varphi(x^\alpha)).$$

Hence

$$\mathrm{Ord}^1\,(\varphi(t)) < \mathrm{Ord}^1\,(t \in \hat{x}^\alpha\varphi(x^\alpha))$$

and

$$\mathrm{Ord}\,(\varphi(t)) < \mathrm{Ord}\,(t \in \hat{x}^\alpha\varphi(x^\alpha)).$$

Theorem 8.10.

$$\mathrm{rank}\,(k_1) < \mathrm{rank}\,(k_2) \to \mathrm{Ord}\,(t \simeq \underset{\sim}{k}_1) < \mathrm{Ord}\,(t \in \underset{\sim}{k}_2).$$

Proof. If $\mathrm{rank}\,(k_1) < \mathrm{rank}\,(k_2)$ then

$$\mathrm{Ord}^1\,(t \simeq \underset{\sim}{k}_1) = \max(g(t), g(\underset{\sim}{k}_1)) \leq \max(g(t), g(\underset{\sim}{k}_2)) = \mathrm{Ord}^1\,(t \in \underset{\sim}{k}_2).$$

If

$$\alpha_0 \overset{\Delta}{=} \mathrm{Ord}^1\,(t \simeq \underset{\sim}{k}_1) = \mathrm{Ord}^1\,(t \in \underset{\sim}{k}_2)$$

then $\alpha_0 = g(t)$ since rank $(k_1) <$ rank (k_2). Therefore

$$\text{Ord}^2\,(t \in \underset{\sim}{k}_2) = 1.$$

But

$$\text{Ord}^2\,(t \simeq \underset{\sim}{k}_1) = 0.$$

Hence

$$\text{Ord}\,(t \simeq \underset{\sim}{k}_1) < \text{Ord}\,(t \in \underset{\sim}{k}_2).$$

Theorem 8.11. rank $(k) \leq \rho(t) \to \text{Ord}\,(t \simeq \underset{\sim}{k}) < \text{Ord}\,(P(t))$.

Proof. $\text{Ord}^1\,(P(t)) = g(t) = \text{Ord}^1\,(t \simeq \underset{\sim}{k})$, since rank $(k) \leq \rho(t)$.

$$\text{Ord}^2\,(P(t)) = 1$$

But

$$\text{Ord}^2\,(t \simeq \underset{\sim}{k}) = 0.$$

Hence

$$\text{Ord}\,(t \simeq \underset{\sim}{k}) < \text{Ord}\,(P(t)).$$

Definition 8.12. If t is a term then

$$\text{Ord}^1\,(t) \triangleq g(t), \qquad \text{Ord}^2\,(t) \triangleq 0, \qquad \text{Ord}^3\,(t) \triangleq 0, \qquad \text{Ord}\,(t) \triangleq \omega^2 \cdot g(t).$$

Theorem 8.13.

1. $\text{Ord}\,(t) < \text{Ord}\,(P(t))$
2. $\max\,(\text{Ord}\,(t_1), \text{Ord}\,(t_2)) < \text{Ord}\,(t_1 \in t_2)$.

Remark. The preceding theorems provide a basis for the definition of a denotation operator D defined on terms and closed limited formulas. The definition is by recursion on $\text{Ord}\,(t)$ and $\text{Ord}\,(\varphi)$.

Definition 8.14

1. $D(\underset{\sim}{k}) \triangleq k,\ k \in K$.
2. $D(\hat{x}^\alpha \varphi(x^\alpha)) \triangleq \{D(t) \mid t \in T_\alpha \wedge D(\varphi(t))\}$.
3. $D(\neg\varphi) \overset{\Delta}{\leftrightarrow} \neg D(\varphi)$.
4. $D(\varphi \wedge \psi) \overset{\Delta}{\leftrightarrow} D(\varphi) \wedge D(\psi)$.
5. $D((\forall x^\alpha)\varphi(x^\alpha)) \overset{\Delta}{\leftrightarrow} (\forall t \in T_\alpha) D(\varphi(t))$.
6. $D(t_1 \in t_2) \overset{\Delta}{\leftrightarrow} D(t_1) \in D(t_2),\ t_1, t_2 \in T$
   $D(K(t)) \overset{\Delta}{\leftrightarrow} D(t) \in K,\ t \in T$
   $D(F(t)) \overset{\Delta}{\leftrightarrow} D(t) \in F,\ t \in T$.

Remark. More exactly D should be defined on $\ulcorner\varphi\urcorner$ and D restricted to closed limited formulas $\ulcorner\varphi\urcorner$ should be regarded as a function onto 2. It should be noted that this recursive definition is permissible since T_α and the class of closed limited formulas φ such that $\text{Ord}\,(\varphi) < \alpha$ are sets.

Theorem 8.15. D is definable in $\mathscr{L}_0(K(\ ), F(\ ))$.

Definition 8.16. D can be extended to an operator $\tilde{D}$ defined for all closed unlimited formulas of $R(K, F)$ by adding

$$\tilde{D}((\forall x)\varphi(x)) \overset{\Delta}{\leftrightarrow} (\forall t \in T)\tilde{D}(\varphi(t)).$$

Remark. Since T is a proper class, $\tilde{D}$ is no longer definable in the language $\mathscr{L}_0(K(\ ), F(\ ))$, simultaneously for all unlimited formulas. As in the case of truth definitions $\tilde{D}(\varphi)$ is definable in $\mathscr{L}_0(K(\ ), F(\ ))$ for any particular formula φ or indeed for any set of formulas φ with less than n quantifiers, n a fixed natural number.

Finally we relate the method of this section with the concepts introduced in §7 by proving the following theorem.

Theorem 8.17. $L[K; F] = \{D(t) \mid t \in T\}$ where $F \subseteq K$ and K is transitive.

Remark. For the proof we need the following.

Definition 8.18. A limited formula φ is of rank $\leq \alpha$ iff every quantifier in φ is of the form $\forall x^\beta$, for some $\beta \leq \alpha$, and every constant term occurring in φ is an element of T_α. We now define an operator D_α for closed limited formulas of rank $\leq \alpha$:

1. $D_\alpha(t_1 \in t_2) \overset{\Delta}{\leftrightarrow} D(t_1) \in D(t_2)$.
2. $D_\alpha(K(t)) \overset{\Delta}{\leftrightarrow} D(t) \in K$.
3. $D_\alpha(F(t)) \overset{\Delta}{\leftrightarrow} D(t) \in F$.
4. $D_\alpha((\forall x^\alpha)\varphi(x^\alpha)) \overset{\Delta}{\leftrightarrow} (\forall x \in T_\alpha)D_\alpha(\varphi(x))$.
5. $D_\alpha(\varphi \wedge \psi) \overset{\Delta}{\leftrightarrow} D_\alpha(\varphi) \wedge D_\alpha(\psi)$.
6. $D_\alpha(\neg\varphi) \overset{\Delta}{\leftrightarrow} \neg D_\alpha(\varphi)$.
7. $D_\alpha((\forall x^\gamma)\varphi(x^\gamma)) \overset{\Delta}{\leftrightarrow} (\forall x \in B'_\gamma)D_\alpha(\varphi(x))$, $\gamma < \alpha$, $B'_\gamma = \{D(t) \mid t \in T_\gamma\}$.

Remark. Then $\bar{K}_\alpha \subseteq B'_\alpha$ and $\alpha < \beta \rightarrow B'_\alpha \in B'_\beta$ since $D(\hat{x}^\alpha(x^\alpha \simeq x^\alpha)) = B'_\alpha$. Set $\mathbf{B}'_\alpha \triangleq \langle B'_\alpha, \bar{K}_\alpha, \bar{F}_\alpha \rangle$.

We then prove by induction on α that

i) $B_\alpha = B'_\alpha$ and
ii) $D(\varphi) \leftrightarrow \mathbf{B}_\alpha \models D_\alpha(\varphi)$ for φ of rank $\leq \alpha$.

We need only consider the case $\alpha \notin K_{\mathrm{II}}$. If i) holds for $\alpha \leq \beta$ and $t \in T_\beta$ then

$$\begin{aligned} D(K(t)) &\leftrightarrow D(t) \in K \\ &\leftrightarrow D(t) \in K \cap B'_\beta \\ &\leftrightarrow D(t) \in \bar{K}_\beta \qquad \text{(by i) for } \alpha = \beta) \\ &\leftrightarrow \mathbf{B}_\beta \models K(D(t)) \end{aligned}$$

Similarly, we can prove

$$D(F(t)) \leftrightarrow \mathbf{B}_\beta \models F(D(t)).$$

We next prove ii) for $\alpha = \beta$ by induction on $\mathrm{Ord}^3(\varphi)$. Since all other cases are trivial or obtained from i) we need only prove

a. $D((\forall x^\beta)\varphi(x^\beta)) \leftrightarrow \mathbf{B}_\beta \models D_\beta((\forall x^\beta)\varphi(x^\beta))$ and
b. $D((\forall x^\gamma)\varphi(x^\gamma)) \leftrightarrow \mathbf{B}_\beta \models D_\beta((\forall x^\gamma)\varphi(x^\gamma)), \gamma < \beta$

assuming ii) holds for all $\varphi(t)$ with $t \in T_\beta$.

$$\begin{aligned} D((\forall x^\beta)\varphi(x^\beta)) &\leftrightarrow (\forall t \in T_\beta) D(\varphi(t)) \\ &\leftrightarrow (\forall t \in T_\beta)[\mathbf{B}_\beta \models D_\beta(\varphi(D(t)))] \quad \text{(by the induction hypothesis)} \\ &\leftrightarrow (\forall a \in B'_\beta)[\mathbf{B}_\beta \models D_\beta(\varphi(a))] \\ &\leftrightarrow (\forall a \in B_\beta)[\mathbf{B}_\beta \models D_\beta(\varphi(a))] \\ &\leftrightarrow \mathbf{B}_\beta \models D_\beta((\forall x^\beta)\varphi(x^\beta)) \end{aligned}$$

b is proved similarly. We now show that $B_{\beta+1} = B'_{\beta+1}$

$$\begin{aligned} D(\hat{x}^\beta\varphi(x^\beta)) &= \{D(t) \mid t \in T_\beta \wedge D(\varphi(t))\} \\ &= \{D(t) \mid t \in T_\beta \wedge \mathbf{B}_\beta \models D_\beta(\varphi(D(t)))\} \\ &= \{a \in B_\beta \mid \mathbf{B}_\beta \models D_\beta(\varphi(a))\}. \end{aligned}$$

Thus if $t = \hat{x}^\gamma\varphi(x^\gamma)$ for some $\gamma \leq \beta$, then

$$D(t) \in B'_{\beta+1} \leftrightarrow D(t) \in Df(\mathbf{B}_\beta).$$

Furthermore

$$\begin{aligned} k = D(k) \in B'_{\beta+1} &\leftrightarrow \text{rank}\,(k) \leq \beta \wedge k \in K \\ &\leftrightarrow k \in K \wedge R(\beta + 1) \\ &\leftrightarrow k \in \bar{K}_{\beta+1}. \end{aligned}$$

Thus $B_{\beta+1} = B'_{\beta+1}$.

Remark. The ramified language and the operator D are very useful in the sense that the definition of D is carried out by using K, F and transfinite recursion i.e., without using any knowledge about V other than the theory of ordinal numbers. Therefore if $On \subseteq V' \subseteq V$ and V' is a standard transitive model of ZF and $F \subseteq K \subseteq V'$ (where K is transitive) then

$$L[K; F]^{V'} = L[K; F].$$

If $\tilde{M}$ is a standard transitive model of ZF, which is a set, and α_0 is the first ordinal not in $\tilde{M}$ i.e., $\alpha_0 = (On)^{\tilde{M}} = On \cap \tilde{M}$ (α_0 is called the order type of $\tilde{M}$), if M is another standard transitive model of ZF such that $\alpha_0 \subseteq M \subseteq \tilde{M}$ and $F \subseteq M$ where F is a class in $\tilde{M}$ i.e., $(\forall x \in \tilde{M})[x \cap F \in \tilde{M}]$ then in $\tilde{M}$, M is a proper class containing all the ordinals of $\tilde{M}$. Therefore we can construct $L[M; F]$ in $\tilde{M}$ and we define this to be $M[F]$. Without knowing $\tilde{M}$, the construction of $M[F]$ can be done using a ramified language where all the ordinals α in x^α, T_α etc. range over α_0 instead of the whole of On. This construction is independent of the choice of $\tilde{M}$ i.e., if $\tilde{M}_1$, $\tilde{M}_2$ are two standard transitive models of ZF with order type α_0, if $M \subseteq \tilde{M}_1$, $M \subseteq \tilde{M}_2$, $F \subseteq M$ and F is a class in $\tilde{M}_1$ and $\tilde{M}_2$, then $L[M; F]$ in $\tilde{M}_1$ and $L[M, F]$ in $\tilde{M}_2$ are the same. Note that we may have $\tilde{M}_1 \nsubseteq \tilde{M}_2$ and $\tilde{M}_2 \nsubseteq \tilde{M}_1$.

Even if there is no such $\tilde{M}$ we can construct $M[F]$ by using ramified language. However in this case $M[F]$ need not be a model of ZF, since we have the following counter example.

Let M be a countable standard transitive model of ZF, $F \subseteq M$ a well ordering of ω whose order type is On^M. A standard transitive model $\tilde{M}$ with the properties described above exists iff $M[F]$ is a standard transitive model of ZF, but $M[F]$ cannot be a model of ZF (since the order-type of $M[F]$ is On^M).

9. Boolean-Valued Relative Constructibility

In this section we will generalize the theory of relative constructibility to Boolean-valued structures for Boolean algebras $\mathbf{B}$ that are sets. Here $\mathscr{L}_0$ will denote the language of the first-order predicate calculus with predicate constants $=$ and $\in$. In addition $\mathscr{L}$ is a first order language that is an extension of $\mathscr{L}_0$. In most applications $\mathscr{L}$ will have only finitely many constants but it may have infinitely many. $\mathbf{M}$ and $\mathbf{M}'$ will be two $\mathbf{B}$-valued structures for the language $\mathscr{L}$. Recall that $\mathbf{M}$ and $\mathbf{M}'$ must each satisfy the Axioms of Equality of Definition 6.5. Also, whenever we consider $[\![\varphi]\!]_{\mathbf{M}}$ we assume that the underlying Boolean algebra is M-complete where $M = |\mathbf{M}|$ i.e., M is the universe of $\mathbf{M}$.

With these conventions we proceed with the task of defining Boolean-valued relative constructibility.

Definition 9.1. $\mathbf{M}$ is a $\mathbf{B}$-valued substructure of $\mathbf{M}'$ iff

1. $M \subseteq M'$,
2. For each n-ary predicate symbol R of $\mathscr{L}$, including $=$ and $\in$,

$$(\forall a_1, \ldots, a_n \in M)[[\![R(a_1, \ldots, a_n)]\!]_{\mathbf{M}} = [\![R(a_1, \ldots, a_n)]\!]_{\mathbf{M}'}]$$

3. $c^{\mathbf{M}} = c^{\mathbf{M}'}$ for each individual constant c of $\mathscr{L}$.

Remark. Most of the conditions 1–3 of page 68 can be easily generalized to the $\mathbf{B}$-valued case. It is, however, more difficult to find an adequate condition corresponding to the requirement that $\mathbf{M}_\alpha$ be transitive.

Definition 9.2. If $\mathbf{M}$ is a $\mathbf{B}$-valued structure for $\mathscr{L}$ and $M' \subseteq M$, then an element $b \in M$ is defined over M' iff

$$(\forall x \in M)\left[[\![x \in b]\!] = \sum_{x' \in M'} [\![x = x']\!][\![x' \in b]\!]\right].$$

Remark. Thus, in order to calculate the value of $[\![x \in b]\!]$, if b is defined over M', we need only know the values $[\![x' \in b]\!]$ for $x' \in M'$.

We now wish to formulate conditions analogous to 1–3 of page 68. Let $\langle \mathbf{M}_\alpha \mid \alpha \in On \rangle$ be a sequence of $\mathbf{B}$-valued structures for the language $\mathscr{L}$ such that M_α is a nonempty set except for M_0,

1. $\mathbf{M}_\alpha$ is a $\mathbf{B}$-valued substructure of $\mathbf{M}_\beta$, for $\alpha < \beta$,

and

2. $M_\alpha = \bigcup_{\beta<\alpha} M_\beta, \alpha \in K_{\mathrm{II}}$.

Then $M \triangleq \bigcup_{\alpha \in On} M_\alpha$.

Again we can define $\mathbf{M}$ such that $|\mathbf{M}| = M$, $\mathbf{M}$ is a $\mathbf{B}$-valued structure and $\mathbf{M}_\alpha$ is a $\mathbf{B}$-valued substructure of $\mathbf{M}$ for all α. $\mathbf{M}$ is uniquely determined by these conditions. $[\![\varphi]\!]$ stands for $[\![\varphi]\!]_{\mathbf{M}}$. Furthermore we require the following conditions.

3. $\mathbf{M}_\alpha$ satisfies the Axiom of Extensionality.
4. For each $b \in M_{\alpha+1}$, b is defined over M_α.
5. For each formula φ of $\mathscr{L}_0$

$$(\forall a_1, \ldots, a_n \in M_\alpha)(\exists b \in M_{\alpha+1})(\forall a \in M_\alpha)[[\![\varphi(a, a_1, \ldots, a_n)]\!]_{\mathbf{M}_\alpha} = [\![a \in b]\!]].$$

Condition 4 replaces the requirement that M_α be transitive for all α in the 2-valued case. Note that 5 is just the condition 3 of page 68 i.e.,

$$Df(\mathbf{M}_\alpha) \subseteq M_{\alpha+1} \quad \text{for} \quad \mathbf{B} = \mathbf{2}.$$

Since $\mathbf{M}_\alpha$ is a $\mathbf{B}$-valued substructure of $\mathbf{M}$

$$(\forall a_1, \ldots, a_n \in M_\alpha)[[\![\varphi(a_1, \ldots, a_n)]\!]_{\mathbf{M}_\alpha} = [\![\varphi(a_1, \ldots, a_n)]\!]]$$

if φ contains no quantifiers.

The following three theorems are proved just as in the case $\mathbf{B} = \mathbf{2}$.

Theorem 9.3. If $\mathbf{A}$ is a $\mathbf{B}$-valued structure for $\mathscr{L}$ and $c^{\mathbf{A}} \in A$ for every individual constant c of $\mathscr{L}$, then there exists a unique $\mathbf{B}$-valued substructure $\mathbf{C}$ of $\mathbf{A}$ such that $|\mathbf{C}| = A$.

Theorem 9.4. If $\mathbf{A}$ is a $\mathbf{B}$-valued structure for $\mathscr{L}$ and $|\mathbf{A}|$ is a set, then there exists a formula Φ of $\mathscr{L}_0$ such that for all closed formulas φ of $\mathscr{L}(C(A))$

$$[\![\varphi]\!]_{\mathbf{A}} = b \leftrightarrow \Phi(\mathbf{A}, \mathbf{B}, \ulcorner\varphi\urcorner, b).$$

Theorem 9.5. If $\mathbf{A}$ is a $\mathbf{B}$-valued structure for $\mathscr{L}$ where $A = |\mathbf{A}|$ may be a proper class then for each formula φ of $\mathscr{L}_0$ there exists a formula ψ of $\mathscr{L}_0$ such that

$$(\forall a_1, \ldots, a_n \in A)[[\![\varphi(a_1, \ldots, a_n)]\!]_{\mathbf{A}} = b \leftrightarrow \psi(\mathbf{A}, \mathbf{B}, a_1, \ldots, a_n, b)].$$

Theorem 9.6. $(\forall a \in M_\alpha)[[\![(\exists x \in a)\varphi(x)]\!] = \sum_{x \in M_\alpha} [\![x \in a]\!]\, [\![\varphi(x)]\!]]$.

Proof. If $a \in M_\alpha$ then $a \in M_{\alpha+1}$ and hence a is defined over M_α.

$$\begin{aligned}[\![(\exists x \in a)\varphi(x)]\!] &= \sum_{x \in M} [\![x \in a]\!]\, [\![\varphi(x)]\!] \\ &= \sum_{x \in M} \sum_{x' \in M_\alpha} [\![x = x']\!]\, [\![x' \in a]\!]\, [\![\varphi(x)]\!]\end{aligned}$$

(Since a is defined over M_α)

$$\leq \sum_{x \in M} \sum_{x' \in M_\alpha} [\![x = x']\!] [\![x' \in a]\!] [\![\varphi(x')]\!] \qquad \text{(Axiom of Equality)}$$

$$\leq \sum_{x' \in M_\alpha} [\![x' \in a]\!] [\![\varphi(x')]\!]$$

$$\leq \sum_{x \in M} [\![x \in a]\!] [\![\varphi(x)]\!]$$

$$= [\![(\exists x \in a)\varphi(x)]\!].$$

Theorem 9.7. $(\forall a \in M_\alpha)[[\![(\forall x \in a)\varphi(x)]\!] = \prod_{x \in M_\alpha} ([\![x \in a]\!] \Rightarrow [\![\varphi(x)]\!])]$.

Remark. Theorem 9.7 follows from duality. The preceding results enable us to cope with bounded quantifiers. As an application we have the following.

Theorem 9.8. $\mathbf{M}$ satisfies the Axiom of Extensionality i.e.,

$$(\forall a, b \in M)[[\![(\forall x)[x \in a \leftrightarrow x \in b] \rightarrow a = b]\!] = \mathbf{1}].$$

Proof. If $a, b \in M$ then $(\exists \alpha)[a \in M_\alpha \wedge b \in M_\alpha]$. Then from Theorem 9.7

$$\begin{aligned} [\![(\forall x)[x \in a \leftrightarrow x \in b]]\!] &= \prod_{x \in M_\alpha} [\![x \in a \leftrightarrow x \in b]\!] \\ &= \prod_{x \in M_\alpha} [\![x \in a \leftrightarrow x \in b]\!]_{\mathbf{M}_\alpha} \\ &\leq [\![a = b]\!]_{\mathbf{M}_\alpha} \qquad \text{(by 3 above)} \\ &= [\![a = b]\!]. \end{aligned}$$

Theorem 9.9. $\mathbf{M}$ satisfies the Axiom of Unions i.e.,

$$(\forall a \in M)[\![(\exists b)(\forall x)[x \in b \leftrightarrow (\exists y \in a)[x \in y]]\!] = \mathbf{1}.$$

Proof. If $a \in M_\alpha$ then $\exists b \in M_{\alpha+1}$ such that

$$(\forall x' \in M_\alpha)[[\![(\exists y \in a)[x' \in y]]\!]_{\mathbf{M}_\alpha} = [\![x' \in b]\!]].$$

Since b is defined over M_α,

$$\begin{aligned} [\![x \in b]\!] &= \sum_{x' \in M_\alpha} [\![x = x']\!] [\![(\exists y \in a)[x' \in y]]\!]_{\mathbf{M}_\alpha} \\ &= \sum_{x' \in M_\alpha} [\![x = x']\!] \sum_{y \in M_\alpha} [\![y \in a]\!] [\![x' \in y]\!] \\ &= \sum_{y \in M_\alpha} [\![y \in a]\!] \sum_{x' \in M_\alpha} [\![x = x']\!] [\![x' \in y]\!] \\ &= \sum_{y \in M_\alpha} [\![y \in a]\!] [\![x \in y]\!] \qquad \text{(Since } y \text{ is defined over } M_\alpha) \\ &= [\![(\exists y \in a)[x \in y]]\!] \qquad \text{(by Theorem 9.6).} \end{aligned}$$

Remark. The Axiom of Pairing is established similarly.

Theorem 9.10. $\mathbf{M}$ satisfies the Axiom of Regularity i.e.,

$$(\forall a \in M)[[\![\exists x \in a \rightarrow (\exists x \in a)(\forall y \in x)[y \notin a]]\!] = \mathbf{1}].$$

Proof. If $a \in M$ we wish to show that

$$[\![\exists x \in a]\!] \leq [\![(\exists x \in a)[a \cap x = 0]]\!].$$

If not then $(\exists x_0 \in M)[[\![x_0 \in a]\!] \nleq [\![(\exists x \in a)[a \cap x = 0]]\!]]$. If

$$\alpha = \mu_\gamma((\exists x_0 \in M_\gamma)[[\![x_0 \in a]\!] \nleq [\![(\exists x \in a)[a \cap x = 0]]\!]]),$$

then $\exists x_0 \in M_\alpha$ such that

$$[\![x_0 \in a]\!] \nleq [\![(\exists x \in a)[a \cap x = 0]]\!].$$

Since $M_0 = 0$, $(\exists \alpha_0)[\alpha = \alpha_0 + 1]$. Then

$$\begin{aligned}
[\![x_0 \cap a \neq 0]\!] &= [\![(\exists y \in x_0)[y \in a]]\!] \\
&= \sum_{y \in M} [\![y \in x_0]\!]\,[\![y \in a]\!] \\
&= \sum_{y \in M} \sum_{y' \in M_{\alpha_0}} [\![y = y']\!]\,[\![y' \in x_0]\!]\,[\![y \in a]\!] \\
&\leq \sum_{y \in M} \sum_{y' \in M_{\alpha_0}} [\![y = y']\!]\,[\![y' \in x_0]\!]\,[\![y' \in a]\!] \\
&\leq \sum_{y \in M} \sum_{y' \in M_{\alpha_0}} [\![y = y']\!]\,[\![y' \in x_0]\!]\,[\![(\exists x \in a)[a \cap x = 0]]\!] \\
&\leq \sum_{y \in M} [\![y \in x_0]\!]\,[\![(\exists x \in a)[a \cap x = 0]]\!].
\end{aligned}$$

Then

$$[\![x_0 \cap a \neq 0]\!] \leq [\![(\exists x \in a)[a \cap x = 0]]\!].$$

Hence

$$\begin{aligned}
\mathbf{1} &= [\![x_0 \cap a = 0]\!] + [\![(\exists x \in a)[a \cap x = 0]]\!] \\
[\![x_0 \in a]\!] &\leq [\![x_0 \in a]\!]\,[\![x_0 \cap a = 0]\!] + [\![(\exists x \in a)[a \cap x = 0]]\!] \\
&\leq [\![(\exists x \in a)[a \cap x = 0]]\!] + [\![(\exists x \in a)[a \cap x = 0]]\!] \\
&= [\![(\exists x \in a)[a \cap x = 0]]\!].
\end{aligned}$$

This contradicts the choice of x_0.

Theorem 9.11. **M** satisfies the Axiom of Infinity.

Proof. Left to the reader.

Remark. We now turn to the proof of the Axiom of Separation.

Theorem 9.12. The function $F: On \to B$ defined by

$$F(\beta) = \sum_{a \in M_\beta} [\![\varphi(a)]\!]$$

is nondecreasing, with respect to the Boolean relation $\leq$ of **B**, and it is continuous.

Proof. Obvious.

Theorem 9.13. If $F: On \to B$ is nondecreasing, then

$$(\exists \beta)(\forall \alpha \geq \beta)[F(\alpha) = F(\beta)]$$

i.e., F is eventually constant.

Proof. If $g(b) = \mu_\beta(b \leq F(\beta))$, $b \in B$ then $g: B \to On$. Since B is a set, $(\exists\beta)[\beta = \sup g``B]$. Then $(\forall\alpha \geq \beta)[F(\alpha) = F(\beta)]$.

Theorem 9.14. $(\exists\beta)[[\![(\exists y)\varphi(y)]\!] = \sum_{a \in M_\beta} [\![\varphi(a)]\!]]$.

Proof. Theorems 9.12 and 9.13.

Corollary 9.15.

$$(\forall\alpha)(\exists\beta)(\forall a_1, \ldots, a_n \in M_\alpha)\left[[\![(\exists y)\varphi(a_1, \ldots, a_n, y)]\!] = \sum_{a \in M_\beta} [\![\varphi(a_1, \ldots, a_n, a)]\!]\right].$$

Theorem 9.16. For each formula φ of $\mathscr{L}$ there exists a semi-normal function F such that

$$(\forall\alpha)(\forall\beta)\left[\beta = F(\alpha) \to (\forall a_1, \ldots, a_n \in M_\alpha)\left[[\![(\exists y)\varphi(a_1, \ldots, a_n, y)]\!] = \sum_{a \in M_\beta} [\![\varphi(a_1, \ldots, a_n, a)]\!]\right]\right].$$

Proof. If

$$F(\alpha) \triangleq \mu_\gamma\left(\gamma \geq \alpha \wedge (\forall a_1, \ldots, a_n \in M_\alpha)\left[[\![(\exists y)\varphi(a_1, \ldots, a_n, y)]\!] = \sum_{a \in M_\gamma} [\![\varphi(a_1, \ldots, a_n, a)]\!]\right]\right),$$

the result then follows from Corollary 9.15.

Corollary 9.17. For each formula φ of $\mathscr{L}$ there exists a semi-normal function F such that

$$(\forall\beta)\left[\beta = F(\beta) \to (\forall a_1, \ldots, a_n \in M_\beta)\left[[\![(\exists y)\varphi(a_1, \ldots, a_n, y)]\!] = \sum_{a \in M_\beta} [\![\varphi(a_1, \ldots, a_n, a)]\!]\right]\right].$$

Theorem 9.18. For each formula φ of $\mathscr{L}$ there exist finitely many semi-normal functions $F_1, \ldots, F_m$ such that

$$(\forall\beta)[\beta = F_1(\beta) = \cdots = F_m(\beta) \to (\forall a_1, \ldots, a_n \in M_\beta)[[\![\varphi(a_1, \ldots, a_n)]\!] = [\![\varphi(a_1, \ldots, a_n)]\!]_{\mathbf{M}_\beta}]].$$

Proof. Left to the reader.

Theorem 9.19. **M** satisfies the Axiom of Separation i.e.,

$$(\forall a_1, \ldots, a_n, a \in M)[[\![(\exists b)(\forall x)[x \in b \leftrightarrow x \in a \wedge \varphi(x, a_1, \ldots, a_n)]\!] = \mathbf{1}].$$

Proof. If $a, a_1, \ldots, a_n \in M$ then there is an α such that

$$[a, a_1, \ldots, a_n \in M_\alpha] \wedge (\forall x' \in M_\alpha)[[\![x' \in a \wedge \varphi(x', a_1, \ldots, a_n)]\!] = [\![x' \in a \wedge \varphi(x', a_1, \ldots, a_n)]\!]_{\mathbf{M}_\alpha}].$$

Therefore $\exists b \in M_{\alpha+1}$ such that

$$(\forall x' \in M_\alpha)[[\![x' \in b]\!] = [\![x' \in a \wedge \varphi(x', a_1, \ldots, a_n)]\!]_{\mathbf{M}_\alpha}].$$

Then

$$\begin{aligned}
[\![x \in b]\!] &= \sum_{x' \in M_\alpha} [\![x = x']\!] [\![x' \in b]\!] \\
&= \sum_{x' \in M_\alpha} [\![x = x']\!] [\![x' \in a]\!] [\![\varphi(x', a_1, \ldots, a_n)]\!]_{\mathbf{M}_\alpha}. \\
&= \sum_{x' \in M_\alpha} [\![x = x']\!] [\![x' \in a]\!] [\![\varphi(x', a_1, \ldots, a_n)]\!] \\
&\leq \sum_{x' \in M_\alpha} [\![x = x']\!] [\![x \in a]\!] [\![\varphi(x, a_1, \ldots, a_n)]\!] \\
&\leq \sum_{x' \in M_\alpha} [\![x \in a]\!] [\![\varphi(x, a_1, \ldots, a_n)]\!] \\
&= [\![x \in a]\!] [\![\varphi(x, a_1, \ldots, a_n)]\!] \\
&= \sum_{x' \in M_\alpha} [\![x = x']\!] [\![x' \in a]\!] [\![\varphi(x, a_1, \ldots, a_n)]\!] \\
&\leq \sum_{x' \in M_\alpha} [\![x = x']\!] [\![x' \in a]\!] [\![\varphi(x', a_1, \ldots, a_n)]\!].
\end{aligned}$$

Then $[\![x \in b]\!] = [\![x \in a]\!] [\![\varphi(x, a_1, \ldots, a_n)]\!]$.

Remark. The problem in the proof of the Axiom of Powers is to find a kind of bound for all b such that $b \subseteq a$.

Theorem 9.20. If $a \in M_{\alpha+1}$ and $[\![(\forall x)[x \in b \leftrightarrow x \in a \wedge \varphi(x, a_1, \ldots, a_n)]]\!] = \mathbf{1}$ then b is defined over M_α, i.e., every definable $\mathbf{B}$-valued subset of $a \in M_{\alpha+1}$ is defined over M_α.

Proof. Under the hypothesis of the proposition

$$\begin{aligned}
[\![c \in b]\!] &= [\![c \in a]\!] [\![\varphi(c, a_1, \ldots, a_n)]\!] \\
&= \sum_{c' \in M_\alpha} [\![c = c']\!] [\![c' \in a]\!] [\![\varphi(c, a_1, \ldots, a_n)]\!] \\
&= \sum_{c' \in M_\alpha} [\![c = c']\!] [\![c' \in a]\!] [\![\varphi(c', a_1, \ldots, a_n)]\!] \\
&= \sum_{c' \in M_\alpha} [\![c = c']\!] [\![c' \in b]\!]
\end{aligned}$$

i.e., b is defined over M_α.

Theorem 9.21. If $b_1, b_2 \in M$ are defined over M_α then

$$(\forall x' \in M_\alpha)[[\![x' \in b_1]\!] = [\![x' \in b_2]\!]] \to [\![b_1 = b_2]\!] = \mathbf{1}.$$

Proof.

$$\begin{aligned}
[\![x \in b_1]\!] &= \sum_{x' \in M_\alpha} [\![x = x']\!] [\![x' \in b_1]\!] \\
&= \sum_{x' \in M_\alpha} [\![x = x']\!] [\![x' \in b_2]\!] \\
&= [\![x \in b_2]\!].
\end{aligned}$$

Thus $[\![\mathrm{b}_1 = b_2]\!] = \mathbf{1}$ by the Axiom of Extensionality.

Theorem 9.22.

$$(\forall\alpha)(\exists\beta)(\forall a \in M)[a \text{ is defined over } M_\alpha \rightarrow (\exists b \in M_\beta)[[\![a = b]\!] = \mathbf{1}]].$$

Proof. B^{M_α} is a set. For $s \in B^{M_\alpha}$ we define

$$f(s) \triangleq \mu_\beta((\exists a \in M_\beta)[a \text{ is defined over } M_\alpha \wedge s = \{\langle x, [\![x \in a]\!]\rangle \mid x \in M_\alpha\}])$$
$$= 0 \text{ if there is no such } \beta.$$

Then

$$(\exists\beta)[\beta \triangleq \sup_{s \in B^{M_\alpha}} f(s)].$$

If a is defined over M_α and

$$s = \{\langle x, [\![x \in a]\!]\rangle \mid x \in M_\alpha\}$$

then $s \in B^{M_\alpha}$ and hence

$$(\exists b \in M_{f(s)})[s = \{\langle x, [\![x \in b]\!]\rangle \mid x \in M_\alpha\}]$$

where b is defined over M_α. Then $[\![a = b]\!] = \mathbf{1}$ and $b \in M_\beta$.

Theorem 9.23. $(\forall\alpha)(\exists b \in M)(\forall a \in M_\alpha)[[\![a \in b]\!] = \mathbf{1}]$.

Proof. $(\exists b \in M_{\alpha+1})(\forall a \in M_\alpha)[[\![a \in b]\!] = [\![a = a]\!]_{\mathbf{M}_\alpha} = \mathbf{1}]$.

Remark. In case $\mathbf{B} = \mathbf{2}$ Theorem 9.23 means that M_α is contained in some $b \in M$.

Theorem 9.24. $\mathbf{M}$ satisfies the Axiom of Powers i.e.,

$$(\forall a \in M)[[\![(\exists x)[x = \mathscr{P}(a)]]\!] = \mathbf{1}].$$

Proof. If $a \in M$ then $(\exists\alpha)[a \in M_\alpha]$. By Theorem 9.22

$$(\exists\beta)(\forall b)[b \text{ is defined over } M_\alpha \rightarrow (\exists b' \in M_\beta)[[\![b = b']\!] = \mathbf{1}]].$$

By Theorem 9.23

$$(\exists c)(\forall b' \in M_\beta)[[\![b' \in c]\!] = \mathbf{1}].$$

It is then sufficient to prove $\mathscr{P}(a) \subseteq c$, i.e.,

$$(\forall x \in M)[[\![x \subseteq a]\!] \leq [\![x \in c]\!]].$$

By the proof of the Axiom of Separation

$$(\exists b \in M)[[\![(\forall y)[y \in b \leftrightarrow y \in x \cap y \in a]]\!] = [\![b = x \cap a]\!] = \mathbf{1}].$$

Then b is defined over M_α. Hence

$$(\exists b' \in M_\beta)[[\![b = b']\!] = \mathbf{1}].$$

Therefore

$$\begin{aligned}[\![x \subseteq a]\!] &= [\![x \subseteq a]\!]\,[\![b = x \cap a]\!] \\ &\leq [\![x = b]\!] \\ &= [\![x = b']\!] \\ &= [\![x = b']\!]\,[\![b' \in c]\!] \\ &\leq [\![x \in c]\!].\end{aligned}$$

Theorem 9.25. **M** satisfies the Axiom Schema of Replacement.

Proof. We take the Axiom of Replacement in the following form:

$$(\exists b)(\forall x \in a)(\exists y \in b)[(\exists y')\varphi(x, y') \rightarrow \varphi(x, y)].$$

We wish to prove

$$(\forall a \in M)[[\![(\exists b)(\forall x \in a)(\exists y \in b)\varphi'(x, y)]\!] = \mathbf{1}],$$

where we abbreviate $(\exists y')\varphi(x, y') \rightarrow \varphi(x, y)$ by $\varphi'(x, y)$. First we note that

$$(\forall x \in M)[[\![(\exists y)\varphi'(x, y)]\!] = \mathbf{1}]$$

If $a \in M$ then $(\exists \alpha)[a \in M_\alpha]$ and since $\sum_{y \in M} [\![\varphi'(x, y)]\!] = \mathbf{1}$.

$$(\exists \beta)(\forall x \in M_\alpha)\left[\sum_{y \in M_\beta} [\![\varphi'(x, y)]\!] = \mathbf{1}\right],$$

by Corollary 9.15. Then, for this β

$$(\exists b)(\forall y \in M_\beta)[[\![y \in b]\!] = \mathbf{1}],$$

by Theorem 9.23. Hence

$$\begin{aligned}\sum_{y \in M} [\![y \in b]\!]\,[\![\varphi'(x, y)]\!] &\geq \sum_{y \in M_\beta} [\![y \in b]\!]\,[\![\varphi'(x, y)]\!] \\ &= \sum_{y \in M_\beta} [\![\varphi'(x, y)]\!].\end{aligned}$$

Therefore

$$(\forall x \in M_\alpha)\left[\sum_{y \in M} [\![y \in b \wedge \varphi'(x, y)]\!] = \mathbf{1}\right]$$

and hence

$$(\forall x \in M_\alpha)[[\![(\exists y \in b)\varphi'(x, y)]\!] = \mathbf{1}].$$

Since a is defined over M_α we have for $x \in M$

$$\begin{aligned}[\![x \in a]\!] &= \sum_{x' \in M_\alpha} [\![x = x']\!]\,[\![x' \in a]\!] \\ &\leq \sum_{x' \in M_\alpha} [\![x = x']\!]\,[\![(\exists y \in b)\varphi'(x', y)]\!] \\ &\leq \sum_{x' \in M_\alpha} [\![(\exists y \in b)\varphi'(x, y)]\!] \\ &= [\![(\exists y \in b)\varphi'(x, y)]\!].\end{aligned}$$

Therefore

$$[\![x \in a]\!] \leq [\![(\exists y \in b)\varphi'(x, y)]\!]$$

hence

$$[\![(\forall x \in a)(\exists y \in b)\varphi'(x, y)]\!] = \mathbf{1}.$$

Remark. We have now proved the following.

Theorem 9.26. If the sequence $\langle \mathbf{M}_\alpha \mid \alpha \in On \rangle$ of $\mathbf{B}$-valued structures for $\mathscr{L}$, satisfies the conditions 1–5 of page 87–88 and $\mathbf{M} = \bigcup_{\alpha \in On} \mathbf{M}_\alpha$ then $\mathbf{M}$ is a $\mathbf{B}$-valued model of *ZF*.

Remark. As an application of the method developed in this section we shall define $V[F]$ by using a ramified language. In §7 $L[K]$ was introduced as the class of all sets that are constructible from K. The class K can be identified with its characteristic function $F: K \to 2$. However we now consider a Boolean-valued set i.e., a function $F: K \to |\mathbf{B}|$ and regard the class of sets constructible from K in the $\mathbf{B}$-valued sense. If we think of V as the class of $\mathbf{2}$-valued sets $V[F]$ becomes an extension of V containing new sets.

Assume $K \in V, f_0: K \to B$ and $\mathscr{L}$ is the language $\mathscr{L}_0(C(V) \cup \{V(\ ), F(\ )\})$ therefore we have individual constants $\underline{k}$ for each $k \in V$.

Definition 9.27. A sequence $\langle \mathbf{T}_\alpha \mid \alpha \in On \rangle$, where

$$\mathbf{T}_\alpha = \langle T_\alpha, \bar{=}, \bar{\in}, \bar{V}_\alpha, \bar{F}_\alpha \rangle,$$

of $\mathbf{B}$-valued structures for the language $\mathscr{L}$ is defined as follows:

T_α is the set of constant terms t with rank $\rho(t) < \alpha$ (Definition 8.2). $[\![\varphi]\!]_{\mathbf{T}_\alpha}$ for a limited formula φ of rank $\leq \alpha$ is defined by recursion on Ord (φ) as follows.

1. $[\![V(t)]\!]_{\mathbf{T}_\alpha} \triangleq \sum_{k \in R(\alpha)} [\![t = \underline{k}]\!]_{\mathbf{T}_\alpha} \qquad t \in T_\alpha.$

2. $[\![F(t)]\!]_{\mathbf{T}_\alpha} \triangleq \sum_{k \in R(\alpha) \cap K} [\![t = \underline{k}]\!]_{\mathbf{T}_\alpha} f_0(k) \qquad t \in T_\alpha.$

3. $[\![\underline{k}_1 \in \underline{k}_2]\!]_{\mathbf{T}_\alpha} \triangleq \mathbf{1}$ if $k_1 \in k_2 \qquad \rho(\underline{k}_1), \rho(\underline{k}_2) < \alpha$

   $\triangleq \mathbf{0}$ if $k_1 \notin k_2$.

4. $[\![t \in \underline{k}]\!]_{\mathbf{T}_\alpha} \triangleq \sum_{k' \in k} [\![t = \underline{k}']\!]_{\mathbf{T}_\alpha} \qquad t \in T_\alpha$, t not an individual constant.

5. $[\![t \in \hat{x}^\beta \varphi(x^\beta)]\!]_{\mathbf{T}_\alpha} \triangleq \sum_{t' \in T_\beta} [\![t = t']\!]_{\mathbf{T}_\alpha} [\![\varphi(t')]\!]_{\mathbf{T}_\alpha} \qquad t \in T_\alpha, \beta < \alpha.$

6. $[\![t_1 = t_2]\!]_{\mathbf{T}_\alpha} \triangleq \prod_{t \in T_\beta} [\![t \in t_1 \leftrightarrow t \in t_2]\!]_{\mathbf{T}_\alpha}, \ t_1, t_2 \in T_\alpha, \beta = \max(\rho(t_1), \rho(t_2)).$

7. $[\![\neg\varphi]\!]_{\mathbf{T}_\alpha} \triangleq -[\![\varphi]\!]_{\mathbf{T}_\alpha}, [\![\varphi_1 \wedge \varphi_2]\!]_{\mathbf{T}_\alpha} = [\![\varphi_1]\!]_{\mathbf{T}_\alpha} [\![\varphi_2]\!]_{\mathbf{T}_\alpha}.$

8. $[\![(\forall x^\beta)\varphi(x^\beta)]\!]_{\mathbf{T}_\alpha} \triangleq \prod_{t \in T_\beta} [\![\varphi(t)]\!]_{\mathbf{T}_\alpha} \qquad \beta \leq \alpha.$

We extend the definition of $[\![\varphi]\!]_{\mathbf{T}_\alpha}$ to unlimited formulas by adding

9. $[\![(\forall x)\varphi(x)]\!]_{\mathbf{T}_\alpha} \triangleq \prod_{t \in T_\alpha} [\![\varphi(t)]\!]_{\mathbf{T}_\alpha}$.

Remark. There is a slight difference in our approach compared with §8 since $=$ is now a predicate constant of $\mathscr{L}_0$ whereas previously we defined $\simeq$. However, the definition of $\simeq$ corresponds to 6 above. Theorems of §8 show that in 1–8 Ord (φ) is always reduced to lower values so that the definition is recursive.

We must now prove that the sequence $\langle \mathbf{T}_\alpha \mid \alpha \in On \rangle$ has the desired properties.

Theorem 9.28. $\forall t_1, t_2 \in T_\alpha$.

1. $[\![t_1 = t_1]\!]_{\mathbf{T}_\alpha} = \mathbf{1}$.
2. $[\![t_1 = t_2]\!]_{\mathbf{T}_\alpha} = [\![t_2 = t_1]\!]_{\mathbf{T}_\alpha}$.

Proof. Obvious from Definition 9.27.

Theorem 9.29. $\forall k_1, k_2 \in R(\alpha)$.

$$k_1 = k_2 \to [\![k_1 = k_2]\!]_{\mathbf{T}_\alpha} = \mathbf{1}$$
$$k_1 \neq k_2 \to [\![k_1 = k_2]\!]_{\mathbf{T}_\alpha} = \mathbf{0}.$$

Proof. If $k_1 \neq k_2$ and $\beta \triangleq \max(\rho(k_1), \rho(k_2))$ then by symmetry we can assume $(\exists k)[k \in k_1 \wedge k \notin k_2]$.

Then $\rho(k) < \beta$, and

$$[\![\underline{k}_1 = \underline{k}_2]\!]_{\mathbf{T}_\alpha} = \prod_{t \in T_\beta} [\![t \in \underline{k}_1 \leftrightarrow t \in \underline{k}_2]\!]_{\mathbf{T}_\alpha}$$
$$\leq [\![\underline{k} \in \underline{k}_1 \leftrightarrow \underline{k} \in \underline{k}_2]\!]_{\mathbf{T}_\alpha} = \mathbf{0}.$$

Then k, for $k \in V$, can be treated as a 2-valued set and Definition 9.27.4 holds for $t = k_1$, too.

Lemma. $\forall t, \underline{k} \in T_\alpha$.

1. $\rho(t) \leq \rho(\underline{k}) \to [\![\underline{k} \in t]\!]_{\mathbf{T}_\alpha} = \mathbf{0}$.
2. $\rho(t) < \rho(\underline{k}) \to [\![\underline{k} = t]\!]_{\mathbf{T}_\alpha} = \mathbf{0}$.

Proof. We prove 1 and 2 simultaneously by induction.

If $\rho(t) \leq \rho(\underline{k})$ and $t = \hat{x}^\beta \varphi(x^\beta)$, then since $\rho(t') < \rho(t) \leq \rho(\underline{k})$

$$[\![\underline{k} \in t]\!]_{\mathbf{T}_\alpha} = \sum_{t' \in T_\beta} [\![\underline{k} = t']\!][\![\varphi(t')]\!] = \mathbf{0}.$$

If $\rho(t) < \rho(k)$ then since $\exists k_1 \in k$, $\rho(t) \leq \rho(\underline{k}_1) < \rho(\underline{k})$

$$[\![t = \underline{k}]\!]_{\mathbf{T}_\alpha} \leq \prod_{k_1 \in k} [\![\underline{k}_1 \in t]\!] = \mathbf{0}.$$

Theorem 9.30. $\forall t_1, t_2, t_3 \in T_\alpha$.

1. $[\![t_1 = t_2]\!]_{\mathbf{T}_\alpha} [\![t_2 = t_3]\!]_{\mathbf{T}_\alpha} \leq [\![t_1 = t_3]\!]_{\mathbf{T}_\alpha}$.

Proof. Let $\alpha_i = \rho(t_i)$, $i = 1, 2, 3$, and let $\beta_1 = \max(\alpha_1, \alpha_2)$, $\beta_2 = \max(\alpha_2, \alpha_3)$, and $\beta_3 = \max(\alpha_3, \alpha_1)$. Writing $[\![\ ]\!]$ for $[\![\ ]\!]_{\mathbf{T}_\alpha}$ we proceed by induction on $\rho(t_3) = \alpha_3$.

Case 1. $\beta_3 \leq \beta_1 \wedge \beta_3 \leq \beta_2$. Obvious from Definition 9.27.6.

Case 2. $\beta_1 < \beta_3 \vee \beta_2 < \beta_3$. From Theorem 9.28, 1 is equivalent to:

2. $[\![t_3 = t_2]\!][\![t_2 = t_1]\!] \leq [\![t_3 = t_1]\!]$.

Consequently, if 1 is proved for $\alpha_1 \leq \alpha_3$ then 1 also holds for $\alpha_3 \leq \alpha_1$. Therefore we may assume $\alpha_1 \leq \alpha_3$ and hence $\beta_2 \not< \beta_3$. Thus we need only consider $\beta_1 < \beta_3$. Then $\beta_1 = \max(\alpha_1, \alpha_2) < \beta_2 = \beta_3 = \alpha_3$.

First of all we assume that t_i is defined over T_α for $i = 1, 2, 3$. (That this is really the case will be shown a little later.)

Let $b = [\![t_1 = t_2]\!][\![t_2 = t_3]\!]$. Then for $t \in T_{\alpha_3} = T_{\beta_3}$

$$\begin{aligned}
b[\![t \in t_1]\!] &= b \sum_{t' \in T_{\alpha_1}} [\![t = t']\!][\![t' \in t_1]\!] \\
&\qquad \text{(since } t_1 \text{ is defined over } T_{\alpha_1}) \\
&= [\![t_2 = t_3]\!] \sum_{t' \in T_{\alpha_1}} [\![t_1 = t_2]\!][\![t = t']\!][\![t' \in t_1]\!] \\
&= [\![t_2 = t_3]\!] \sum_{t' \in T_\alpha} \prod_{s \in T_{\beta_1}} [\![s \in t_1 \leftrightarrow s \in t_2]\!][\![t = t']\!][\![t' \in t_1]\!] \\
&\leq \sum_{t' \in T_{\alpha_1}} [\![t_2 = t_3]\!][\![t' \in t_2]\!][\![t = t']\!], \\
&\qquad \text{(since we can take } s = t') \\
&\leq \sum_{t' \in T_{\alpha_1}} [\![t' \in t_3]\!][\![t = t']\!] \\
&\qquad \text{(by the same technique as above)} \\
&\leq \sum_{t' \in T_{\alpha_3}} [\![t = t']\!][\![t' \in t_3]\!] \\
&= [\![t \in t_3]\!] \qquad \text{(since } t_3 \text{ is defined over } T_{\alpha_3}).
\end{aligned}$$

Therefore $b \leq [\![t \in t_1 \rightarrow t \in t_3]\!]$ for $t \in T_{\beta_3}$. On the other hand for $t \in T_{\beta_3}$

$$\begin{aligned}
b[\![t \in t_3]\!] &= [\![t_1 = t_2]\!] \sum_{t' \in T_{\alpha_3}} [\![t_2 = t_3]\!][\![t = t']\!][\![t' \in t_3]\!] \\
&\leq [\![t_1 = t_2]\!] \sum_{t' \in T_{\alpha_3}} [\![t' \in t_2]\!][\![t = t']\!] \\
&\qquad \text{(by the same technique as above)} \\
&= \sum_{t' \in T_{\alpha_3}} [\![t = t']\!] \sum_{t'' \in T_{\alpha_2}} [\![t_1 = t_2]\!][\![t' = t'']\!][\![t'' \in t_2]\!] \\
&\leq \sum_{t' \in T_{\alpha_3}} [\![t = t']\!] \sum_{t'' \in T_{\alpha_2}} [\![t' = t'']\!][\![t'' \in t_1]\!] \\
&\qquad \text{(by the same technique as above)} \\
&\leq \sum_{t'' \in T_{\alpha_2}} [\![t = t'']\!][\![t'' \in t_1]\!] \\
&\qquad \text{(by our induction hypothesis, since } \rho(t'') < \rho(t_3)) \\
&= \sum_{t'' \in T_{\alpha_2}} [\![t = t'']\!] \sum_{t' \in T_{\alpha_1}} [\![t'' = t']\!][\![t' \in t_1]\!] \\
&\leq \sum_{t' \in T_{\alpha_1}} [\![t = t']\!][\![t' \in t_1]\!] \\
&\qquad \text{(by the induction hypothesis, since } \rho(t') < \rho(t_3)) \\
&= [\![t \in t_1]\!].
\end{aligned}$$

Therefore $b \leq [\![t \in t_1 \leftrightarrow t \in t_3]\!]$ for all $t \in T_{\beta_3}$.

We next wish to show that if $\beta_1 < \beta_3$ then t_i is defined over T_α. We first prove that if $\rho(\underline{k}) \leq \rho(t_3)$ then k is defined over $T_{\rho(\underline{k})}$.

$$\begin{aligned}
[\![t \in \underline{k}]\!] &= \sum_{k' \in k} [\![t = \underline{k}']\!][\![\underline{k}' \in \underline{k}]\!] \\
&\leq \sum_{t' \in T_{\rho(\underline{k})}} [\![t = t']\!][\![t' \in \underline{k}]\!] \qquad (\text{since } k \subseteq T_{\rho(\underline{k})}) \\
&= \sum_{t' \in T_{\rho(\underline{k})}} [\![t = t']\!] \sum_{k' \in k} [\![t' = \underline{k}']\!] \\
&\leq \sum_{t' \in T_{\rho(\underline{k})}} \sum_{k' \in k} [\![t = \underline{k}']\!] \qquad (\text{by the induction hypothesis}) \\
&= \sum_{k' \in k} [\![t = \underline{k}']\!] = [\![t \in \underline{k}]\!].
\end{aligned}$$

i.e., $[\![t \in \underline{k}]\!] = \sum_{t' \in T_{\rho(\underline{k})}} [\![t = t']\!][\![t' \in \underline{k}]\!]$.

Next we show that if $t = \hat{x}^\beta \varphi(x^\beta)$ for some $\beta \leq \alpha_3$ then t is defined over T_β.

$$\begin{aligned}
[\![t_0 \in t]\!] &= \sum_{t' \in T_\beta} [\![t_0 = t']\!][\![\varphi(t')]\!] \\
&\leq \sum_{t' \in T_\beta} [\![t_0 = t']\!] \sum_{t'' \in T_\beta} [\![t_0 = t'']\!][\![\varphi(t'')]\!] \\
&\leq \sum_{t' \in T_\beta} [\![t_0 = t']\!] \sum_{t'' \in T_\beta} [\![t' = t'']\!][\![\varphi(t'')]\!], \qquad (\text{by the induction hypothesis}) \\
&= \sum_{t' \in T_\beta} [\![t_0 = t']\!][\![t' = t]\!] \\
&\leq \sum_{t' \in T_\beta} [\![t_0 = t']\!] \sum_{t'' \in T_\beta} [\![t' = t'']\!][\![\varphi(t'')]\!] \\
&\leq \sum_{t'' \in T_\beta} [\![t_0 = t'']\!][\![\varphi(t'')]\!], \qquad (\text{by the induction hypothesis}) \\
&= [\![t_0 \in t]\!].
\end{aligned}$$

Since $\alpha_1 \leq \alpha_2 < \alpha_3$ or $\alpha_2 \leq \alpha_1 < \alpha_3$ it follows that t_i is defined over T_{α_i}.

Remark. We have also proved that if $t \in T_\alpha$, then t is defined over $T_{\rho(t)}$. The same argument can be used to prove the following result.

Theorem 9.31. If $t \in T_\alpha$ then t is defined over T_α.

Remark. We now return to the remaining Axioms of Equality.

Theorem 9.32. $\forall t_1, t_2, t_3 \in T_\alpha$

1. $[\![t_1 = t_2]\!]_{\mathbf{T}_\alpha}[\![t_2 \in t_3]\!]_{\mathbf{T}_\alpha} \leq [\![t_1 \in t_3]\!]_{\mathbf{T}_\alpha}$.
2. $[\![t_1 = t_2]\!]_{\mathbf{T}_\alpha}[\![t_3 \in t_1]\!]_{\mathbf{T}_\alpha} \leq [\![t_3 \in t_2]\!]_{\mathbf{T}_\alpha}$.
3. $[\![t_1 = t_2]\!]_{\mathbf{T}_\alpha}[\![V(t_1)]\!]_{\mathbf{T}_\alpha} \leq [\![V(t_2)]\!]_{\mathbf{T}_\alpha}$.
4. $[\![t_1 = t_2]\!]_{\mathbf{T}_\alpha}[\![F(t_1)]\!]_{\mathbf{T}_\alpha} \leq [\![F(t_2)]\!]_{\mathbf{T}_\alpha}$.

Proof.

1. $$[\![t_1 = t_2]\!][\![t_2 \in t_3]\!] \le \sum_{t \in T_\alpha} [\![t_1 = t_2]\!][\![t = t_2]\!][\![t \in t_3]\!]$$
$$\le \sum_{t \in T_\alpha} [\![t = t_1]\!][\![t \in t_3]\!]$$
$$= [\![t_1 \in t_3]\!] \qquad \text{(by Theorem 9.31).}$$

2. $$[\![t_1 = t_2]\!][\![t_3 \in t_1]\!] = [\![t_1 = t_2]\!] \sum_{t \in T_{\rho(t_1)}} [\![t = t_3]\!][\![t \in t_1]\!]$$
$$= \sum_{t \in T_{\rho(t_1)}} [\![t = t_3]\!] \prod_{s \in T_{\beta_1}} [\![s \in t_1 \leftrightarrow s \in t_2]\!][\![t \in t_1]\!],$$
$$\text{where } \beta_1 = \max(\rho(t_1), \rho(t_2))$$
$$\le \sum_{t \in T_{\rho(t_1)}} [\![t = t_3]\!][\![t \in t_2]\!]$$
$$\le \sum_{t \in T_\alpha} [\![t = t_3]\!][\![t \in t_2]\!], \qquad (\text{since } \rho(t_1) < \alpha)$$
$$= [\![t_3 \in t_2]\!].$$

3 and 4 follow from Theorem 9.30.

Remark. We have now proved that $\mathbf{T}_\alpha$ is a $\mathbf{B}$-valued structure. (See Definition 6.5)

Theorem 9.33. $\mathbf{T}_\alpha$ satisfies the Axiom of Extensionality.

Proof. Obvious from Definition 9.27.6.

Theorem 9.34. If $t \in T_{\alpha+1}$ then t is defined over T_α.

Proof. See remark following Theorem 9.30.

Theorem 9.35. If φ is a formula of $\mathscr{L}$ then

$$(\forall t_1, \ldots, t_n \in T_\alpha)(\exists t' \in T_{\alpha+1})(\forall t \in T_\alpha)[[\![\varphi(t, t_1, \ldots, t_n)]\!]_{\mathbf{T}_\alpha} = [\![t \in t']\!]_{\mathbf{T}_\alpha}].$$

Proof. If $t' \triangleq \hat{x}^\alpha \varphi^\alpha(x^\alpha, t_1, \ldots, t_n)$ where φ^α is the formula obtained from φ by replacing $\forall x$ by $\forall x^\alpha$, then $t' \in T_{\alpha+1}$ and for $t \in T_\alpha$

$$[\![t \in \hat{x}^\alpha \varphi^\alpha(x^\alpha, t_1, \ldots, t_n)]\!]_{\mathbf{T}_{\alpha+1}} = \sum_{t'' \in T_\alpha} [\![t = t'']\!]_{\mathbf{T}_{\alpha+1}} [\![\varphi^\alpha(t'', t_1, \ldots, t_n)]\!]_{\mathbf{T}_{\alpha+1}}$$
$$= [\![\varphi^\alpha(t, t_1, \ldots, t_n)]\!]_{\mathbf{T}_{\alpha+1}}$$
$$= [\![\varphi^\alpha(t, t_1, \ldots, t_n)]\!]_{\mathbf{T}_\alpha}$$
$$= [\![\varphi(t, t_1, \ldots, t_n)]\!]_{\mathbf{T}_\alpha}.$$

Remark. Thus we see that $\{\mathbf{T}_\alpha \mid \alpha \in On\}$ satisfies the conditions 1–5 on p. 87–88. Therefore if $\mathbf{T} = \langle T, \bar{=}, \bar{\in}, \bar{V}, \bar{F}\rangle$ is defined from $\langle \mathbf{T}_\alpha \mid \alpha \in On\rangle$ as $\mathbf{M}$ from $\langle \mathbf{M}_\alpha \mid \alpha \in On\rangle$ then $\mathbf{T}$ is a $\mathbf{B}$-valued model of ZF. We can also define a denotation operator D as in §8 and put $V[f_0] = \{D(t) \mid t \in T\}$. However we are more interested in standard $\mathbf{2}$-valued models. In order to obtain a suitable homomorphism of $V[f_0]$ onto a $\mathbf{2}$-valued model we relativize our results to some M. Let M be a transitive model of ZF, and let $\mathbf{B}$ be an M-complete Boolean algebra with $\mathbf{B} \in M$. Furthermore assume that $K \in M$ and $f_0: K \to B$, $f_0 \in M$. By relativizing our previous definition of $\mathbf{T}$ to M we obtain $(V[f_0])^M$. Let $h: |\mathbf{B}| \to |\mathbf{2}|$ be an M-complete homomorphism. (If M is countable there are such homomorphisms by the Rasiowa-Sikorski Theorem.) Then we can pass to a $\mathbf{2}$-valued standard model.

Definition 9.36. $F_0 \triangleq \{k \in K \mid h(f_0(k)) = \mathbf{1}\}$.
For t, t_1, t_2 constant terms,

1. $D(t_1 \in t_2) \overset{\Delta}{\leftrightarrow} D(t_1) \in D(t_2)$.
2. $D(V(t)) \overset{\Delta}{\leftrightarrow} D(t) \in M$.
3. $D(F(t) \overset{\Delta}{\leftrightarrow} D(t) \in F_0$.

The remainder of the definition is the same as in Definition 8.14.
Let $M[F_0] \triangleq M[h] \triangleq \{D(t) \mid t \in T\}$, where $T = \{T_\alpha{}^M \mid \alpha \in M\}$.

When we wish to identify the particular denotation operator associated with a particular $M[F_0]$ we will write $D_{M[F_0]}$ instead of D.

4. $D'(V(t)) \overset{\Delta}{\leftrightarrow} (\exists k \in M) D'(t = \underline{k})$.
5. $D'(F(t)) \overset{\Delta}{\leftrightarrow} (\exists k \in F_0) D'(t = \underline{k})$.
6. $D'(\underline{k}_1 \in \underline{k}_2) \overset{\Delta}{\leftrightarrow} k_1 \in k_2$, $\quad k_1, k_2 \in K$.
7. $D'(t \in \underline{k}) \overset{\Delta}{\leftrightarrow} (\exists k' \in K) D'(t = \underline{k}')$.
8. $D'(t \in \hat{x}^\beta \varphi(x^\beta)) \overset{\Delta}{\leftrightarrow} (\exists t' \in T_\beta)[D'(t = t') \wedge D'(\varphi(t'))]$, $\beta \in M$.
9. $D'(t_1 = t_2) \overset{\Delta}{\leftrightarrow} (\forall t \in T_\beta) D'(t \in t_1 \leftrightarrow t \in t_2)$, $\beta = \max(\rho(t_1), \rho(t_2))$.
10. $D'(\neg\varphi) \overset{\Delta}{\leftrightarrow} \neg D'(\varphi)$, $D'(\varphi \wedge \psi) \leftrightarrow D'(\varphi) \wedge D'(\psi)$.
11. $D'((\forall x^\beta)\varphi(x^\beta)) \overset{\Delta}{\leftrightarrow} (\forall t \in T_\beta) D'(\varphi(t))$, $\beta \in M$.

Remark. It is easy to see that D is equivalent to D' and the following theorem holds:

Theorem 9.37. $M[F_0] \models \varphi \leftrightarrow h(\llbracket\varphi\rrbracket) = \mathbf{1}$.

Remark. Since $\mathbf{T}$ is a $\mathbf{B}$-valued model of ZF, we have $\llbracket\varphi\rrbracket = \mathbf{1}$ for each axiom φ. Consequently we have the following result.

Theorem 9.38. $M[F_0]$ is a standard transitive model of ZF. If M satisfies the AC so does $M[F_0]$.

Proof. To show that $M[F_0]$ satisfies the AC if M does we note that if M satisfies the AC then since $K \in M$, K is well ordered and, since $F_0 \subseteq K$, F_0 is also well ordered in $M[F_0]$. Hence $M[F_0]$ satisfies the AC.

Remark. Comparing the results of Theorem 9.38 with those discussed at the end of §8 note that we did not require the existence of a model $\tilde{M}$ of ZF with the same order type as M such that F_0 is a class in $\tilde{M}$ but instead F_0 must satisfy certain requirements to ensure that $M[F_0]$ be a model of ZF. Defining $M[F_0]$ by considering $\mathbf{B}$-valued relative constructibility has many advantages as will become clear from the applications in the next sections. We can however give an application of this method now:

Suppose that M is a countable standard transitive model of ZF, $\mathbf{B} \in M$ and $\mathbf{B}$ is M-complete, also $K \in M$ and $(V[f_0])^M$ is defined as before. Now assume that there is some sentence φ such that $[\![\varphi]\!] \neq \mathbf{0}$. Then there is a homomorphism $h\colon |\mathbf{B}| \to |\mathbf{2}|$ that is M-complete, sends $[\![\varphi]\!]$ to $\mathbf{1}$, and from which we get a standard $\mathbf{2}$-valued model $M[F_0]$ of ZF in which φ is true in the ordinary sense of $\mathbf{2}$-valued logic.

10. Forcing

As an application of the general theory developed in the previous sections we give a definition of "forcing" and derive its elementary properties. Throughout this section, M denotes a standard transitive model of ZF, $\mathbf{P} \in M$ is a partial order structure, and $\mathbf{B}$ is the corresponding M-complete Boolean algebra of regular open sets of $\mathbf{P}$ in the relative sense of M. Furthermore we have

h an M-complete homomorphism of $\mathbf{B}$ into $\mathbf{2}$,
F an M-complete ultrafilter for $\mathbf{B}$, and
G a set that is $\mathbf{P}$-generic over M,

such that h, F, and G are related to each other as described in §2. Thus one of them may be given and the remaining sets are obtained from it as in §2.

We now specialize the construction of $M[F_0]$ to one of the following cases:

1. $K = B$ and $f_0: B \to B$ is the identity on B or
2. $K = P$ and $f_0: P \to B$ is defined by

$$f_0(p) = [p]^{-0}, \qquad p \in P.$$

In case 1

$$\begin{aligned} F_0 &= \{b \in B \mid h(f_0(b)) = \mathbf{1}\} \\ &= \{b \in B \mid h(b) = \mathbf{1}\} \\ &= F. \end{aligned}$$

In case 2

$$\begin{aligned} F_0 &= \{p \in P \mid h([p]^{-0}) = \mathbf{1}\} \\ &= \{p \in P \mid [p]^{-0} \in F\} \\ &= G. \end{aligned}$$

Since h, F, G are obtainable from each other in a simple way, we have in both cases $M[F_0] = M[h] = M[G] = M[F]$ and Theorem 10.1 follows.

Theorem 10.1. If G is $\mathbf{P}$-generic over M then $M[G]$ is a standard transitive model of ZF that has the same order type as M. For any formula φ of $\mathscr{L}_0$

$$\begin{aligned} M[G] \models \varphi &\leftrightarrow h([\![\varphi]\!]) = \mathbf{1} \\ &\leftrightarrow [\![\varphi]\!] \in F \\ &\leftrightarrow [\![\varphi]\!] \cap G \neq 0. \end{aligned}$$

Furthermore if M satisfies the AC so does $M[G]$.

Remark. $[\![\varphi]\!] = b$ is definable in $\mathscr{L}_0$ from $\mathbf{P}$ and $\ulcorner\varphi\urcorner$ (uniformly in $\ulcorner\varphi\urcorner$ if φ ranges over limited formulas only).

Definition 10.2. If $p \in P$ and φ is a formula, limited or unlimited, then

$$p \Vdash \varphi \overset{\Delta}{\leftrightarrow} p \in [\![\varphi]\!].$$

Theorem 10.3. $p \Vdash \varphi$ is definable in $\mathscr{L}_0$ from $\ulcorner\varphi\urcorner$, b and $\mathbf{P}$ (uniformly in $\ulcorner\varphi\urcorner$ if φ ranges over limited formulas only).

Remark. As can be seen from Theorem 10.1 there is a close relationship between satisfaction in $M[G]$ and the notion of forcing. In particular the forcing relation satisfies certain recursive conditions similar to the notion of satisfaction in $M[G]$:

Theorem 10.4. Let $k, k_1, k_2 \in V$ and t, t_1, t_2 be constant terms.

1. $p \Vdash \neg\varphi \leftrightarrow (\forall q \le p) \neg (q \Vdash \varphi)$.
2. $p \Vdash \varphi_1 \wedge \varphi_2 \leftrightarrow p \Vdash \varphi_1 \wedge p \Vdash \varphi_2$.
3. $p \Vdash (\forall x)\varphi(x) \leftrightarrow (\forall t \in T)[p \Vdash \varphi(t)]$.
4. $p \Vdash (\forall x^\alpha)\varphi(x^\alpha) \leftrightarrow (\forall q \le p)(\exists q' \le q)(\forall t \in T_\alpha)[q' \Vdash \varphi(t)]$.
5. $p \Vdash V(t) \leftrightarrow (\forall q \le p)(\exists q' \le q)(\exists k)[q' \Vdash t = \underline{k}]$ in particular $p \Vdash V(\underline{k})$.
6. $p \Vdash F(t) \leftrightarrow (\forall q \le p)(\exists q' \le q)(\exists b \in B)[b \in F_0 \wedge q' \Vdash t = \underline{k}]$.
7. $p \Vdash \underline{k}_1 \in \underline{k}_2 \leftrightarrow k_1 \in k_2$.
8. $p \Vdash t \in \underline{k} \leftrightarrow (\forall q \le p)(\exists q' \le q)(\exists k' \in k)[q' \Vdash t = \underline{k}']$.
9. $p \Vdash t \in \hat{x}^\beta\varphi(x^\beta) \leftrightarrow (\forall q \le p)(\exists q' \le q)(\exists t' \in T_\beta)[q' \Vdash t = t' \wedge q' \Vdash \varphi(t')]$.
10. $p \Vdash t_1 = t_2 \leftrightarrow p \Vdash (\forall x^\beta)[x^\beta \in t_1 \leftrightarrow x^\beta \in t_2]$ where $\beta = \max(\rho(t_1), \rho(t_2))$.

Proof. The proofs of most of these statements are obvious from the definition:

1. $p \Vdash \neg\varphi \leftrightarrow p \in {}^{-}[\![\varphi]\!]$
$\quad\leftrightarrow (\forall q \le p)[q \notin [\![\varphi]\!]]$
$\quad\leftrightarrow (\forall q \le p) \neg [q \Vdash \varphi]$.
2. $p \Vdash \varphi_1 \wedge \varphi_2 \leftrightarrow p \in [\![\varphi_1 \wedge \varphi_2]\!]$
$\quad\leftrightarrow p \in [\![\varphi_1]\!] \wedge p \in [\![\varphi_2]\!]$
$\quad\leftrightarrow p \Vdash \varphi_1 \wedge p \Vdash \varphi_2$.
3. $p \Vdash (\forall x)\varphi(x) \leftrightarrow p \in \left(\bigcap_{t \in T} [\![\varphi(t)]\!]\right)^{-0}$
$\quad\leftrightarrow p \in \bigcap_{t \in T} [\![\varphi(t)]\!]$ by Theorem 1.35
$\quad\leftrightarrow (\forall t \in T)[p \Vdash \varphi(t)]$.

4–10. The proofs are left to the reader.

Remark. Note that in order to define forcing and prove Theorem 10.4 we need not assume that M is countable. However, in order to prove the existence of an M-complete homomorphism of $\mathbf{B}$ into $\mathbf{2}$, or equivalently the existence of a set G that is $\mathbf{P}$-generic over M, we need some further conditions on M. We collect all these requirements in the following definition.

Definition 10.5. $\langle M, \mathbf{P}\rangle$ is a setting for forcing iff

1. M is a standard transitive model of ZF,
2. $\mathbf{P}$ is a partially ordered structure and $\mathbf{P} \in M$,
3. M is countable.

Remark. Under these assumptions we know that for each $p \in P$ there is a G that is $\mathbf{P}$-generic over M and $p \in G$. In fact 1 and 3 could be weakened. In particular, it would be sufficient to require instead of 3

3′. $\mathscr{P}(\mathbf{P}) \cap M$ is countable.

The following theorem is a kind of completeness theorem for forcing.

Theorem 10.6. If $\langle M, \mathbf{P}\rangle$ is a setting for forcing then

$$p \Vdash \varphi \leftrightarrow (\forall G')[G' \text{ is } \mathbf{P}\text{-generic over } M \wedge p \in G' \to M[G'] \vDash \varphi].$$

Proof. Using the one-to-one correspondence between $\mathbf{P}$-generic sets over M and M-complete homomorphisms from $\mathbf{B}$ into $\mathbf{2}$ we need only show:

$$p \Vdash \varphi \leftrightarrow (\forall h')[h': |\mathbf{B}| \to |\mathbf{2}| \text{ is an } M\text{-complete homomorphism} \wedge h'([p]^{-0}) = \mathbf{1} \to h'(\llbracket\varphi\rrbracket) = \mathbf{1}].$$

But the right-hand side is equivalent to

$$[p]^{-0} \leq \llbracket\varphi\rrbracket$$

which in turn is equivalent to each of the following:

$$p \in \llbracket\varphi\rrbracket$$
$$p \Vdash \varphi.$$

Remark. We could also define forcing either (1) by using the recursive conditions of Theorem 10.4 or (2) by the equivalence of Theorem 10.6. On the other hand, the definition of forcing by (1) allows us to define a $\mathbf{B}$-valued interpretation $\llbracket\ \rrbracket$ by

$$\llbracket\varphi\rrbracket = \{p \in P \mid p \Vdash \varphi\}.$$

Theorem 10.7. $q \leq p \wedge p \Vdash \varphi \to q \Vdash \varphi$.

Proof. Obvious from the definition.

Corollary 10.8. $\neg[p \Vdash \varphi \wedge p \Vdash \neg\varphi]$.

Remark. On the other hand, we need not have $p \Vdash \varphi \vee p \Vdash \neg\varphi$.

Theorem 10.9. If G is $\mathbf{P}$-generic over M and $S = \{p \in P \mid p \Vdash \varphi\}$ is dense then $M[G] \vDash \varphi$.

Proof. $S = \llbracket\varphi\rrbracket$ is regular open in $\mathbf{P}$ since $\llbracket\varphi\rrbracket \in B$. Then, since S is dense, $S = S^{-0} = \mathbf{1}$. Therefore $\llbracket\varphi\rrbracket = \mathbf{1}$.

Definition 10.10. For each $S \subseteq P$, S is dense beneath p iff $[p] \subseteq S^{-}$.

Theorem 10.11. If G is **P**-generic over M, if $p \in G$ and if $S \in M$ is dense beneath p then $G \cap S \neq 0$.

Proof. Under the given hypothesis if

$$S' = S \cup \{q \in P \mid \neg \text{Comp}(p, q)\}$$

then $S' \in M$ and S' is dense, hence $G \cap S' \neq 0$. But any two elements of G are compatible, hence $G \cap S \neq 0$.

Theorem 10.12. If G is **P**-generic over M and $p \in G$ then

$$p \Vdash (\exists x)\varphi(x) \rightarrow (\exists q \leq p)(\exists t \in T)(q \in G \wedge q \Vdash \varphi(t)).$$

Proof.

$$p \Vdash (\exists x)\varphi(x) \leftrightarrow p \in \sum_{t \in T} [\![\varphi(t)]\!] = \left(\bigcup_{t \in T} [\![\varphi(t)]\!]\right)^{-0}$$

$$\leftrightarrow [p] \subseteq \left(\bigcup_{t \in T} [\![\varphi(t)]\!]\right)^{-}.$$

So $p \Vdash (\exists x)\varphi(x)$ implies that $\bigcup_{t \in T} [\![\varphi(t)]\!]$ is dense beneath p, and the same holds for $S' = [p] \cap \bigcup_{t \in T} [\![\varphi(t)]\!]$. Also $S' \in M$ since $B \in M$. Therefore by Theorem 10.11

$$p \Vdash (\exists x)\varphi(x) \rightarrow (\exists q \in G)\left[q \leq p \wedge q \in \bigcup_{t \in T} [\![\varphi(t)]\!]\right]$$

$$\rightarrow (\exists q)(\exists t \in T)[q \leq p \wedge q \in G \wedge q \Vdash \varphi(t)].$$

11. The Independence of $V = L$ and the CH

Cohen's technique of forcing was created for the specific purpose of proving the independence of several axioms of set theory from those of general set theory. In this section we will use Cohen's method to prove the independence of $V = L$ and the CH from the axioms of ZF.

Let M be a countable standard transitive model of $ZF + V = L$.

Definition 11.1.

$P \triangleq \{\langle p_1, p_2\rangle \mid p_1 \subseteq \omega \wedge p_2 \subseteq \omega \wedge \bar{\bar{p}}_1 < \omega \wedge \bar{\bar{p}}_2 < \omega \wedge p_1 \cap p_2 = 0\}.$
$\langle p_1, p_2\rangle \leq \langle p_1', p_2'\rangle \overset{\Delta}{\leftrightarrow} p_1' \subseteq p_1 \wedge p_2' \subseteq p_2.$
$\mathbf{P} \triangleq \langle P, \leq\rangle.$
$\forall G \subseteq P, \tilde{a}(G) \triangleq \{n \in \omega \mid (\exists p_1, p_2)[n \in p_1 \wedge \langle p_1, p_2\rangle \in G]\}.$
$\forall a \subseteq \omega, \tilde{G}(a) \triangleq \{\langle p_1, p_2\rangle \mid p_1 \subseteq a \wedge p_2 \subseteq \omega - a \wedge \langle p_1, p_2\rangle \in P\}.$

Exercise. Prove that the partial order structure $\mathbf{P}$ is fine in the sense of Definition 5.21.

Remark. $P \in M$, $\tilde{a}(G) \subseteq \omega$ and $\tilde{G}(a) \subseteq P$.

Lemma 1. $a_1 \subseteq \omega \wedge a_2 \subseteq \omega \wedge a_1 \neq a_2 \rightarrow \tilde{G}(a_1) \neq \tilde{G}(a_2)$.

Proof. Without loss of generality we may assume

$$(\exists n \in \omega)[n \in a_1 \wedge n \notin a_2].$$

Then $\langle\{n\}, 0\rangle \in \tilde{G}(a_1)$ but $\langle\{n\}, 0\rangle \notin \tilde{G}(a_2)$.

Lemma 2. If G is $\mathbf{P}$-generic over M then $\tilde{G}(\tilde{a}(G)) = G$.

Proof. If G is $\mathbf{P}$-generic over M and $p = \langle p_1, p_2\rangle \in G$ then $p_1 \subseteq \tilde{a}(G)$ and $p_2 \subseteq \omega - \tilde{a}(G)$.

For if $n \in p_2$ and $n \in \tilde{a}(G)$ then $\exists q = \langle q_1, q_2\rangle \in G$, $n \in q_1$. Since $p, q \in G$, $\exists r = \langle r_1, r_2\rangle \in G$, $r \leq p \wedge r \leq q$. Since $r \leq p \wedge n \in p_2$ we have $n \in r_2$. But also $n \in r_1$ since $r \leq q \wedge n \in q_1$. But $r \in P$ and hence $r_1 \cap r_2 = 0$. This is a contradiction. Hence $G \subseteq \tilde{G}(\tilde{a}(G))$.

If $p = \langle p_1, p_2\rangle \in \tilde{G}(\tilde{a}(G))$ then $p_1 \subseteq \tilde{a}(G) \wedge p_2 \subseteq \omega - \tilde{a}(G)$. If

$$p_1 = \{n_1, \ldots, n_k\}$$

then since $p_1 \subseteq \tilde{a}(G)$, $\exists q^i \in G$, $q^i = \langle q_1{}^i, q_2{}^i\rangle$, $i = 1, 2$, such that

$$n_1 \in q_1^1 \wedge n_2 \in q_1^2.$$

$\exists r \in G, r \leq q^1 \wedge r \leq q^2, r = \langle r_1, r_2 \rangle$. Then $n_1, n_2 \in r_1$. Thus, by induction $\exists q = \langle q_1, q_2 \rangle \in G$, $p_1 \subseteq q_1$, i.e., $q \leq \langle p_1, 0 \rangle$. Let $p_2 = \{m_1, \ldots, m_l\}$ and let $S = \bigcup_{i=1}^{l} [\langle \{m_i\}, 0 \rangle] \cup [\langle 0, p_2 \rangle]$. Then S is dense and hence $S \cap G \neq 0$. Let q' be in $S \cap G$. Since $p_2 \subset \omega - \tilde{a}(G), q_1' \cap p_2 = 0$ where $q' = \langle q_1', q_2' \rangle$. Therefore $q' \leq \langle 0, p_2 \rangle$. Since $q, q' \in G, \exists r \in G, r \leq q \wedge r \leq q'$. So $r \leq \langle p_1, p_2 \rangle$ since $q \leq \langle p_1, 0 \rangle$ and $q' \leq \langle 0, p_2 \rangle$. Hence $p = \langle p_1, p_2 \rangle \in G$. Therefore $\tilde{G}(\tilde{a}(G)) \subseteq G$.

Lemma 3. If G_1, G_2 are each **P**-generic over M then

$$\tilde{a}(G_1) = \tilde{a}(G_2) \leftrightarrow G_1 = G_2.$$

Proof. Lemmas 1 and 2.

Remark. Thus $\tilde{a}$ is a one-to-one correspondence between **P**-generic sets over M and certain subsets of ω. Also $M[\tilde{G}(a)] = M[a]$ and $M[G] = M[\tilde{a}(G)]$ for $a \subseteq \omega$ and $G \subseteq P$.

Theorem 11.2. If $a \in M$ and $a \subseteq \omega$ then $\tilde{G}(a)$ is not **P**-generic over M.

Proof. If $a \in M$ and $a \subseteq \omega$ then $\tilde{G}(a) \in M$. If $S = P - \tilde{G}(a)$ then $S \in M$. For each $p = \langle p_1, p_2 \rangle \in P$ there exists an $n \in \omega$ such that $n \notin p_1$ and $n \notin p_2$. Let

$$\begin{aligned} p' &= \langle p_1 \cup \{n\}, p_2 \rangle \quad \text{if} \quad n \notin a \\ &= \langle p_1, p_2 \cup \{n\} \rangle \quad \text{if} \quad n \in a. \end{aligned}$$

Then $p' \leq p$ and $p' \in S$. Therefore S is dense. But $S \cap \tilde{G}(a) = 0$, consequently $\tilde{G}(a)$ is not **P**-generic over M.

Theorem 11.3. If G is **P**-generic over M then $M[G]$ is a standard transitive model of $ZF + AC + GCH + V \neq L$.

Proof. If $a = \tilde{a}(G)$ then $M[G] = M[a]$ and hence, by Theorem 11.2 and Lemma 2, $a \notin M$. Therefore $M[a]$ is not a model of $V = L$.

Since $a \subseteq \omega \subseteq L$, $L_a = L[a]$ and hence

1. $V = L[a] \rightarrow GCH$.

But $L[a]$ relativized to $M[a]$ is just $M[a]$. Therefore the relativization of 1 to $M[a]$ gives the GCH in $M[a]$.

Corollary 11.4. If there exists a standard transitive model of ZF then there exists a standard transitive model of $ZF + AC + GCH + V \neq L$.

Exercises. If **P** is as defined above

1. Prove that $^-[\langle \{n\}, 0 \rangle]^{-0} = [\langle 0, \{n\} \rangle]^{-0}$.
2. Calculate
   a. $^-[\langle p_1, p_2 \rangle]^{-0}$.
   b. $[\langle p_1, p_2 \rangle]^{-0} \cdot [\langle q_1, q_2 \rangle]^{-0}$.
   c. $[\langle p_1, p_2 \rangle]^{-0} + [\langle q_1, q_2 \rangle]^{-0}$.

Remark. Let $\langle M, \mathbf{P} \rangle$ be any setting for forcing and $\pi \in M$ be an automorphism of $\mathbf{P}$ ($\pi \in \text{Aut}(\mathbf{P})$). Then π induces an automorphism $\tilde{\pi} \in M$ of $\mathbf{B}$, the Boolean algebra of regular open sets of $\mathbf{P}$ in M. Let G be $\mathbf{P}$-generic over M. Let F and h be the M-complete ultrafilter for $\mathbf{B}$ and the M-complete homomorphism of $\mathbf{B}$ into $\mathbf{2}$ respectively, obtained from G. The following theorem shows how G, F, and h transform under π.

Theorem 11.5. $\pi``G$ is $\mathbf{P}$-generic over $M$, $\tilde{\pi}``F$ is the M-complete ultrafilter for $\mathbf{B}$ and $h \circ \tilde{\pi}^{-1}$ is the M-complete homomorphism of $\mathbf{B}$ into $\mathbf{2}$ corresponding to $\pi``G$. Furthermore $M[G] = M[F] = M[h] = M[\pi``G] = M[\tilde{\pi}``F] = M[h \circ \tilde{\pi}^{-1}]$.

Proof. $S \subseteq P$ is dense iff $\pi``S$ is dense. Therefore

$$S \cap \pi``G \neq 0 \leftrightarrow \pi^{-1}(S) \cap G \neq 0.$$

Thus $\pi``G$ is $\mathbf{P}$-generic over $M$. The ultrafilter corresponding to $\pi``G$ is

$$\begin{aligned} \{b \in B \mid b \cap \pi``G \neq 0\} &= \{b \in B \mid \pi^{-1}(b) \cap G \neq 0\} \\ &= \{b \in B \mid \tilde{\pi}^{-1}(b) \in F\} \\ &= \tilde{\pi}``F. \end{aligned}$$

Finally

$$\begin{aligned} b \in \tilde{\pi}``F &\leftrightarrow \tilde{\pi}^{-1}(b) \in F \\ &\leftrightarrow h \circ \tilde{\pi}^{-1}(b) = \mathbf{1}. \end{aligned}$$

Therefore $h \circ \tilde{\pi}^{-1}$ is the M-complete homomorphism of $\mathbf{B}$ into $\mathbf{2}$ corresponding to $\tilde{\pi}``F$.

Theorem 11.6. If

1. $(\forall p, q \in P)(\exists \pi \in \text{Aut}(\mathbf{P}))[\pi \in M \wedge \text{Comp}(\pi(p), q)]$, and
2. G_1 and G_2 are $\mathbf{P}$-generic over M,

then $M[G_1]$ and $M[G_2]$ are elementarily equivalent in the language

$$\mathscr{L}_0(C(M)).$$

Proof. If φ is a closed formula of $\mathscr{L}_0(C(M))$ and if

$$M[G_1] \models \varphi \quad \text{and} \quad M[G_2] \models \neg\varphi$$

then $\exists p_1, p_2$

$$p_1 \in G \wedge p_1 \Vdash \varphi \wedge p_2 \in G_2 \wedge p_2 \Vdash \neg\varphi.$$

By 1 $(\exists \pi \in \text{Aut}(\mathbf{P}))[\pi \in M \wedge (\exists p \leq \pi(p_1))[p \leq p_2]]$. Let G be $\mathbf{P}$-generic over M and such that $p \in G$. (Such a G exists.) Then since $p \leq p_2$, $p_2 \in G$ and hence, by Theorem 10.6, $M[G] \models \neg\varphi$. But also $p_1 \geqslant \pi^{-1}(p) \in (\pi^{-1})``G$ hence $\pi^{-1}(p) \Vdash \varphi$. Therefore $M[(\pi^{-1})``G] \models \varphi$. But $M[G] = M[(\pi^{-1})``G]$. This is a contradiction.

Exercise. Check that the partial order structure $\mathbf{P}$ of Definition 11.1 satisfies condition 1 of Theorem 11.6.

Remark. Let **P** be the partial order structure of Definition 11.1 and used for the proof of the independence of $V = L$. For $k \subseteq \omega$ with $\bar{\bar{k}} < \omega$ we define an automorphism π_k of **P** as follows:

$$\pi_k(\langle p_1, p_2 \rangle) = \langle q_1, q_2 \rangle \quad \text{where} \quad \begin{aligned} q_1 &= (p_1 - k) \cup (p_2 \cap k) \\ q_2 &= (p_2 - k) \cup (p_1 \cap k). \end{aligned}$$

Obviously $\pi_k \in M$. We then obtain the following strengthening of Theorem 11.3.

Theorem 11.7. If G is **P**-generic over M then in $M[G]$ there is no well-ordering of $\mathscr{P}(\omega)$ definable in $\mathscr{L}_0(C(M))$.

Proof. If φ is a formula of $\mathscr{L}_0(C(M))$ defining a well-ordering of $\mathscr{P}(\omega)$ in $M[G]$, then

$$\exists p \in G, \qquad p \Vdash \text{“}\varphi \text{ well-orders } \mathscr{P}(\omega)\text{”}.$$

Since $\tilde{a}(G) \in M[G]$, $\tilde{a}(G) = D_{M[G]}(t_0)$ for some term t_0 of the ramified language. Then since $\tilde{a}(G) \subseteq \omega$, $\exists p \in G$,

$$p \Vdash \text{“}\{t_0 \,\Delta\, k \mid k \subseteq \omega \wedge \bar{\bar{k}} < \omega\} \text{ has a } \varphi\text{-first element”}$$

where $a_1 \,\Delta\, a_2 \overset{\Delta}{=} (a_1 \cup a_2) - (a_1 \cap a_2)$ is the symmetric difference of a_1 and a_2.
Furthermore it is easy to check that

$$\tilde{a}(\pi_k``(G)) = \tilde{a}(G) \,\Delta\, k \quad \text{for} \quad k \subseteq \omega \quad \text{and} \quad \bar{\bar{k}} < \omega.$$

Then $(\exists q \in G)(\exists k_0)[k_0 \subseteq \omega \wedge \bar{\bar{k}}_0 < \omega \wedge q \Vdash$ “$t_0 \,\Delta\, k_0$ is the φ-first element of $\{t_0 \,\Delta\, k \mid k \subseteq \omega \wedge \bar{\bar{k}} < \omega\}$”].
Hence in $M[G]$, $\tilde{a}(G) \,\Delta\, k_0$ is the φ-first element of

$$\{\tilde{a}(G) \,\Delta\, k \mid k \subseteq \omega \wedge \bar{\bar{k}} < \omega\}.$$

If $q = \langle q_1, q_2 \rangle$ then there exists a $k_1 \neq 0$ such that $k_1 \subseteq \omega$, $\bar{\bar{k}}_1 < \omega$, $k_1 \cap k_0 = 0$, $k_1 \cap q_1 = 0$, and $k_1 \cap q_2 = 0$. If

$$H \overset{\Delta}{=} \pi_{k_1}``G$$

then H is **P**-generic over M and $\tilde{a}(H) = \tilde{a}(G) \,\Delta\, k_0$. Since $q \in G$ and $\pi_k(q) = q$, $q \in H$. Therefore, by Theorem 11.6, in $M[H]$, $\tilde{a}(H) \,\Delta\, k_0$ is the φ-first element of $\{\tilde{a}(H) \,\Delta\, k \mid k \subseteq \omega \wedge \bar{\bar{k}} < \omega\}$. Since $\pi_{k_1} \in M$, $M[G] = M[H]$ and

$$\begin{aligned} \{\pi_k``G \mid k \subseteq \omega \wedge \bar{\bar{k}} < \omega\} &= \{\pi_k``H \mid k \subseteq \omega \wedge \bar{\bar{k}} < \omega\} \\ \{\tilde{a}(\pi_k``G) \mid k \subseteq \omega \wedge \bar{\bar{k}} < \omega\} &= \{\tilde{a}(\pi_k``H) \mid k \subseteq \omega \wedge \bar{\bar{k}} < \omega\} \\ \{\tilde{a}(G) \,\Delta\, k \mid k \subseteq \omega \wedge \bar{\bar{k}} < \omega\} &= \{\tilde{a}(H) \,\Delta\, k \mid k \subseteq \omega \wedge \bar{\bar{k}} < \omega\}. \end{aligned}$$

Thus $\tilde{a}(G) \,\Delta\, k_0$ is also the φ-first element of

$$\{\tilde{a}(H) \,\Delta\, k \mid k \subseteq \omega \wedge \bar{\bar{k}} < \omega\} \quad \text{in} \quad M[H].$$

Therefore $\tilde{a}(G) \,\Delta\, k_0 = \tilde{a}(H) \,\Delta\, k_0$, but

$$\tilde{a}(H) = \tilde{a}(G) \,\Delta\, k_1 \quad \text{and} \quad k_1 \cap k_0 = 0.$$

This is a contradiction.

Remark. We now return to the general case.

Theorem 11.8. If $\langle M, \mathbf{P}\rangle$ is a setting for forcing, if M is a model of the AC that satisfies the countable chain condition on $\mathbf{P}$ i.e.,

$$(\forall S \subseteq P)[(\forall p_1, p_2 \in S)[p_1 \neq p_2 \rightarrow \neg \text{Comp}(p_1, p_2)] \rightarrow \bar{\bar{S}} < \aleph_0]$$

and if G is $\mathbf{P}$-generic over M then the cardinals in M and $M[G]$ respectively are the same i.e., every cardinal in M is a cardinal in $M[G]$ and vice versa.

Proof. Since $M \subseteq M[G]$ and $On^M = On^{M[G]}$, every cardinal in $M[G]$ is a cardinal in M because "a is not a cardinal" $\leftrightarrow (\exists f)\varphi(f, a)$, for some φ that is absolute with respect to transitive models.

Conversely if there is a cardinal in M that is not a cardinal in $M[G]$ then there is a cardinal λ in $M[G]$, and hence in M, but

$$\gamma = (\lambda^+)^M$$

is not a cardinal in $M[G]$. Let λ be the smallest such cardinal. Then

$$\lambda = \bar{\bar{\gamma}}^{M[G]} < \gamma = (\lambda^+)^M.$$

Hence

$$(\exists f \in M[G])[f \colon \lambda \xrightarrow{\text{onto}} \gamma].$$

Furthermore f is denoted by a term t of the ramified language and

$$(\exists p \in P)[p \in G \wedge p \Vdash t \colon \lambda \xrightarrow{\text{onto}} \gamma].$$

Abbreviating $q \Vdash \langle \alpha, \beta\rangle \in t$ by $\varphi(q, \alpha, \beta)$, φ is M-definable and furthermore

$$(\forall \beta, \beta' < \gamma)(\forall q \leq p)(\forall q' \leq p)$$
$$[\beta \neq \beta' \wedge q \Vdash \langle \alpha, \beta\rangle \in t \wedge q' \Vdash \langle \alpha, \beta'\rangle \in t \rightarrow \neg \text{Comp}(q, q')]$$

for otherwise there exists a $q'' \leq q$, such that $q'' \leq q'$, $q'' \Vdash \langle \alpha, \beta\rangle \in t$, $q'' \Vdash \langle \alpha, \beta'\rangle \in t$, and $q'' \Vdash t \colon \lambda \xrightarrow{\text{onto}} \gamma$. This is a contradiction. Then if

$$S_\alpha = \{\beta < \gamma \mid (\exists q \leq p)\varphi(q, \alpha, \beta)\}, \qquad \alpha < \lambda,$$

we have $S_\alpha \in M$ and $\bar{\bar{S}}_\alpha{}^M \leq \aleph_0$. Since f is a function onto γ

$$\gamma \subseteq \bigcup_{\alpha < \lambda} S_\alpha$$

and $\langle S_\alpha \mid \alpha < \lambda\rangle \in M$. Therefore $\bar{\bar{\gamma}}^M \leq \overline{\overline{\lambda \times \omega}}^M = \bar{\bar{\lambda}}^M < \gamma = \bar{\bar{\gamma}}^M$. This is a contradiction.

Remark. Let $\mathbf{P}$ be a partial order structure and let $\mathbf{B}$ be the associated Boolean algebra of regular open subsets of P. Then P satisfies the c.c.c. iff B satisfies the c.c.c.: ($\Rightarrow$). Suppose $S \subseteq B$ and $(\forall b_1)(\forall b_2)[b_1 \cdot b_2 \in S \wedge b_1 \neq b_2 \rightarrow b_1 \cdot b_2 = 0]$. By the AC we choose a p from each $b \in S$. Let S' consist of such p's. Then any two elements of S' are incompatible. Therefore, $\overline{\overline{S'}} \leq \omega$, and hence $\bar{\bar{S}} \leq \omega$. Conversely, let $S' \subseteq P$ and suppose any two elements of S' are incompatible. Let $S = \{[p]^{-0} \mid p \in S'\}$. Then $S \subseteq B$ and any two elements of S are disjoint. This follows from the following fact (See Theorem 1.29.2.):

$$(\forall p, q \in P)[[p] \cap [q] = 0 \leftrightarrow [p]^{-0} \cap [q]^{-0} = 0].$$

Corollary 11.9. If $\langle M, \mathbf{P}\rangle$ is a setting for forcing, if G is $\mathbf{P}$-generic over M, if M satisfies

$$(\forall S \subseteq P)[(\forall p_1, p_2 \in S)[p_1 \neq p_2 \rightarrow \neg \mathrm{Comp}\,(p_1, p_2)] \rightarrow \bar{\bar{S}} < \lambda]$$

and if λ is a regular cardinal in M, then the sets of cardinals $\geq \lambda$ in M and $M[G]$ respectively are the same.

Proof. We first show that λ is a cardinal in $M[G]$. Otherwise

$$(\exists f \in M[G])(\exists \lambda_0 < \lambda)[f\colon \lambda_0 \xrightarrow{\text{onto}} \lambda].$$

Then, as in the previous proof, we obtain

$$\lambda \leq \bigcup_{\alpha < \lambda_0} S_\alpha$$

which contradicts the assumption that λ is regular in M. Using the argument of Theorem 11.8 it follows that if $\mu \geq \lambda$ is a cardinal in $M[G]$ then $(\mu^+)^M$ is a cardinal in $M[G]$.

Remark. Next we will prove the independence of the CH from the axioms of $ZF + AC$. The idea of the proof is the following. Choosing some suitable $\mathbf{P} \in M$ which satisfies the countable chain condition in M we adjoin α-many subsets of ω. If α is a cardinal $> \aleph_0$, in M, then the CH is violated in the resulting model $M[G]$ since cardinals are preserved by passing from M to $M[G]$. The formal proof proceeds as follows: We define $\mathbf{P}$ by

$$P \overset{\Delta}{=} \{p \mid (\exists d)[d \subseteq \alpha \times \omega \wedge \bar{\bar{d}} < \omega \wedge p\colon d \rightarrow 2]\}$$
$$p_1 \leq p_2 \overset{\Delta}{\leftrightarrow} p_2 \subseteq p_1, \quad p_1, p_2 \in P$$

Let $\langle M, \mathbf{P}\rangle$ be a setting for forcing such that M satisfies the AC.

Theorem 11.10. P satisfies the c.c.c. in M.

Proof. We show by induction on n that

(i) $S \subseteq P \wedge S \in M \wedge (\forall p_1, p_2 \in S)[p_1 \neq p_2 \rightarrow \neg \mathrm{Comp}\,(p_1, p_2)]$
$\wedge\ (\forall p \in S)[\overline{\overline{\mathscr{D}(p)}}^* = n] \rightarrow \bar{\bar{S}}^M \leq \aleph_0.$

From this the theorem follows by defining

$$S^n = \{p \in S \mid \overline{\overline{\mathscr{D}(p)}} = n\}.$$

Then $S = \bigcup_{n\in\omega} S^n$ is countable in M by (i).

To prove (i) we assume $S \neq 0$. Then $\exists p_1 \in S$, and

$$(\forall p \in S)[p_1 \neq p \rightarrow (\exists \delta < \alpha)(\exists m \in \omega)[\langle \delta, m\rangle \in \mathscr{D}(p) \cap \mathscr{D}(p_1) \wedge p(\delta, m) \neq p_1(\delta, m)]$$

* $\mathscr{D}(p) \overset{\Delta}{=} \{x \mid (\exists y)[\langle x, y\rangle \in p]\}$

If $\mathscr{D}(p_1)$ is $\{\langle\delta_1, m_1\rangle, \ldots, \langle\delta_n, m_n\rangle\}$

$$S_{i0} \triangleq \{p \in S \mid \langle\delta_i, m_i\rangle \in \mathscr{D}(p) \wedge p(\delta_i, m_i) = 0\}, \qquad i = 1, \ldots, n.$$
$$S_{i1} \triangleq \{p \in S \mid \langle\delta_i, m_i\rangle \in \mathscr{D}(p) \wedge p(\delta_i, m_i) = 1\}, \qquad i = 1, \ldots, n.$$

Then $S = S_{10} \cup S_{11} \cup \cdots \cup S_{n0} \cup S_{n1}$. If

$$S'_{i0} \triangleq \{p - \{\langle\langle\delta_i, m_i\rangle, 0\rangle\} \mid p \in S_{i0}\}, \qquad i = 1, \ldots, n$$
$$S'_{i1} \triangleq \{p - \{\langle\langle\delta_i, m_i\rangle, 1\rangle\} \mid p \in S_{i1}\}, \qquad i = 1, \ldots, n.$$

Then $\overline{\overline{S}}{}_{i0}^{M} = \overline{\overline{S}}{}_{i0}^{\prime M}$ and $\overline{\overline{S}}{}_{i1}^{M} = \overline{\overline{S}}{}_{i1}^{\prime M}$. But, by the induction hypothesis for $n - 1$.

$$\overline{\overline{S}}{}_{i0}^{\prime M} \leq \aleph_0 \wedge \overline{\overline{S}}{}_{i1}^{\prime M} \leq \aleph_0.$$

Therefore $\overline{\overline{S}}{}^{M} \leq \aleph_0$.

Remark. If G is **P**-generic over M, then for each $\delta < \alpha$ we define

$$a_\delta(G) \triangleq \{n \in \omega \mid (\exists p \in G)[p(\delta, n) = 1]\}.$$

Claim: $(\forall\delta, \delta' < \alpha)[\delta \neq \delta' \rightarrow a_\delta(G) \neq a_{\delta'}(G)]$.

Let f be the function defined on α by

$$f(\delta) = a_\delta(G),\ \delta < \alpha.$$

It can be proved that $f \in M[G]$. Let t be a term in the ramified language that denotes f. Suppose $a_\delta(G) = a_{\delta'}(G)$ i.e. $f(\delta) = f(\delta')$. Then

$$M[G] \models [t(\delta) = t(\delta')]$$

and hence

1. $(\exists q \in G)[q \Vdash t(\delta) = t(\delta')]$.

We choose n such that

$$(\forall\delta'' < \alpha)[\langle\delta'', n\rangle \notin \mathscr{D}(q)].$$

Since $\delta \neq \delta'$

2. $(\exists q' \leq q)[q'(\delta, n) = 1 \wedge q'(\delta', n) = 0]$.

Now choose a G' that is **P**-generic over M and such that $q' \in G'$. Then

$$n \in a_\delta(G') \wedge n \notin a_{\delta'}(G').$$

Hence

$$a_\delta(G') \neq a_{\delta'}(G').$$

But $q' \in G'$ and by 1 and 2 $q' \Vdash t(\delta) = t(\delta')$. Thus $M[G'] \Vdash t(\delta) = t(\delta')$. Therefore $a_\delta(G') = a_{\delta'}(G')$. This is a contradiction.

We have thus established that

$$(\forall\delta < \alpha)[a_\delta(G) \subseteq \omega] \quad \text{and} \quad (\forall\delta, \delta' < \alpha)[\delta \neq \delta' \rightarrow a_\delta(G) \neq a_{\delta'}(G)].$$

Therefore

$$\overline{\overline{P(\omega)}}{}^{M[G]} \geq \alpha.$$

Starting with some $\alpha > \omega_1{}^M$ we have $\alpha > \omega_1{}^{M[G]}$. This proves the following:

Theorem 11.11. If G is **P**-generic over M (where **P** and M are as specified above) then $M[G]$ is a standard transitive model of $ZF + AC + \neg CH$. Furthermore, for any given cardinal $\alpha \in M$ we can find a G and a **P** such that

$$\overline{\overline{P(\omega)}} \geq \alpha \quad \text{in} \quad M[G].$$

12. The Independence of the AC

In order to prove the independence of the Axiom of Choice from the axioms of ZF we cannot use the models which were employed in the previous section, since if M is a model of $ZF + AC$ and G is $\mathbf{P}$-generic over M, then $M[G]$ also satisfies the AC. Yet the model N which we shall construct and which violates the AC is of the form $M[G]$. The corresponding language will have countably many symbols and we shall add to M countably many generic sets together with a set containing all these generic sets. In order to deal with this new situation we introduce the following:

Definition 12.1. Let $\mathbf{P}_i = \langle P_i, \leq \rangle$, $i \in I$, be a family of partial order structures with I a set. Then

$$\mathbf{P} = \prod_{i \in I}^{s} \mathbf{P}_i$$

(the *strong product* of the $\mathbf{P}_i$'s) is defined to be the partial order structure $\langle P, \leq \rangle$ where

$$P \triangleq \prod_{i \in I} P_i$$

and

$$p_1 \leq p_2 \overset{\Delta}{\leftrightarrow} (\forall i \in I)[p_1(i) \leq p_2(i)] \quad \text{for} \quad p_1, p_2 \in P.$$

If each P_i has a greatest element 1_i, $i \in I$, then

$$\mathbf{P} = \prod_{i \in I}^{w} \mathbf{P}_i$$

(the *weak product* of the $\mathbf{P}_i$'s) is defined to be the partial order structure $\langle P_0, \leq \rangle$ where

$$P_0 \triangleq \left\{ p \mid p \in \prod_{i \in I} P_i \wedge (\exists F \subseteq I)[\overline{\overline{F}} < \omega \wedge (\forall i \in I - F)[p(i) = 1_i]] \right\}$$

i.e., P_0 is the set of $p \in P$ where $p(i) = 1_i$ for all but finitely many $i \in I$, and $\leq$ is $\leq$ of $\mathbf{P}$ restricted to P_0.

Remark. The topology of $\prod_{i \in I}^{s} \mathbf{P}_i$ is the strong topology of $\prod_{i \in I} P_i$ and the topology of $\prod_{i \in I}^{w} \mathbf{P}_i$ is the relative topology of both the weak topology of $\prod_{i \in I} P_i$ and the strong product topology of $\prod_{i \in I} P_i$. P_0 is dense in $\prod_{i \in I} P_i$ with the weak topology, but not necessarily dense in $\prod_{i \in I} P_i$ with the strong product topology.

Problem. Is the complete Boolean algebra of regular open sets of $\prod_{i\in I}^{s} \mathbf{P}_i$ or $\prod_{i\in I}^{w} \mathbf{P}_i$ determined by the complete Boolean algebras of regular open sets of $\mathbf{P}_i$ $(i \in I)$?

Now consider $\prod_{i\in I}^{w} P_i$. An element $p \in \prod_{i\in I}^{w} P_i$ is sometimes denoted by $\{\dots p_{i_1}, \dots, p_{i_n} \dots\}$ where

$$p_{i_1} = p(i_1) \wedge \cdots \wedge p_{i_n} = p(i_n) \wedge (\forall i \in I)[i \neq i_1 \wedge \cdots \wedge i \neq i_n \rightarrow p(i) = 1_i].$$

In this case we also write

$$[\dots p_{i_1}, \dots, p_{i_n} \dots] \quad \text{for} \quad [p].$$

Let $\mathbf{P}$ be the partial order structure of Definition 11.1 that was used for the proof of the independence of $V = L$. Let $\mathbf{P}_i$, $i \in \omega$, be isomorphic copies of $\mathbf{P}$ and let j_i be an isomorphism of $\mathbf{P}$ onto $\mathbf{P}_i$. Note that $\mathbf{P}$ has a greatest element $\langle 0, 0 \rangle$.

Consider the language $\mathscr{L}_0(\{V(\ )\} \cup \{F_i(\ ) \mid i \in \omega\} \cup \{S(\ )\})$ and the corresponding ramified language $R(\{V(\ )\} \cup \{F_i(\ ) \mid i \in \omega\} \cup \{S(\ )\})$. Let $\mathbf{B}$ be the complete Boolean algebra of all regular open sets in $\prod_{i\in\omega}^{w} \mathbf{P}_i$ and $f_i\colon P \rightarrow B$ be defined by $f_i(p) = [\dots p_i \dots]^{-0}$ where $p_i = j_i(p)$.

We define $\mathbf{T}_\alpha = \langle T_\alpha, \bar{=}, \bar{\in}, \bar{V}_\alpha, \bar{F}_{0\alpha}, \dots, \bar{F}_{i\alpha}, \dots, \bar{S}_\alpha \rangle$ by

$$\llbracket V(t) \rrbracket_{\mathbf{T}_\alpha} = \sum_{k\in R(\alpha)} \llbracket t = \underline{k} \rrbracket_{\mathbf{T}_\alpha}$$

$$\llbracket F_i(t) \rrbracket_{\mathbf{T}_\alpha} = \sum_{k\in R(\alpha)\cap P} \llbracket t = \underline{k} \rrbracket_{\mathbf{T}_\alpha} \cdot f_i(k), \qquad i \in \omega,$$

$$\llbracket S(t) \rrbracket_{\mathbf{T}_\alpha} = \sum_{i<\omega} \llbracket t = \hat{x}^\omega F_i(x^\omega) \rrbracket_{\mathbf{T}_\alpha} \quad \text{if} \quad \alpha > \omega$$

$$= \mathbf{0} \quad \text{otherwise}$$

and the remaining conditions are those of Definition 9.27.3–9 except that K is replaced by P.

Let $\mathbf{G}$ be the group of all permutations of ω such that $\pi(n) \neq n$ for only finitely many $n \in \omega$, and let $\mathbf{G}_n$, $n \in \omega$, be the subgroup of $\mathbf{G}$ consisting of all $\pi \in \mathbf{G}$ such that $(\forall m \leq n)[\pi(m) = m]$.

We extend $\pi \in \mathbf{G}$ to terms and formulas of our ramified language as follows:

1. $\pi(\underline{k}) = \underline{k}$.
2. $\pi(t_1 \in t_2) \leftrightarrow \pi(t_1) \in \pi(t_2)$.
3. $\pi(t_1 = t_2) \leftrightarrow \pi(t_1) = \pi(t_2)$.
4. $\pi(\neg\varphi) \leftrightarrow \neg\pi(\varphi)$, $\pi(\varphi \wedge \psi) \leftrightarrow \pi(\varphi) \wedge \pi(\psi)$.
5. $\pi((\forall x^\alpha)\varphi(x^\alpha)) \leftrightarrow (\forall x^\alpha)\pi(\varphi(x^\alpha))$.
   $\pi((\forall x)\varphi(x)) \leftrightarrow (\forall x)\pi(\varphi(x))$.
6. $\pi(V(t)) \leftrightarrow V(\pi(t))$.
   $\pi(S(t)) \leftrightarrow S(\pi(t))$.
7. $\pi(F_i(t)) \leftrightarrow F_{\pi(i)}(\pi(t))$, $\qquad i \in \omega$.

So it is only by condition 7 that π is in general not the identity on the terms and formulas.

Finally, let $p^1, \ldots, p^n$ be elements of P. Then for

$$p = \{\ldots p_{i_1}{}^1, \ldots, p_{i_n}{}^n \ldots\} \in \prod_{i \in \omega}^{w} P_i$$

where $p_{i_k}{}^k = j_{i_k}(p^k)$, $k = 1, 2, \ldots, n$, we define $\pi(p)$ to be

$$\{\ldots p^1_{\pi(i_1)}, \ldots, p^n_{\pi(i_n)} \ldots\}.$$

That is, $\pi(p)$ is an element of $\prod_{i \in \omega}^{w} P_i$ whose $\pi(i_k)$th coordinate is the counterpart of p^k in $P_{\pi(i_k)}$, $k = 1, \ldots, n$, and whose mth coordinate is 1 if $m \neq \pi(i_k)$ for all $k = 1, 2, \ldots, n$. Then one can prove by transfinite induction on Ord (φ):

Theorem 12.2. Let $p \in \prod_{i \in \omega}^{w} P_i$. Then $p \Vdash \varphi \leftrightarrow \pi(p) \Vdash \pi(\varphi)$.

Theorem 12.3. For every formula φ,

$$(\forall t \in T)\left(\forall p \in \prod_{i \in \omega}^{w} P_i\right)(\exists m)(\forall \pi)[\pi \in \mathbf{G}_m \rightarrow \pi(t) = t \wedge \pi(\varphi) = \varphi \wedge \pi(p) = p].$$

Proof. Take m to be the maximum of all $i \in \omega$ for which F_i occurs in φ or t or p_i is not 1_i. Then m has the required properties.

Now let M be a countable standard transitive model of ZF and $h\colon |\mathbf{B}| \rightarrow |\mathbf{2}|$ be an M-complete homomorphism. Then $N = M[h]$ is defined at the end of §9: $N = \{D(t) \mid t \in T\}$ where D is defined as in Definition 9.36 except that now

$$F_i \triangleq \{p \in P \mid h([\![F_i(\underline{p})]\!]) = \mathbf{1}\}$$

and

$$\begin{aligned} &D(F_i(t)) \leftrightarrow D(t) \in F_i, \qquad i \in \omega, \\ &D(S(t)) \leftrightarrow (\exists i < \omega) D(t = \hat{x}_n{}^{\omega} F_i(x_n{}^{\omega})). \end{aligned}$$

As in §9 N is a standard transitive model of ZF. Defining $S \triangleq D(\hat{x}_n{}^{\omega+1} S(x_n{}^{\omega+1}))$ we obtain:

$$S = \{F_i \mid i \in \omega\} \quad \text{and} \quad S \in N.$$

Furthermore, since $[\![F_i(\underline{p})]\!] = f_i(p) = [\ldots p_i \ldots]^{-0}$ for $p \in P$, F_i is $\mathbf{P}$-generic over M.

Let a_i be $\tilde{a}(F_i)$ i.e.,

$$a_i \triangleq \{n \in \omega \mid (\exists p_1, p_2)[n \in p_1 \wedge \langle p_1, p_2 \rangle \in F_i]\}, \qquad i \in \omega.$$

Clearly, $F_i \in N$ and $a_i \in N$ for $i \in \omega$ and $S \in N$. With this notation we have the following:

Theorem 12.4. In N, S is an infinite subset of $P(\omega)$, and yet S contains no countable subset. In particular, $P(\omega)$ is not well-ordered in N and hence the AC does not hold in N.

Proof. Recall that by Lemmas 2 and 3, pages 106–107,

$$a_i = a_j \leftrightarrow F_i = F_j \quad \text{for} \quad i, j < \omega,$$

and

$$\tilde{G}(a_i) = F_i \quad \text{for} \quad i \in \omega.$$

First we prove that S is infinite by showing that

1. $i, j < \omega \wedge i \neq j \rightarrow F_i \neq F_j$.

Suppose $F_i = F_j$ for some $i, j \in \omega$ with $i \neq j$. Then since $F_i = D(\hat{x}_n{}^{\omega} F_i(x_n{}^{\omega}))$,

$$p \Vdash \hat{x}_n{}^{\omega} F_i(x_n{}^{\omega}) = \hat{x}_n{}^{\omega} F_j(x_n{}^{\omega})$$

for some p such that $h([p]^{-0}) = \mathbf{1}$.

We can assume that $p = \{\ldots p_i, \ldots, p_j \ldots\}$. Let $p_i = \langle p_i{}^1, p_i{}^2 \rangle$ and $p_j = \langle p_j{}^1, p_j{}^2 \rangle$, and choose $n \in \omega$ such that $n \notin p_i{}^1 \cup p_i{}^2 \cup p_j{}^1 \cup p_j{}^2$. Let q be

$$\{\ldots q_i, \ldots, q_j \ldots\}$$

where

$$q_i = \langle p_i{}^1 \cup \{n\}, p_i{}^2 \rangle$$
$$q_j = \langle p_j{}^1, p_j{}^2 \cup \{n\} \rangle$$

(and $q_k = 1_k = \langle 0, 0 \rangle$ for $k \neq i, j$).

Since $h([p]^{-0}) = \mathbf{1}$, we have $p_i \in F_i$ and $p_j \in F_j$.

Case 1. $n \in a_i$. Then $q_i \in \tilde{G}(\tilde{a}(F_i)) = F_i$ since $q_i \leq p_i \in F_i$. But $F_i = F_j$ in N, implies $[\![F_i(p^1)]\!] = [\![F_j(p^1)]\!]$ for all $p^1 \in P$. Hence $q_j \in F_j$, and by definition of q_j, $n \notin a_j$. Thus $a_i \neq a_j$ contrary to our assumption that $F_i = F_j$.

Case 2. $n \notin a_i$. Then $q_i \notin F_i$, so $q_j \notin F_j$ as above. Consequently since $F_j = \tilde{G}(a_j)$, $q_j{}^1 \nsubseteq a_j$ or $q_j{}^2 \nsubseteq \omega - a_j$. But $p_j{}^1 \subseteq a_j$ and $p_j{}^2 \subseteq \omega - a_j$, so $n \in a_j$. Again this is a contradiction.

This proves 1. So it remains to show that

2. "S contains no countable subset"

holds in N.

Assume, in N, that S contains a countable subset. Then, by Theorem 10.12, there exists a term $t \in T$ such that for some p

$$h([p]^{-0}) = \mathbf{1} \wedge p \Vdash t \colon \underline{\omega} \xrightarrow{1:1} \hat{x}_n{}^{\omega+1} S(x_n{}^{\omega+1}).$$

Throughout this proof we say "F_i appears in t" iff p_i (the ith component of p) is not $1_i = \langle 0, 0 \rangle$. Choose $n \in \omega$ such that $j < n$ whenever F_j appears in t and

$$(\forall \pi \in \mathbf{G}_n)[\pi(t) = t].$$

By Theorem 10.12, there exists a $p' \leq p$, a $k \in \omega$ and some $m > n$ such that

$$p' \Vdash t(\underline{k}) = \hat{x}_n{}^{\omega} F_m(x_n{}^{\omega}).$$

Pick some $i > m$ such that F_i does not appear in p', and some $\pi \in \mathbf{G}_n$ which permutes i and m, and let $p'' = \pi(p')$. Then $\pi(t) = t$ and, by Theorem 12.2, $p'' \Vdash t(\underline{k}) = \hat{x}_n{}^{\omega} F_i(x_n{}^{\omega})$. There exists a q such that $q \leq p'$ and $q \leq p''$.

Then

$$q \Vdash t(\underline{k}) = \hat{x}_n{}^{\omega} F_m(x_n{}^{\omega})$$

and

$$q \Vdash t(\underline{k}) = \hat{x}_n{}^{\omega} F_i(x_n{}^{\omega}).$$

This is a contradiction.

Remark. We conclude this section with some results which are useful for certain applications. Returning to the general case, let $\langle M, \mathbf{P} \rangle$ be a setting for forcing where M is a standard transitive model of $ZF + AC$, and let $\mathbf{B}$ be the Boolean algebra of regular open sets of $\mathbf{P}$ in M.

Theorem 12.5. If G is $\mathbf{P}$-generic over M and

$$(\forall S \subseteq P)[\bar{\bar{S}} \leq \omega \wedge \text{Comp}\,(S) \rightarrow (\exists p \in P)(\forall q \in S)[p \leq q]]$$

(i.e., every countable compatible subset of $\mathbf{P}$ has a lower bound) holds in M, then every ω-sequence of ordinals in $M[G]$ is already in M.

Proof. Let $D_{M[G]}(t)$ be an ω-sequence of ordinals in $M[G]$. By Theorem 10.9, it suffices to show that $\{p \in P \mid p \Vdash V(t)\}$ is dense, i.e.,

$$(\forall p \in P)(\exists q \leq p)[q \Vdash V(t)]$$

or by Theorem 10.4.5 relativized to M. Recall that the interpretation of $V(t)$ is $t \in M$.

1. $(\forall p \in P)(\exists q \leq p)(\exists s \in M)[q \Vdash t = \underline{s}]$.

To prove 1, let $p \in P$ and $p \Vdash$ "t is an ω-sequence of ordinals." Using the AC in M, define in M a descending sequence $\langle p_i \mid i \in \omega \rangle$ and a sequence $s = \langle s_i \mid i \in \omega \rangle$ such that

$$p_0 = p,\ p_{i+1} \leq p_i \text{ and } p_{i+1} \Vdash t(\underline{i}) = \underline{s}_i.$$

Now let q be a lower bound of all p_i, $i \in \omega$. Then $q \leq p$ and $q \Vdash t = \underline{s}$.

Remark. Let $\mathbf{P}_1$ and $\mathbf{P}_2$ be two partial order structures in M where M is a countable standard transitive model of $ZF + AC$. Let $\mathbf{P} = \mathbf{P}_1 \times \mathbf{P}_2$ and assume that $\mathbf{P}_1$ and $\mathbf{P}_2$ both have a greatest element 1. There is a simple relationship between generic sets with respect to $\mathbf{P}$ on the one hand and the factors $\mathbf{P}_1$ and $\mathbf{P}_2$ on the other hand:

Theorem 12.6. If G_1 is $\mathbf{P}_1$-generic over M and G_2 is $\mathbf{P}_2$-generic over $M[G_1]$, then $G_1 \times G_2$ is $\mathbf{P}$-generic over M.

Proof. Assume the hypothesis of the theorem and let S be an element of M that is dense in $\mathbf{P}$. Define

$$S_2 \triangleq \{p_2 \in P_2 \mid (G_1 \times \{p_2\}) \cap S \neq 0\}.$$

Claim: S_2 is dense in P_2.

Let q_2 be any element of P_2 and define

$$S_1 \triangleq \{p_1 \in P_1 \mid (\{p_1\} \times [q_2]) \cap S \neq 0\}.$$

Then since S is dense in $\mathbf{P}$

$$(\forall q_1 \in P_1)(\exists p_1, p_2)[\langle p_1, p_2 \rangle \in S \wedge p_1 \leq q_1 \wedge p_2 \leq q_2].$$

Therefore S_1 is dense in $\mathbf{P}_1$ and hence $G_1 \cap S_1 \neq 0$. This implies, by definition of S_1, that

$$(G_1 \times [q_2]) \cap S \neq 0.$$

Thus S_2 is dense in $\mathbf{P}_2$. Since $S_2 \in M[G_1]$ and G_2 is $\mathbf{P}_2$-generic over $M[G_1]$, $G_2 \cap S_2 \neq 0$ which means, by definition of S_2, that

$$(G_1 \times G_2) \cap S \neq 0.$$

Therefore $G_1 \times G_2$ is $\mathbf{P}$-generic over M.

Theorem 12.7. If G is $\mathbf{P}$-generic over M, then there exists a G_1 which is $\mathbf{P}_1$-generic over M and a G_2 which is $\mathbf{P}_2$-generic over $M[G_1]$ such that $G = G_1 \times G_2$.

Proof. Let G be $\mathbf{P}$-generic over M and define

$$G_1 \triangleq \{p_1 \in P_1 \mid \langle p_1, 1 \rangle \in G\},$$

$$G_2 \triangleq \{p_2 \in P_2 \mid \langle 1, p_2 \rangle \in G\}.$$

Then $G = G_1 \times G_2$ ($G \subseteq G_1 \times G_2$ is obvious. To prove $G_1 \times G_2 \subseteq G$ use the proof method of Theorem 2.4).

1. G_1 is $\mathbf{P}_1$-generic over M.

Let $S \in M$ be dense in $\mathbf{P}_1$. Then $S \times P_2 \in M$ and it is dense in $\mathbf{P}$, hence $G \cap (S \times P_2) \neq 0$ and therefore $G_1 \cap S \neq 0$.

2. G_2 is $\mathbf{P}_2$-generic over $M[G_1]$.

Let $S \in M[G_1]$ be dense in $\mathbf{P}_2$. Then $S = D_{M[G_1]}(t)$ for some term t, and for some $p_1 \in G_1$, $p_1 \Vdash'$ "t is dense in $\mathbf{P}_2$" (where $\Vdash'$ refers to $\mathbf{P}_1$).

Define

$$E \triangleq \{\langle q_1, q_2 \rangle \in P \mid q_1 \leq p_1 \wedge q_1 \Vdash' \underline{q_2} \in t\},$$
$$p \triangleq \langle p_1, 1 \rangle.$$

Claim: E is dense beneath p, i.e.,

$$(\forall r \leq p)(\exists q \leq r)[q \in E].$$

Take any $r \leq p$, $r = \langle r_1, r_2 \rangle$. There exists an H_1 such that $r_1 \in H_1$ and H_1 is $\mathbf{P}_1$-generic over M. Consider $M[H_1]$; "$D_{M[H_1]}(t)$ is dense in $\mathbf{P}_2$" holds in

$M[H_1]$. There exists a $q_2 \leq r_2$ such that $q_2 \in D_{M[H_1]}(t)$ and hence there exists $q_1 \leq r_1$ such that $q_1 \Vdash' q_2 \in t$. Then $q = \langle q_1, q_2 \rangle \leq r$ and $q \in E$.

Since $E \in M$ and $p \in G$, we have, by Theorem 10.11,

$$G \cap E \neq 0.$$

Let $q = \langle q_1, q_2 \rangle \in G \cap E$. Then

$$M[G_1] \models q_2 \in S.$$

Therefore $G_2 \cap S \neq 0$ and 2 is proved.

13. Boolean-Valued Set Theory

The use of ramified language in Cohen-type independence proofs often requires proofs by induction which may become rather cumbersome in special cases. A different though essentially equivalent approach which avoids ramified language is provided by the theory of Boolean-valued models as developed by Scott and Solovay.

The analogue of the recursive definition of $R(\alpha)$ we define in the following way:

Definition 13.1. Let $\mathbf{B} = \langle B, +, \cdot, ^{-}, \mathbf{0}, \mathbf{1} \rangle$ be a complete Boolean algebra. Then $V_\alpha^{(\mathbf{B})}$ is defined by recursion with respect to α as follows:

$$V_0^{(\mathbf{B})} \triangleq 0.$$

$$V_\alpha^{(\mathbf{B})} \triangleq \{u \mid [u: \mathscr{D}(u) \to B] \wedge (\exists \xi < \alpha)[\mathscr{D}(u) \subseteq V_\xi^{(\mathbf{B})}]\}, \qquad \alpha > 0.$$

$$V^{(\mathbf{B})} \triangleq \bigcup_{\alpha \in On} V_\alpha^{(\mathbf{B})}.$$

Remark. Elements of $V^{(\mathbf{B})}$ are called **B**-valued sets, these are functions u from their domain, $\mathscr{D}(u)$, into B where $\mathscr{D}(u)$ itself consists of **B**-valued sets.

Theorem 13.2. $\alpha \in K_{\text{II}} \to V_\alpha^{(\mathbf{B})} = \bigcup_{\xi < \alpha} V_\xi^{(\mathbf{B})}$.

Remark. In order to obtain a **B**-valued structure $\mathbf{V}^{(\mathbf{B})} = \langle V^{(\mathbf{B})}, \overline{=}, \overline{\in} \rangle$, we define $\overline{=}$ and $\overline{\in}$ in the following way.

Definition 13.3. For $u, v \in V^{(\mathbf{B})}$,

1. $[\![u \mathbin{\overline{\in}} v]\!] \triangleq \sum_{y \in \mathscr{D}(v)} (v(y) \cdot [\![u = y]\!])$.

2. $[\![u \mathbin{\overline{=}} v]\!] \triangleq \prod_{x \in \mathscr{D}(u)} [u(x) \Rightarrow [\![x \in v]\!]] \cdot \prod_{y \in \mathscr{D}(v)} [v(y) \Rightarrow [\![y \in u]\!]]$.

Remark. Thus $\overline{\in}$ and $\overline{=}$ are defined simultaneously by recursion. Hereafter we will write $=$ and $\in$ for $\overline{=}$ and $\overline{\in}$ respectively.

There are several ways to check that 1 and 2 really constitute a definition by recursion:

1. The definition of $[\![u \in v]\!]$ and $[\![u = v]\!]$ is recursive with respect to the well-founded relation

$$\{\langle\langle u, v\rangle, \langle u', v'\rangle\rangle \mid \operatorname{rank}(u) \# \operatorname{rank}(v) < \operatorname{rank}(u') \# \operatorname{rank}(v')\}$$

where $\alpha \# \beta$ is the natural sum of α and β. (For a definition and elementary properties of the natural sum of ordinals the reader may consult one of the following monographs: H. Bachmann: Transfinite Zahlers, Ergebnisse elev. Math. Vol. 1 (1955), pp. 102f, or A. A. Fraenkel: Abstract Set Theory (1953), p. 297.)

2. Alternatively, we would use Gödel's pairing function J_0 which is a one-to-one correspondence between $On \times On$ and On with the following property.

$$\begin{aligned} J_0(\alpha, \beta) < J_0(\alpha', \beta') \leftrightarrow & \max(\alpha, \beta) < \max(\alpha', \beta') \\ & \vee [[\max(\alpha, \beta) = \max(\alpha', \beta')] \wedge [\beta < \beta' \vee [\beta = \beta' \wedge \alpha < \alpha']]]]. \end{aligned}$$

If we assign to $[\![u \in v]\!]$ the ordinal $J_0(\text{rank}(v), \text{rank}(u))$ and to $[\![u = v]\!]$ the ordinal $\max(J_0(\text{rank}(u), \text{rank}(v)), J_0(\text{rank}(v), \text{rank}(u)))$, it is easy to see that $[\![u \in v]\!]$ and $[\![u = v]\!]$ in 1 and 2 respectively are reduced to $[\![u = v']\!]$ and $[\![u' \in v']\!]$ in such a way that the associated ordinals are reduced to lower ordinals.

3. We would also eliminate $\in$ in 2 by substituting the definition 1:

$$\begin{aligned} [\![u = v]\!] = & \prod_{x \in \mathscr{D}(u)} \left[u(x) \Rightarrow \sum_{y \in \mathscr{D}(v)} [v(y) \cdot [\![x = y]\!]] \right] \\ & \cdot \prod_{y \in \mathscr{D}(v)} \left[v(y) \Rightarrow \sum_{x \in \mathscr{D}(u)} [u(x) \cdot [\![y = x]\!]] \right] \end{aligned}$$

which is a definition by recursion with respect to the well-founded relation $\{\langle\langle u, v\rangle, \langle u', v'\rangle\rangle \mid \max(\text{rank}(u), \text{rank}(v)) < \max(\text{rank}(u'), \text{rank}(v'))\}$. Then 1 becomes an explicit definition in terms of $=$.

Next we prove that the Axioms of Equality hold in $\mathbf{V}^{(\mathbf{B})}$ (see Definition 6.5).

Theorem 13.4. For $u, v \in V^{(\mathbf{B})}$,

1. $[\![u = v]\!] = [\![v = u]\!]$.
2. $[\![u = u]\!] = \mathbf{1}$.
3. $x \in \mathscr{D}(u) \rightarrow u(x) \leq [\![x \in u]\!]$.

Proof. 1. The definition of $[\![u = v]\!]$ is symmetric in u and v.

2 and 3 are proved by induction on rank (u). Let $x \in \mathscr{D}(u)$. Then

$$[\![x \in u]\!] = \sum_{y \in \mathscr{D}(u)} u(y) \cdot [\![x = y]\!],$$

hence

$$\begin{aligned} u(x) \cdot [\![x = x]\!] \leq [\![x \in u]\!] \quad & \text{if} \quad x \in \mathscr{D}(u) \\ u(x) \leq [\![x \in u]\!] \quad & \text{by the induction hypothesis} \quad [\![x = x]\!] = \mathbf{1}. \end{aligned}$$

Therefore

$$(\forall x \in \mathscr{D}(u))[[u(x) \Rightarrow [\![x \in u]\!]] = \mathbf{1}]$$

and

$$[\![u = u]\!] = \prod_{x \in \mathscr{D}(u)} [u(x) \Rightarrow [\![x \in u]\!]] = \mathbf{1}.$$

Theorem 13.5. Let $u, u', v, v', w \in V^{(\mathbf{B})}$. Then for each α

1. $[\operatorname{rank}(u) < \alpha] \wedge [\operatorname{rank}(u') < \alpha] \wedge [\operatorname{rank}(v) \leq \alpha]$
$\rightarrow [\![u = u']\!] \cdot [\![u \in v]\!] \leq [\![u' \in v]\!]$.
2. $[\operatorname{rank}(u) < \alpha] \wedge [\operatorname{rank}(v) \leq \alpha] \wedge [\operatorname{rank}(v') \leq \alpha]$
$\rightarrow [\![u \in v]\!] \cdot [\![v = v']\!] \leq [\![u \in v']\!]$.
3. $[\operatorname{rank}(u) \leq \alpha] \wedge [\operatorname{rank}(v) \leq \alpha] \wedge [\operatorname{rank}(w) \leq \alpha]$
$\rightarrow [\![u = v]\!] \cdot [\![v = w]\!] \leq [\![u = w]\!]$.

Proof. (By induction on α).

1. If $\operatorname{rank}(u) < \alpha$, $\operatorname{rank}(u') < \alpha$, and $\operatorname{rank}(v) \leq \alpha$, then

$$
\begin{aligned}
[\![u = u']\!] \cdot [\![u \in v]\!] &= \sum_{y \in \mathscr{D}(v)} v(y) \cdot [\![y = u]\!] \cdot [\![u = u']\!] \\
&\leq \sum_{y \in \mathscr{D}(v)} v(y) \cdot [\![y = u']\!] \qquad \text{by the induction hypothesis for 3} \\
&= [\![u' \in v]\!].
\end{aligned}
$$

2. If $\operatorname{rank}(u) < \alpha$, $\operatorname{rank}(v) \leq \alpha$, and $\operatorname{rank}(v') \leq \alpha$, then for $y \in \mathscr{D}(v)$

$$
\begin{aligned}
[\![u = y]\!] \cdot v(y) \cdot [\![v = v']\!] &\leq [\![u = y]\!] \cdot v(y)(v(y) \Rightarrow [\![y \in v']\!]) \\
&\leq [\![u = y]\!] [\![y \in v']\!] \\
&\leq [\![u \in v']\!] \qquad \text{by 1.}
\end{aligned}
$$

Therefore taking the sup over all $y \in \mathscr{D}(v)$

$$[\![u \in v]\!] \cdot [\![v = v']\!] \leq [\![u \in v']\!].$$

3. If $\operatorname{rank}(u) \leq \alpha$, $\operatorname{rank}(v) \leq \alpha$, and $\operatorname{rank}(w) \leq \alpha$, then for $x \in \mathscr{D}(u)$

$$
\begin{aligned}
[\![u = v]\!] \cdot [\![v = w]\!] \cdot u(x) &\leq [\![u = v]\!] \cdot [\![v = w]\!] \cdot [\![x \in u]\!] \qquad \text{by Theorem 13.4(3)} \\
&\leq [\![x \in v]\!] \cdot [\![v = w]\!] \qquad \text{by 2} \\
&\leq [\![x \in w]\!] \qquad \text{by 2.}
\end{aligned}
$$

Hence

$$[\![u = v]\!] \cdot [\![v = w]\!] \leq \prod_{x \in \mathscr{D}(w)} [u(x) \Rightarrow [\![x \in w]\!]]$$

and by symmetry,

$$[\![u = v]\!] \cdot [\![v = w]\!] \leq \prod_{x \in \mathscr{D}(w)} [w(x) \Rightarrow [\![x \in u]\!]].$$

Hence by Definition 13.3.2.

$$[\![u = v]\!] \cdot [\![v = w]\!] \leq [\![u = w]\!].$$

Corollary 13.6. For $u, u', v, v', w \in V^{(\mathbf{B})}$,

1. $[\![u = u']\!] \cdot [\![u \in v]\!] \leq [\![u' \in v]\!]$.
2. $[\![u \in v]\!] \cdot [\![v = v']\!] \leq [\![u \in v']\!]$.
3. $[\![u = v]\!] \cdot [\![v = w]\!] \leq [\![u = w]\!]$.

Corollary 13.7. For $u_1, \ldots, u_n, u_1', \ldots, u_n' \in V^{(\mathbf{B})}$,

$$[\![u_1 = u_1']\!] \cdots [\![u_n = u_n']\!] [\![\varphi(u_1, \ldots, u_n)]\!] \leq [\![\varphi(u_1', \ldots, u_n')]\!].$$

Remark. Therefore $\mathbf{V}^{(\mathbf{B})} = \langle V^{(\mathbf{B})}, \bar{=}, \bar{\in} \rangle$ is a $\mathbf{B}$-valued structure for the language $\mathscr{L}_0$. Even more, $\mathbf{V}^{(\mathbf{B})}$ is a $\mathbf{B}$-valued model of ZF i.e. $\mathbf{V}^{(\mathbf{B})}$ satisfies each axiom of ZF. Here a formula φ of the language $\mathscr{L}_0$ is satisfied by $\mathbf{V}^{(\mathbf{B})}$ iff $[\![\varphi]\!] = \mathbf{1}$ interpreting $\in$ (of $\mathscr{L}_0$) by $\bar{\in}$. We shall not give a direct proof of this statement but use the results of §9.

Definition 13.8. $\mathbf{V}_\alpha^{(\mathbf{B})} = \langle V_\alpha^{(\mathbf{B})}, \bar{=}, \bar{\in} \rangle$ is defined by

$$[\![u = v]\!]_\alpha \overset{\Delta}{=} [\![u = v]\!]$$
$$[\![u \in v]\!]_\alpha \overset{\Delta}{=} [\![u \in v]\!]$$

for $u, v \in V_\alpha^{(\mathbf{B})}$. (We write $[\![\ \ ]\!]_\alpha$ for $[\![\ \ ]\!]_{V_\alpha^{(\mathbf{B})}}$.)

Remark. Thus $\mathbf{V}_\alpha^{(\mathbf{B})}$ is a $\mathbf{B}$-valued structure for $\mathscr{L}_0$. Next we shall prove that this sequence of structures satisfies the conditions specified in §9. (See Remark following Definition 9.2) Obviously $\mathbf{V}_\alpha^{(\mathbf{B})}$ satisfies 1 and 2. We will now show that $\mathbf{V}_\alpha^{(\mathbf{B})}$ also satisfies 3, 4, and 5.

Theorem 13.9. $V_\alpha^{(\mathbf{B})}$ satisfies the Axiom of Extensionality.

Proof. Let $u, v \in V_\alpha^{(\mathbf{B})}$. Then

$$\begin{aligned}
[\![(\forall x)[x \in u \leftrightarrow x \in v]]\!]_\alpha &= \prod_{x \in V_\alpha^{(\mathbf{B})}} [\![x \in u \rightarrow x \in v]\!]_\alpha \cdot \prod_{x \in V_\alpha^{(\mathbf{B})}} [\![x \in v \rightarrow x \in u]\!]_\alpha \\
&\leq \prod_{x \in \mathscr{D}(u)} (u(x) \Rightarrow [\![x \in v]\!]) \prod_{x \in \mathscr{D}(v)} (v(x) \Rightarrow [\![x \in u]\!]) \\
&\qquad\qquad \text{by Theorem 13.4(3)} \\
&= [\![u = v]\!] = [\![u = v]\!]_\alpha.
\end{aligned}$$

Theorem 13.10. If $u \in V_{\alpha+1}^{(\mathbf{B})}$ then u is defined over $V_\alpha^{(\mathbf{B})}$, i.e.,

$$(\forall v \in V_{\alpha+1}^{(\mathbf{B})})\left[[\![v \in u]\!] = \sum_{x \in V_\alpha^{(\mathbf{B})}} [\![v = x]\!][\![x \in u]\!]\right].$$

Proof. Let u and v be in $V_{\alpha+1}^{(\mathbf{B})}$.

$$\begin{aligned}
[\![v \in u]\!] &= \sum_{x \in \mathscr{D}(u)} u(x) \cdot [\![v = x]\!] \\
&\leq \sum_{x \in \mathscr{D}(u)} [\![x \in u]\!] \cdot [\![x = v]\!] && \text{by Theorem 13.4(3)} \\
&\leq \sum_{x \in V_\alpha^{(\mathbf{B})}} [\![x \in u]\!] \cdot [\![x = v]\!] && \text{since } \mathscr{D}(u) \subseteq V_\alpha^{(\mathbf{B})} \\
&\leq [\![v \in u]\!] && \text{by Theorem 13.6.}
\end{aligned}$$

Therefore

$$[\![v \in u]\!] = \sum_{x \in V_\alpha^{(\mathbf{B})}} [\![v = x]\!] \cdot [\![x \in u]\!].$$

Theorem 13.11. For every formula φ of $\mathscr{L}_0$,

$$(\forall a_1, \ldots, a_n \in V_\alpha^{(\mathbf{B})})(\exists b \in V_{\alpha+1}^{(\mathbf{B})})(\forall a \in V_\alpha^{(\mathbf{B})})[[\![\varphi(a, a_1, \ldots, a_n)]\!]_\alpha = [\![a \in b]\!]].$$

Proof. Let $a_1, \ldots, a_n \in V_\alpha^{(\mathbf{B})}$ and define $b: V_\alpha^{(\mathbf{B})} \to B$ by

$$b(a) = [\![\varphi(a, a_1, \ldots, a_n)]\!]_\alpha \quad \text{for} \quad a \in V_\alpha^{(\mathbf{B})}.$$

Then $b \in V_{\alpha+1}^{(\mathbf{B})}$ and

$$\begin{aligned}[\![a \in b]\!] &= \sum_{a' \in V_\alpha^{(\mathbf{B})}} [\![\varphi(a', a_1, \ldots, a_n)]\!]_\alpha \cdot [\![a' = a]\!] \\ &\leq [\![\varphi(a, a_1, \ldots, a_n)]\!]_\alpha \qquad \text{by the Axioms of Equality.}\end{aligned}$$

On the other hand, for $a \in V_\alpha^{(\mathbf{B})}$

$$[\![a \in b]\!] \geq [\![\varphi(a, a_1, \ldots, a_n)]\!]_\alpha$$

by Theorem 13.4(3).

Remark. Since the conditions of §9 are satisfied by $\langle \mathbf{V}_\alpha^{(\mathbf{B})} \mid \alpha \in On \rangle$, we have, by Theorem 9.26, the following result:

Theorem 13.12. $\mathbf{V}^{(\mathbf{B})}$ is a **B**-valued model of ZF.

Remark. The following theorem is very useful.

Theorem 13.13. For $u \in V^{(\mathbf{B})}$,

1. $[\![(\exists x \in u)\varphi(x)]\!] = \sum_{x \in \mathscr{D}(u)} u(x) \cdot [\![\varphi(x)]\!]$.
2. $[\![(\forall x \in u)\varphi(x)]\!] = \prod_{x \in \mathscr{D}(u)} (u(x) \Rightarrow [\![\varphi(x)]\!])$.

Proof. For $u \in V^{(\mathbf{B})}$,

$$\begin{aligned}[\![(\exists x \in u)\varphi(x)]\!] &= \sum_{x' \in V^{(\mathbf{B})}} [\![x' \in u]\!] \cdot [\![\varphi(x')]\!] \\ &= \sum_{x' \in V^{(\mathbf{B})}} \sum_{x \in \mathscr{D}(u)} u(x) \cdot [\![x = x']\!] \cdot [\![\varphi(x')]\!] \\ &\leq \sum_{x \in \mathscr{D}(u)} u(x) \cdot [\![\varphi(x)]\!] \qquad \text{by the Axioms of Equality} \\ &\leq \sum_{x' \in V^{(\mathbf{B})}} [\![x' \in u]\!] \cdot [\![\varphi(x')]\!] \qquad \text{by Theorem 13.4(3).}\end{aligned}$$

This proves 1, and 2 follows by duality.

Definition 13.14. Let $\mathbf{B}'$ be a complete Boolean algebra. Then **B** is a complete subalgebra of $\mathbf{B}'$ iff **B** is a subalgebra of $\mathbf{B}'$, **B** is complete, but in addition, for each $A \subseteq |\mathbf{B}|$

$$\prod\nolimits^{\mathbf{B}} A = \prod\nolimits^{\mathbf{B}'} A$$

and

$$\sum\nolimits^{\mathbf{B}} A = \sum\nolimits^{\mathbf{B}'} A,$$

that is, a class $A \subseteq |\mathbf{B}|$ has the same sup and inf relative to **B** that it has relative to $\mathbf{B}'$.

Remark. Next we shall show how V can be embedded in $V^{(\mathbf{B})}$. As preparation we prove the following.

Theorem 13.15. Let $\mathbf{B}$ be a complete subalgebra of the complete Boolean algebra $\mathbf{B}'$. Then

1. $V^{(\mathbf{B})} \subseteq V^{(\mathbf{B}')}$.
2. $u, v \in V^{(\mathbf{B})} \to [\![u \in v]\!]^{(\mathbf{B})} = [\![u \in v]\!]^{(\mathbf{B}')} \wedge [\![u = v]\!]^{(\mathbf{B})} = [\![u = v]\!]^{(\mathbf{B}')}$.

Proof. (By induction)

1. Obvious, since any function into B is also a function into B'.
2. Follows from the fact that $\prod$ and $\sum$, over values in $\mathbf{B}$, are the same in $\mathbf{B}$ and $\mathbf{B}'$ respectively.

Remark. Since any (standard) set u may be identified with the function f_u having domain u and assuming the constant value $\mathbf{1}$ on u, we expect that V can be identified with some part of $V^{(\mathbf{B})}$. The corresponding mapping is defined in the following way.

Definition 13.16. For $y \in V$, $\check{y} \triangleq \{\langle \check{x}, \mathbf{1}\rangle \mid x \in y\}$ is defined by recursion with respect to the well-founded $\in$-relation.

Remark. Obviously, $\check{y} \in V^{(\mathbf{2})}$.

Theorem 13.17. For $x, y \in V$,

1. $x \in y \leftrightarrow [\![\check{x} \in \check{y}]\!] = \mathbf{1} \wedge x \notin y \leftrightarrow [\![\check{x} \in \check{y}]\!] = \mathbf{0}$,
2. $x = y \leftrightarrow [\![\check{x} = \check{y}]\!] = \mathbf{1} \wedge x \neq y \leftrightarrow [\![\check{x} = \check{y}]\!] = \mathbf{0}$,
3. $(\forall u \in V^{(\mathbf{2})})(\exists! v \in V)[[\![u = \check{v}]\!] = \mathbf{1}]$.

Proof. 1 and 2 are proved simultaneously by induction from Definition 13.3. Proving this is in fact a very good exercise that we leave to the reader. In order to prove 3, let $u \in V^{(\mathbf{2})}$ and assume as induction hypothesis

(i) $(\forall x \in \mathscr{D}(u))(\exists! z \in V)[[\![x = \check{z}]\!] = \mathbf{1}]$.

(Note that $u \in V^{(\mathbf{2})} \to \mathscr{D}(u) \subseteq V^{(\mathbf{2})}$.)

Let y be $\{z \in V \mid (\exists x \in \mathscr{D}(u))[[\![x = \check{z}]\!] = \mathbf{1}] \wedge [\![x \in u]\!] = \mathbf{1}]\}$ (which is a set by (i), since $\mathscr{D}(u)$ is a set). Obviously, $[\![u = \check{y}]\!] = \mathbf{1}$ (using (i)). The uniqueness of y follows from 2 and the Axioms of Equality for $V^{(\mathbf{2})}$.

Therefore, identifying V with the indicated part of $V^{(\mathbf{2})}$ we obtain an embedding of V in $V^{(\mathbf{B})}$.

Exercise. Define $v \in V^{(\mathbf{B})}$ as follows. Let $b \in B$ and $\mathscr{D}(v) = \{0, \check{1}\}$, $v(0) = b$, $v(\check{1}) = {}^{-}b$. Then

1. $[\![v \subseteq \check{2}]\!] = \mathbf{1}$.
2. $[\![0 \in v]\!] = b \wedge [\![\check{1} \in v]\!] = {}^{-}b$.
3. $[\![v \in \check{3}]\!] = b$.

Theorem 13.18. Let **B** be a complete subalgebra of the complete Boolean algebra **B**′ and $\varphi(u_1, \ldots, u_n)$ be a formula in which every quantifier is bounded (i.e., of the form $\exists x \in y$ or $\forall x \in y$). Then for $u_1, \ldots, u_n \in V^{(\mathbf{B})}$,

(i) $[\![\varphi(u_1, \ldots, u_n)]\!]^{(\mathbf{B})} = [\![\varphi(u_1, \ldots, u_n)]\!]^{(\mathbf{B}')}$.

Proof. (By induction on the number of logical symbols in φ.) If φ is atomic, (i) is true by Theorem 13.15. The only nontrivial case is

$$\varphi(u, u_1, \ldots, u_n) = (\exists x \in u)\psi(x, u, u_1, \ldots, u_n).$$

Then for $u, u_1, \ldots, u_n \in V^{(\mathbf{B})}$,

$$\text{(ii)}\ [\![\varphi(u, u_1, \ldots, u_n)]\!]^{(\mathbf{B})} = \sum_{x \in \mathscr{D}(u)}^{(\mathbf{B})} u(x) \cdot [\![\psi(x, u, u_1, \ldots, u_n)]\!]^{(\mathbf{B})}$$

by Theorem 13.13

$$= \sum_{x \in \mathscr{D}(u)}^{(\mathbf{B})} u(x) \cdot [\![\psi(x, u, u_1, \ldots, u_n)]\!]^{(\mathbf{B}')}$$

by the induction hypothesis

$$= [\![\varphi(u, u_1, \ldots, u_n)]\!]^{(\mathbf{B}')}$$

by Theorem 13.13,

since $\sum_{x \in \mathscr{D}(u)}$ in (ii) is the same in **B** and **B**′ (note that $\mathscr{D}(u) \subseteq V^{(\mathbf{B})}$ by assumption).

Corollary 13.19. If $\phi(u_1, \ldots, u_n)$ is a bounded formula (i.e., a formula containing only bounded quantifiers), then for $u_1, \ldots, u_n \in V$,

$$\phi(u_1, \ldots, u_n) \leftrightarrow [\![\phi(\check{u}_1, \ldots, \check{u}_n)]\!]^{(\mathbf{B})} = \mathbf{1}.$$

Proof. Apply Theorem 13.18 to the Boolean algebra **2** which is a complete subalgebra of each Boolean algebra **B** and use Theorem 13.17.

Remark. As an application of Corollary 13.19 we give a direct proof of the following theorem.

Theorem 13.20. $\mathbf{V}^{(\mathbf{B})}$ satisfies the Axiom of Infinity.

Proof. We have $\check{\omega} \in V^{(\mathbf{2})} \subseteq V^{(\mathbf{B})}$ and

$$(\exists x)[x \in \omega] \wedge (\forall x \in \omega)(\exists y \in \omega)[x \in y]$$

is a bounded formula which is provable in *ZF*. Hence by Corollary 13.19

$$[\![(\exists x)[x \in \check{\omega}] \wedge (\forall x \in \check{\omega})(\exists y \in \check{\omega})[x \in y]]\!] = \mathbf{1}$$

which is one form of the Axiom of Infinity.

Remark. Another formula with bounded quantifiers is Ord (α) which expresses "α is an ordinal," hence by another application of Corollary 13.19 we obtain the following.

Theorem 13.21. $[\![\text{Ord}\,(\check{\alpha})]\!] = \mathbf{1}$ for each $\alpha \in On$.

Remark. On the other hand, we have the following result.

Theorem 13.22. For $u \in V^{(\mathbf{B})}$, $[\![\mathrm{Ord}(u)]\!] = \sum_{\alpha \in On} [\![u = \check{\alpha}]\!]$.

Proof. $[\![u = \check{\alpha}]\!] = [\![u = \check{\alpha}]\!] \cdot [\![\mathrm{Ord}(\check{\alpha})]\!] \leq [\![\mathrm{Ord}(u)]\!]$ by Corollary 13.7. Therefore $\sum_{\alpha \in On} [\![u = \check{\alpha}]\!] \leq [\![\mathrm{Ord}(u)]\!]$. In order to prove $[\![\mathrm{Ord}(u)]\!] \leq \sum_{\alpha \in On} [\![u = \check{\alpha}]\!]$, note, from Theorem 13.17, that

$$\alpha \neq \beta \rightarrow [\![x = \check{\alpha}]\!] \cdot [\![x = \check{\beta}]\!] \leq [\![\check{\alpha} = \check{\beta}]\!] = \mathbf{0}.$$

Therefore, for each $x \in V^{(\mathbf{B})}$,

$$D_x \triangleq \{\xi \mid [\![x = \check{\xi}]\!] > 0\}$$

is a set (using the fact that the mapping $\xi \rightarrow [\![x = \check{\xi}]\!]$ is a one-to-one function on D_x and $[\![x = \check{\alpha}]\!]$ ranges over the set B). Thus $D = \bigcup_{x \in \mathscr{D}(u)} D_x$ is a set, and taking an ordinal α greater than sup D we obtain

$$(\forall \beta \geq \alpha)(\forall x \in \mathscr{D}(u))[[\![x = \check{\beta}]\!] = \mathbf{0}].$$

Therefore

$$\text{(i)}\ [\![\check{\alpha} \in u]\!] = \sum_{x \in \mathscr{D}(u)} u(x) \cdot [\![x = \check{\alpha}]\!] = \mathbf{0}.$$

Since $V^{(\mathbf{B})}$ is a model of ZF, $[\![\mathrm{Ord}(\check{\alpha})]\!] = \mathbf{1}$ and

$$[\![\mathrm{Ord}(u) \rightarrow u \in \check{\alpha} \vee u = \check{\alpha} \vee \check{\alpha} \in u]\!] = \mathbf{1}$$

i.e.,

$$\begin{aligned} [\![\mathrm{Ord}(u)]\!] &\leq [\![u \in \check{\alpha}]\!] + [\![u = \check{\alpha}]\!] + [\![\check{\alpha} \in u]\!] = [\![u \in \check{\alpha}]\!] + [\![u = \check{\alpha}]\!] \qquad \text{by (i)} \\ &= \sum_{\xi < \alpha} [\![u = \check{\xi}]\!] + [\![u = \check{\alpha}]\!] \qquad \text{since } \check{\alpha}(\check{\xi}) = \mathbf{1} \\ &= \sum_{\xi \leq \alpha} [\![u = \check{\xi}]\!] \leq \sum_{\xi \in On} [\![u = \check{\xi}]\!]. \end{aligned}$$

Corollary 13.23.

1. $[\![(\exists u)[\mathrm{Ord}(u) \wedge \phi(u)]]\!] = \sum_{\alpha \in On} [\![\phi(\check{\alpha})]\!]$.

2. $[\![(\forall u)[\mathrm{Ord}(u) \rightarrow \phi(u)]]\!] = \prod_{\alpha \in On} [\![\phi(\check{\alpha})]\!]$.

Proof.

$$\begin{aligned} [\![(\exists u)[\mathrm{Ord}(u) \wedge \phi(u)]]\!] &= \sum_{u \in V^{(\mathbf{B})}} [\![\mathrm{Ord}(u)]\!] \cdot [\![\phi(u)]\!] \\ &\leq \sum_{u \in V^{(\mathbf{B})}} \sum_{\alpha \in On} [\![u = \check{\alpha}]\!] [\![\phi(u)]\!] \qquad \text{by Theorem 13.22} \\ &\leq \sum_{\alpha \in On} [\![\phi(\check{\alpha})]\!] \\ &\leq \sum_{u \in V^{(\mathbf{B})}} [\![\mathrm{Ord}(u)]\!] \cdot [\![\phi(u)]\!] \qquad \text{by Theorem 13.21.} \end{aligned}$$

Therefore quantification over ordinals can be replaced by quantification (in the Boolean sense) over the standard ordinals. In view of Theorem 13.21, this result corresponds to the fact that M and $M[G]$ have the same ordinals.

In order to help the reader in getting more familiar with the Boolean-valued model $\mathbf{V}^{(\mathbf{B})}$ we conclude this section with some examples.

1. $[\![0 = \check{0}]\!] = \mathbf{1}$.

Proof. Note that 0 in $\check{0}$ is the empty set in V whereas 0 on the left-hand side of $0 = \check{0}$ is the empty set in $V^{(\mathbf{B})}$, i.e., $0 = a$ is to be replaced by its defining formula $(\forall x)[x \notin a]$. Now for $x \in V^{(\mathbf{B})}$,

$$[\![x \in \check{0}]\!] = \sum_{v \in \mathscr{D}(\check{0})} [\![x = \check{v}]\!] = \mathbf{0} \quad \text{since} \quad \mathscr{D}(\check{0}) = 0 \qquad \text{(empty set in } V\text{)}.$$

Therefore

$$[\![(\forall x)[x \notin \check{0}]]\!] = \prod_{x \in V^{(\mathbf{B})}} [\![x \notin \check{0}]\!] = \mathbf{1}.$$

2. $[\![\check{\alpha} + 1 = (\alpha + 1)^{\vee}]\!] = \mathbf{1}$.

Proof. The meaning of 2 is

$$[\![(\forall \gamma)[\gamma \in \check{\alpha} \vee \gamma = \check{\alpha} \leftrightarrow \gamma \in (\alpha + 1)^{\vee}]\!] = \mathbf{1}.$$

To prove this we first note, from Definition 13.3, that

$$\begin{aligned} [\![\check{\gamma} \in \check{\alpha} \vee \check{\gamma} = \check{\alpha}]\!] &= \left(\sum_{\delta \in \alpha} [\![\check{\delta} = \check{\gamma}]\!]\right) + [\![\check{\alpha} = \check{\gamma}]\!] \\ &= \sum_{\delta \in (\alpha + 1)} [\![\check{\delta} = \check{\gamma}]\!] \\ &= [\![\check{\gamma} \in (\alpha + 1)^{\vee}]\!]. \end{aligned}$$

Then by Corollary 13.23

$$\begin{aligned} [\![(\forall \gamma)[\gamma \in \check{\alpha} \vee \gamma = \check{\alpha} \leftrightarrow \gamma \in (\alpha + 1)^{\vee}]\!] &= \prod_{\delta \in On} [\![\check{\gamma} \in \check{\alpha} \vee \check{\gamma} = \check{\alpha} \leftrightarrow \check{\gamma} \in (\alpha + 1)^{\vee}]\!] \\ &= \mathbf{1} \end{aligned}$$

3. $[\![0 \in \check{\omega} \wedge (\forall x \in \check{\omega})[x + 1 \in \check{\omega}]]\!] = \mathbf{1}$.

Proof.

$$\begin{aligned} [\![(\forall x \in \check{\omega})[x + 1 \in \check{\omega}]]\!] &= \prod_{n \in \omega} [\![\check{n} + 1 \in \check{\omega}]\!] \\ &= \prod_{n \in \omega} [\![(n + 1)^{\vee} \in \check{\omega}]\!] \qquad \text{by 2} \\ &= \mathbf{1}. \end{aligned}$$

3 also follows directly from Corollary 13.19.

4. For $a \in V^{(\mathbf{B})}$, $n \in \omega \rightarrow [\![0 \in a \wedge (\forall x \in a)[x + 1 \in a] \rightarrow \check{n} \in a]\!] = \mathbf{1}$.

Proof. (By induction on n.) The case $n = 0$ follows from 1.

$$[\![(\forall x \in a)[x + 1 \in a]]\!] \cdot [\![\check{n} \in a]\!] \leq [\![\check{n} + 1 \in a]\!]$$

hence, by the induction hypothesis and 2,

$[\![0 \in a \wedge (\forall x \in a)[x + 1 \in a]]\!] \leq [\![(n + 1)^{\vee} \in a]\!]$.

5. $[\![0 \in a \wedge (\forall x \in a)[x + 1 \in a] \rightarrow \check{\omega} \subseteq a]\!] = \mathbf{1}$ for $a \in V^{(\mathbf{B})}$.

Proof.

$$[\![0 \in a \wedge (\forall x \in a)[x + 1 \in a]]\!] \leq \prod_{n \in \omega} [\![\check{n} \in a]\!] \qquad \text{by 4}$$

$$= [\![\check{\omega} \subseteq a]\!].$$

6. $[\![\omega = \check{\omega}]\!] = \mathbf{1}$.

Proof. $\omega = a \leftrightarrow \varphi(a)$ where $\varphi(a)$ is

$$0 \in a \wedge (\forall x \in a)[x + 1 \in a] \wedge (\forall y)[0 \in y \wedge (\forall x \in y)[x + 1 \in y] \rightarrow a \subseteq y].$$

Thus, from 3 and 5, $[\![\omega = \check{\omega}]\!] = \mathbf{1}$. Therefore, ω in $\mathbf{V}^{(\mathbf{B})}$ is $\check{\omega}$.

14. Another Interpretation of $\mathbf{V}^{(\mathbf{B})}$

The aim of this section is to prove that "M is a standard transitive model of ZF containing all the ordinals" and "$V = M[F]$" hold in $\mathbf{V}^{(\mathbf{B})}$ for suitable M and F (Theorems 14.21 and 14.24).

We introduce a new unary predicate constant $M(\ )$ and extend our former structure $\mathbf{V}_\alpha^{(\mathbf{B})} = \langle V_\alpha^{(\mathbf{B})}, \bar{=}, \bar{\in}\rangle$ to a $\mathbf{B}$-valued structure (denoted by the same symbol) $\mathbf{V}_\alpha^{(\mathbf{B})} = \langle V_\alpha^{(\mathbf{B})}, \bar{=}, \bar{\in}, \overline{M}_\alpha\rangle$ by defining

$$[\![M(u)]\!]_\alpha = \sum_{k \in R(\alpha)} [\![u = \check{k}]\!] \quad \text{for} \quad u \in V_\alpha^{(\mathbf{B})}.$$

In order to show that the new extended structure $\mathbf{V}_\alpha^{(\mathbf{B})}$ is in fact a $\mathbf{B}$-valued structure, we have to show that the Axioms of Equality remain valid.

Theorem 14.1. $u, v \in V_\alpha^{(\mathbf{B})} \rightarrow [\![u = v]\!] \cdot [\![M(u)]\!]_\alpha \leq [\![M(v)]\!]_\alpha$.

Proof. For $u, v \in V_\alpha^{(\mathbf{B})}$,

$$\begin{aligned}[\![u = v]\!] \cdot [\![M(u)]\!]_\alpha &= \sum_{k \in R(\alpha)} [\![u = v]\!] [\![u = \check{k}]\!] \\ &\leq \sum_{k \in R(\alpha)} [\![v = \check{k}]\!] = [\![M(v)]\!]_\alpha.\end{aligned}$$

Theorem 14.2. Let $u \in V_{\alpha+1}^{(\mathbf{B})}$ and $k \in V$. Then

1. $\alpha \leq \text{rank}(k) \rightarrow [\![\check{k} \in u]\!] = \mathbf{0}$.
2. $\alpha < \text{rank}(k) \rightarrow [\![\check{k} = u]\!] = \mathbf{0}$.

Proof. (By induction on α.)

1. Let $\alpha \leq \text{rank}(k)$. Since $\mathscr{D}(u) \subseteq V_\alpha^{(\mathbf{B})}$

$$[\![\check{k} \in u]\!] = \sum_{x \in \mathscr{D}(u)} u(x) \cdot [\![x = \check{k}]\!] = \mathbf{0}$$

by the induction hypothesis for 2.

2. Let $\alpha < \text{rank}(k)$. Then $(\exists k_1 \in k)[\alpha \leq \text{rank}(k_1)]$ and hence by 1

$$[\![\check{k} = u]\!] \leq \prod_{k_1 \in k} [\![\check{k}_1 \in u]\!] = \mathbf{0}.$$

Remark. From Theorem 14.2 the following result is easily proved.

Theorem 14.3. If $u \in V_\alpha^{(\mathbf{B})}$ and $\alpha \leq \beta$, then $[\![M(u)]\!]_\alpha = [\![M(u)]\!]_\beta$.

Remark. Therefore the new structures $\mathbf{V}_\alpha^{(\mathbf{B})} = \langle V_\alpha^{(\mathbf{B})}, \bar{=}, \bar{\in}, \overline{M}_\alpha \rangle$ satisfy the conditions of §9 (p. 87–88) and hence by Theorem 9.26 we obtain the following.

Theorem 14.4. $\mathbf{V} = \langle V^{(\mathbf{B})}, \bar{=}, \bar{\in}, \overline{M} \rangle$ is a **B**-valued model of *ZF*.

Remark. Note that $M(u)$ is a Boolean expression of "u is a member of $\check{V}$" where $\check{V} = \{\check{x} \mid x \in V\}$ is a Boolean representation of $V^{(\mathbf{2})}$ which is isomorphic to V. Therefore we can talk about $\check{V}$ in $V^{(\mathbf{B})}$ by using the predicate M.

Definition 14.5. An element b of a Boolean algebra **B** (which need not be complete) is called an *atom* iff $b \neq \mathbf{0}$ and

$$(\forall b' \in |\mathbf{B}|)[b' \leq b \rightarrow b' = \mathbf{0} \vee b' = b]$$

i.e., iff b is a minimal element of $\mathbf{B} - \{\mathbf{0}\}$. A Boolean algebra **B** is called *nonatomic* iff **B** has no atoms.

Remark. A complete nonatomic Boolean algebra **B** does not have any complete ultrafilter F, since otherwise $\prod_{b \in F} b$ would be an atom of **B**.

For the following, **B** always denotes a complete Boolean algebra.

Theorem 14.6. If **B** is nonatomic and $S \subseteq B = |\mathbf{B}|$, then

$$\prod_{b \in S} b \cdot \prod_{b \in B - S} (^-b) = \mathbf{0}.$$

Proof. Suppose $\prod_{b \in S} b \cdot \prod_{b \in B - S} (^-b) \neq \mathbf{0}$. Then

$$(\forall b \in B) \neg [b \in S \wedge {}^-b \in S]$$

and

$$(\forall b \in B) \neg [b \in B - S \wedge {}^-b \in B - S]$$

hence

$$B - S = \{^-b \mid b \in S\}.$$

Let $b_0 = \prod_{b \in S} b$. If $b_0 \in S$, S is a complete ultrafilter contrary to the assumption that **B** is nonatomic. On the other hand, if $b_0 \notin S$, then $b_0 \in B - S$, hence

$$\prod_{b \in S} b \cdot \prod_{b \in B - S} (^-b) \leq b_0(^-b_0) = \mathbf{0}.$$

This is a contradiction.

Remark. For the remaining part of this section, let

$$\tilde{B} = \{\check{b} \mid b \in B\} \quad \text{and} \quad F\colon \tilde{B} \rightarrow B$$

be defined by

$$F(\check{b}) = b \quad \text{for} \quad b \in B.$$

Obviously, $F \in V^{(\mathbf{B})}$.

Theorem 14.7. If b_0 is an atom in $\mathbf{B}$ and $S = \{b \in B \mid b_0 \leq b\}$, then $[\![F = \check{S}]\!] = b_0$.

Proof. From the definition of S and the fact that b_0 is an atom it is easily shown that

$$b \in S \leftrightarrow {}^{-}b \in B - S.$$

Since, by Theorem 13.16, $[\![\check{x} = \check{b}]\!] = \mathbf{0}$ unless $x = b$ it follows from Theorem 13.2 that $[\![\check{b} \in F]\!] = F(\check{b})$. Then

$$\begin{aligned} [\![F = \check{S}]\!] &= \prod_{b \in B} [F(\check{b}) \Rightarrow [\![\check{b} \in \check{S}]\!]] \cdot \prod_{b \in S} F(\check{b}) \\ &= \prod_{b \in B - S} ({}^{-}b) \cdot b_0 = \prod_{b \in S} b \cdot b_0 \\ &= b_0. \end{aligned}$$

Theorem 14.8. $[\![F \subseteq \check{B}]\!] = \mathbf{1}$.

Proof. $[\![F \subseteq \check{B}]\!] = \prod_{b \in B} (F(\check{b}) \Rightarrow [\![\check{b} \in \check{B}]\!]) = \mathbf{1}$.

Remark. If $\mathbf{B}$ has an atom, $[\![M(F)]\!] > \mathbf{0}$ by Theorem 14.7. On the other hand, if $\mathbf{B}$ is nonatomic $[\![M(F)]\!] = \mathbf{0}$:

Theorem 14.9. If $\mathbf{B}$ is nonatomic, then $[\![M(F)]\!] = \mathbf{0}$.

Proof. $[\![M(F)]\!] = \sum_{k \in V} [\![F = \check{k}]\!]$

$$\begin{aligned} [\![F = \check{k}]\!] \neq \mathbf{0} &\rightarrow \mathbf{0} < [\![F = \check{k}]\!] \cdot [\![F \subseteq \check{B}]\!] && \text{by Theorem 14.8} \\ &\rightarrow \mathbf{0} < [\![\check{k} \subseteq \check{B}]\!] = \prod_{x \in \mathscr{D}(\check{k})} [\![\check{x} \in \check{B}]\!] = \mathbf{1} \\ &\rightarrow k \subseteq B && \text{by Corollary 13.18.} \end{aligned}$$

Therefore $[\![M(F)]\!] = \sum_{S \subseteq B} [\![F = \check{S}]\!]$. But for $S \subseteq B$,

$$\begin{aligned} [\![F = \check{S}]\!] &= \prod_{b \in B} (F(\check{b}) \Rightarrow [\![\check{b} \in \check{S}]\!]) \cdot \prod_{b \in S} F(\check{b}) \\ &= \prod_{b \in B - S} {}^{-}b \cdot \prod_{b \in S} b = \mathbf{0} && \text{by Theorem 14.6.} \end{aligned}$$

Hence $[\![M(F)]\!] = \mathbf{0}$.

Remark. Since $M(u)$ is the Boolean expression of "$u \in \check{V}$" we might expect that M is a model of ZF in the Boolean sense, i.e., $[\![\phi^M]\!] = \mathbf{1}$ for every axiom ϕ of ZF where ϕ^M denotes ϕ with all quantifiers restricted to M. We also write $a \in M$ for $M(a)$.

Theorem 14.10. M is transitive in $V^{(\mathbf{B})}$, i.e.,

$$(\forall u \in V^{(\mathbf{B})})[[\![(\forall x \in u)[M(u) \rightarrow M(x)]]\!] = \mathbf{1}].$$

Proof. Let $u \in V^{(\mathbf{B})}$. Because of Theorem 13.13.2 it suffices to show that

$$\prod_{x \in \mathscr{D}(u)} (u(x) \Rightarrow [\![M(u) \rightarrow M(x)]\!]) = \mathbf{1}$$

i.e.,

(i) $(\forall x \in \mathscr{D}(u))[u(x) \cdot [\![M(u)]\!] \leq [\![M(x)]\!]]$.

Therefore, if $x \in \mathscr{D}(u)$,

$$[\![u = \check{k}]\!] \leq (u(x) \Rightarrow [\![x \in \check{k}]\!]).$$

(ii) $u(x) \cdot [\![u = \check{k}]\!] \leq [\![x \in \check{k}]\!]$ and

(iii) $[\![x \in \check{k}]\!] = \sum_{k_1 \in k} [\![x = \check{k}_1]\!] \leq \sum_{k_1 \in V} [\![x = \check{k}_1]\!] = [\![M(x)]\!]$.

Combining (ii) and (iii),

$$u(x) \cdot \sum_{k \in V} [\![u = \check{k}]\!] \leq \sum_{k \in V} [\![x \in \check{k}]\!] \leq [\![M(x)]\!].$$

This proves (i).

Theorem 14.11. $(\forall x \in V^{(\mathbf{B})})(\exists k \in V)[[\![M(x)]\!] \leq [\![x \in \check{k}]\!]]$.

Proof. Let $x \in V^{(\mathbf{B})}$. Then $x \in V^{(\mathbf{B})}_{\alpha+1}$ for some α. Choose $k = R(\alpha + 1)$. Then

$$[\![M(x)]\!] = \sum_{k_0 \in V} [\![x = \check{k}_0]\!] = \sum_{k_0 \in R(\alpha+1)} [\![x = \check{k}_0]\!] \qquad \text{by Theorem 14.2.2}$$

$$= \sum_{k_0 \in k} [\![x = \check{k}_0]\!] [\![\check{k}_0 \in \check{k}]\!] \leq [\![x \in \check{k}]\!].$$

Remark. In fact, in Theorem 14.11 we have $=$ instead of $\leq$ i.e.,

$$[\![x \in \check{k}]\!] = \sum_{a \in k} [\![x = \check{a}]\!] \leq [\![M(x)]\!].$$

Theorem 14.12. M is almost universal in $V^{(\mathbf{B})}$, i.e.,

$$(\forall u \in V^{(\mathbf{B})})[[\![u \subseteq M \rightarrow (\exists y \in M)[u \subseteq y]]\!] = \mathbf{1}].$$

Proof. Let $u \in V^{(\mathbf{B})}$. By Theorem 14.11, for each $x \in \mathscr{D}(u)$ there exists a k_x such that

$$[\![M(x)]\!] \leq [\![x \in \check{k}_x]\!].$$

(From the proof of Theorem 14.11 we see that we can take

$$k_x = R(\alpha(x)) \quad \text{where} \quad \alpha(x) = \mu_\beta(x \in V^{(\mathbf{B})}_{\beta+1}).$$

Since $\mathscr{D}(u)$ is a set we can then define $\{k_x \mid x \in \mathscr{D}(u)\}$ without using the Axiom of Choice.)

Let $k = \bigcup_{x \in \mathscr{D}(u)} k_x$. Then

$$[\![u \subseteq \check{k}]\!] = \prod_{x \in \mathscr{D}(u)} (u(x) \Rightarrow [\![x \in \check{k}]\!])$$

$$\geq \prod_{x \in \mathscr{D}(u)} (u(x) \Rightarrow [\![M(x)]\!]).$$

Therefore,

$$[\![u \subseteq M]\!] \leq \prod_{x \in \mathscr{D}(u)} (u(x) \Rightarrow [\![M(x)]\!]) \leq \sum_{y \in V^{(\mathbf{B})}} [\![u \subseteq \check{y}]\!] [\![M(\check{y})]\!]$$

$$\leq \sum_{y \in V^{(\mathbf{B})}} [\![M(y)]\!] [\![u \subseteq y]\!] = [\![(\exists y \in M)[u \subseteq y]]\!].$$

Remark. We know that $\mathbf{V}^{(\mathbf{B})}$ satisfies the Axiom of Pairing. Boolean-valued pairsets and ordered pairs are defined in the following way.

Definition 14.13. For $u, v \in V^{(\mathbf{B})}$:

$$\{u, v\}^{(\mathbf{B})} \triangleq \{u, v\} \times \{\mathbf{1}\}$$
$$\{u\}^{(\mathbf{B})} \triangleq \{u, u\} \times \{\mathbf{1}\} = \{u\} \times \{\mathbf{1}\}$$
$$\langle u, v\rangle^{(\mathbf{B})} \triangleq \{\{u\}^{(\mathbf{B})}, \{u, v\}^{(\mathbf{B})}\}^{(\mathbf{B})}.$$

These definitions are justified by the following theorem.

Theorem 14.14. For $u, v \in V^{(\mathbf{B})}$,

1. $[\![\{u, v\}^{(\mathbf{B})} = \{u, v\}]\!] = \mathbf{1}$.
2. $[\![\{u\}^{(\mathbf{B})} = \{u\}]\!] = \mathbf{1}$.
3. $[\![\langle u, v\rangle^{(\mathbf{B})} = \langle u, v\rangle]\!] = \mathbf{1}$.

Proof. 1. It is sufficient to prove

$$[\![(\forall x)[x = u \vee x = v \leftrightarrow x \in \{u, v\}^{(\mathbf{B})}]]\!] = \mathbf{1}.$$

But this follows from the fact that for all $x \in V^{(\mathbf{B})}$

$$[\![x \in \{u, v\}^{(\mathbf{B})}]\!] = [\![x = u]\!] + [\![x = v]\!]$$
$$= [\![x = u \vee x = v]\!].$$

The arguments for 2 and 3 are similar.

Theorem 14.15. $\{\check{k}_1, \check{k}_2\}^{(\mathbf{B})} = \{k_1, k_2\}^{\vee}$ and hence

$$[\![\{\check{k}_1, \check{k}_2\} = \{k_1, k_2\}^{\vee}]\!] = \mathbf{1}.$$

Proof. Obvious from the definitions and Theorem 14.14.1.

Theorem 14.16.

1. $[\![(\exists x)(\exists y)[\{x, y\}^{(\mathbf{B})} = \check{k}_1 \wedge \varphi(x, y)]]\!] = \sum_{\{k_2, k_3\} = k_1} [\![\varphi(\check{k}_2, \check{k}_3)]\!]$.

2. $[\![(\exists x)(\exists y)[\{x, y\}^{(\mathbf{B})} \in \check{k}_1 \wedge \varphi(x, y)]]\!] = \sum_{\{k_2, k_3\} \in k_1} [\![\varphi(\check{k}_2, \check{k}_3)]\!]$.

Proof. 1. In *ZF* we have

$$(\exists x)(\exists y)[\{x, y\} = \check{k}_1 \wedge \varphi(x, y)] \leftrightarrow (\exists x \in \check{k}_1)(\exists y \in \check{k}_1)[\{x, y\} = \check{k}_1 \wedge \varphi(x, y)].$$

Hence by Theorem 13.13.1

$$[\![(\exists x)(\exists y)[\{x, y\} = \check{k}_1 \wedge \varphi(x, y)]]\!] = \sum_{\{k_2, k_3\} \in k_1} [\![\{\check{k}_2, \check{k}_3\} = \check{k}_1]\!] [\![\varphi(\check{k}_2, \check{k}_3)]\!]$$

$$= \sum_{\{k_2, k_3\} = k_1} [\![\varphi(\check{k}_2, \check{k}_3)]\!] \qquad \text{by Theorem 14.15.}$$

2. Similarly, in ZF

$$(\exists x)(\exists y)[\{x, y\} \in \check{k}_1 \wedge \varphi(x, y)] \leftrightarrow (\exists z \in \check{k}_1)(\exists x)(\exists y)[\{x, y\} = z \wedge z \in \check{k}_1 \wedge \varphi(x, y)]$$

therefore

$$[\![(\exists x)(\exists y)[\{x, y\} \in \check{k}_1 \wedge \varphi(x, y)]]\!] = \sum_{k \in k_1} [\![(\exists x)(\exists y)[\{x, y\} = \check{k} \wedge \varphi(x, y)]]\!]$$

$$= \sum_{k \in k_1} \sum_{\{k_2, k_3\} = k} [\![\varphi(\check{k}_2, \check{k}_3)]\!] \qquad \text{by 1}$$

$$= \sum_{\{k_2, k_3\} \in k_1} [\![\varphi(\check{k}_2, \check{k}_3)]\!].$$

Remark. By the same method we can prove the following:

Theorem 14.17.

1. $[\![(\exists x)(\exists y)[\langle x, y\rangle = \check{k}_1 \wedge \varphi(x, y)]]\!] = \sum_{\langle k_2, k_3\rangle = k_1} [\![\varphi(\check{k}_2, \check{k}_3)]\!].$

2. $[\![(\exists x)(\exists y)[\langle x, y\rangle \in \check{k}_1 \wedge \varphi(x, y)]]\!] = \sum_{\langle k_2, k_3\rangle \in k_1} [\![\varphi(\check{k}_2, \check{k}_3)]\!].$

Theorem 14.18. $[\![On \subseteq M]\!] = \mathbf{1}.$

Proof. Let $u \in V^{(\mathbf{B})}$. Then

$$[\![\mathrm{Ord}\,(u)]\!] = \sum_{\alpha \in On} [\![u = \check{\alpha}]\!] \qquad \text{by Theorem 13.22}$$

$$\leq \sum_{k \in V} [\![u = \check{k}]\!] = [\![M(u)]\!].$$

Remark. Before proving that (in the Boolean sense) M is closed under Gödel's eight fundamental operations $\mathscr{F}_1 \cdots \mathscr{F}_8$, (See Definition 14.2 Introduction to Axiomatic Set Theory) we prove the following absoluteness property for the $\mathscr{F}_i$'s:

Theorem 14.19. $[\![\mathscr{F}_i(\check{k}_1, \check{k}_2) = \mathscr{F}_i(k_1, k_2)^{\vee}]\!] = \mathbf{1}$ for $i = 1, \ldots, 8$.

Proof. By Theorem 14.15 (for $i = 1$) and the following lemmas.

Lemma 1. $[\![\check{k}_1 \cap E = (k_1 \cap E)^{\vee}]\!] = \mathbf{1}$ where $E = \{\langle x, y\rangle \mid x \in y\}$.

Proof. Let $u \in V^{(\mathbf{B})}$. Then

$$[\![u \in (\check{k}_1 \cap E)]\!] = [\![(\exists x)(\exists y)[\langle x, y\rangle \in \check{k}_1 \wedge \langle x, y\rangle = u \wedge x \in y]]\!]$$

$$= \sum_{\langle k_2, k_3\rangle \in k_1} [\![u = \langle \check{k}_2, \check{k}_3\rangle \wedge \check{k}_2 \in \check{k}_3]\!] \qquad \text{by Theorem 14.17.2}$$

$$= \sum_{k_0 \in k_1 \cap E} [\![u = \check{k}_0]\!] = [\![u \in (k_1 \cap E)^{\vee}]\!].$$

Lemma 2. $[\![(\check{k}_1 - \check{k}_2) = (k_1 - k_2)^{\vee}]\!] = \mathbf{1}.$

Proof.

$$\begin{aligned}\check{k}_1 - \check{k}_2 &= \{\langle \check{k}, \mathbf{1}\rangle \mid k \in k_1\} - \{\langle \check{k}, \mathbf{1}\rangle \mid k \in k_2\} \\ &= \{\langle \check{k}, \mathbf{1}\rangle \mid k \in k_1 - k_2\} = (k_1 - k_2)^{\vee}.\end{aligned}$$

Lemma 3. $[\![(\check{k}_1 \ulcorner \check{k}_2) = (k_1 \ulcorner k_2)^{\vee}]\!] = \mathbf{1}.$

Proof. For $u \in V^{(\mathbf{B})}$,

$$\begin{aligned}[\![u \in (\check{k}_1 \ulcorner \check{k}_2)]\!] &= [\![(\exists x)(\exists y)[\langle x, y\rangle \in \check{k}_1 \wedge u = \langle x, y\rangle \wedge x \in \check{k}_2]]\!] \\ &= \sum_{\langle k_3, k_4\rangle \in k_1} [\![u = \langle \check{k}_3, \check{k}_4\rangle \wedge \check{k}_3 \in \check{k}_2]\!] \qquad \text{by Theorem 14.17.2} \\ &= \sum_{\substack{\langle k_3, k_4\rangle \in k_1 \\ k_3 \in k_2}} [\![u = \langle \check{k}_3, \check{k}_4\rangle]\!] \\ &= \sum_{k \in (k_1 \ulcorner k_2)} [\![u = \check{k}]\!] \qquad \text{by Theorem 14.15} \\ &= [\![u \in (k_1 \ulcorner k_2)^{\vee}]\!].\end{aligned}$$

Lemma 4. $[\![\check{k}_1 \cap \check{k}_2 = (k_1 \cap k_2)^{\vee}]\!] = \mathbf{1}.$

Proof.

$$\begin{aligned}\check{k}_1 \cap \check{k}_2 &= \{\langle \check{k}, \mathbf{1}\rangle \mid k \in k_1\} \cap \{\langle \check{k}, \mathbf{1}\rangle \mid k \in k_2\} \\ &= \{\langle \check{k}, \mathbf{1}\rangle \mid k \in k_1 \cap k_2\} \\ &= (k_1 \cap k_2)^{\vee}.\end{aligned}$$

Remark. Therefore we need not consider the intersection with k_1 in $\mathscr{F}_i(k_1, k_2)$ for $i = 5, \ldots, 8$.

Lemma 5. $[\![\mathscr{D}(\check{k}_1) = \mathscr{D}(k_1)^{\vee}]\!] = \mathbf{1}.$

Proof. Let $u \in V^{(\mathbf{B})}$.

$$\begin{aligned}[\![u \in \mathscr{D}(\check{k}_1)]\!] &= [\![(\exists x)(\exists y)[\langle x, y\rangle \in \check{k}_1 \wedge u = x]]\!] \\ &= \sum_{\langle k_3, k_4\rangle \in k_1} [\![u = \check{k}_3]\!] \qquad \text{by Theorem 14.17.2} \\ &= \sum_{k_3 \in \mathscr{D}(k_1)} [\![u = \check{k}_3]\!] = [\![u \in \mathscr{D}(k_1)^{\vee}]\!].\end{aligned}$$

Lemma 6. $[\![(\check{k}_1)^{-1} = (k_1{}^{-1})^{\vee}]\!] = \mathbf{1}.$

Proof. For $u \in V^{(\mathbf{B})}$,

$$\begin{aligned}[\![u \in (\check{k}_1)^{-1}]\!] &= [\![(\exists x)(\exists y)[\langle x, y\rangle \in \check{k}_1 \wedge u = \langle y, x\rangle]]\!] \\ &= \sum_{\langle k_2, k_3\rangle \in k_1} [\![u = \langle k_3, k_2\rangle^{\vee}]\!] \\ &= [\![u \in (k_1{}^{-1})^{\vee}]\!].\end{aligned}$$

Lemma 7.

$$[\![\{\langle x, y, z\rangle \mid \langle x, z, y\rangle \in \check{k}_1\} = \{\langle x, y, z\rangle \mid \langle x, z, y\rangle \in k_1\}^{\vee}]\!] = \mathbf{1}.$$

Proof. (By the same method as before.) Let $u \in V^{(\mathbf{B})}$. Then

$$
\begin{aligned}
&[\![u \in \{\langle x, y, z\rangle \mid \langle x, z, y\rangle \in \check{k}_1\}]\!] \\
&\quad = [\![(\exists x, w)[\langle x, w\rangle \in \check{k}_1 \wedge (\exists y, z)[w = \langle z, y\rangle \wedge u = \langle x, y, z\rangle]]]\!] \\
&\quad = \sum_{\langle k_3, k_4\rangle \in k_1} [\![(\exists y)(\exists z)[\check{k}_4 = \langle z, y\rangle \wedge u = \langle \check{k}_3, y, z\rangle]]\!] \\
&\quad = \sum_{\langle k_3, k_4\rangle \in k_1} \sum_{\langle k_6, k_5\rangle = k_4} [\![u = \langle \check{k}_3, \check{k}_5, \check{k}_6\rangle]\!] \\
&\quad = \sum_{\langle k_3, k_4\rangle \in k_1} \sum_{\langle k_6, k_5\rangle = k_4} [\![u = \langle k_3, k_5, k_6\rangle^{\vee}]\!] \\
&\quad = \sum_{k \in \{\langle k_3, k_5, k_6\rangle \mid \langle k_3, k_6, k_5\rangle \in k_1\}} [\![u = \check{k}]\!] \\
&\quad = [\![u \in \{\langle x, y, z\rangle \mid \langle x, z, y\rangle \in k_1\}^{\vee}]\!].
\end{aligned}
$$

Remark. Finally, by the same method, we can prove the following.

Lemma 8. $[\![\mathscr{F}_8(\check{k}_1, \check{k}_2) = \mathscr{F}_8(k_1, k_2)^{\vee}]\!] = \mathbf{1}$.

Remark. This completes the proof of Theorem 14.19. From Theorem 14.19 we can easily prove the following result.

Theorem 14.20. For $i = 1, \ldots, 8$,

$$(\forall u, v \in V^{(\mathbf{B})})[[\![M(u) \wedge M(v) \rightarrow M(\mathscr{F}_i(u, v))]\!] = \mathbf{1}].$$

Proof. Let $u, v \in V^{(\mathbf{B})}$. Then

$$
\begin{aligned}
[\![u = \check{k}_1]\!] \cdot [\![u = \check{k}_2]\!] &\leq [\![\mathscr{F}_i(u, v) = \mathscr{F}_i(\check{k}_1, \check{k}_2)]\!] && \text{by Corollary 13.7} \\
&\leq [\![\mathscr{F}_i(u, v) = \mathscr{F}_i(k_1, k_2)^{\vee}]\!] && \text{by Theorem 14.19} \\
&\leq [\![M(F_i(u, v))]\!].
\end{aligned}
$$

Therefore,

$$[\![M(u)]\!] \cdot [\![M(v)]\!] \leq [\![M(\mathscr{F}_i(u, v))]\!].$$

Remark. As a consequence of Theorems 14.10, 14.12, 14.18, and 14.20 we have the following.

Theorem 14.21. In $\mathbf{V}^{(\mathbf{B})}$, M is a standard transitive model of ZF containing all the ordinals, i.e.,

$$[\![(\forall u, v)[M(u) \wedge v \in u \rightarrow M(v)]]\!] = \mathbf{1}$$

$$[\![(\forall u)[\mathrm{Ord}\,(u) \rightarrow M(u)]]\!] = \mathbf{1}.$$

$[\![\phi^M]\!] = \mathbf{1}$ for each axiom ϕ of ZF where ϕ^M is obtained from ϕ by relativizing all the quantifiers occurring in ϕ to M.

Remark. There is an easier way to show $[\![\phi^M]\!] = \mathbf{1}$ for every axiom ϕ of ZF than the one stated above. Let ϕ be $(\forall x)\psi(x)$. Then

$$[\![\phi^M]\!] = [\![(\forall x)[M(x) \to \psi(x)]]\!] = \prod_{u \in V^{(\mathbf{B})}} ([\![M(u)]\!] \Rightarrow [\![\psi(u)]\!])$$

$$= \prod_{u \in V^{(\mathbf{B})}} \left(\sum_{k \in V} [\![u = \check{k}]\!] \Rightarrow [\![\psi(u)]\!]\right)$$

$$= \prod_{u \in V^{(\mathbf{B})}} \prod_{k \in V} ([\![u = \check{k}]\!] \Rightarrow [\![\psi(u)]\!])$$

$$= \prod_{u \in V^{(\mathbf{B})}} \prod_{k \in V} ([\![u = \check{k}]\!] \Rightarrow [\![\psi(\check{k})]\!])$$

$$= \prod_{k \in V} [\![\psi(\check{k})]\!].$$

Now if ϕ is an axiom of ZF, then since V is a model of ZF, $\psi(k)$ is true for each $k \in V$. Hence, $(\forall k \in V)[[\![\psi(\check{k})]\!] = \mathbf{1}]$, because $\mathbf{V}^{(\mathbf{B})} \models ZF$.

A similar argument shows that if ϕ is $(\exists x)\psi(x)$, then

$$[\![\phi^{\mathrm{M}}]\!] = \sum_{k \in V} [\![\psi(\check{k})]\!].$$

Since V is a model of ZF, $(\exists k_0 \in V)[\psi(k_0)]$, hence $[\![\psi(\check{k}_0)]\!] = \mathbf{1}$.

Since also $[\![F \subseteq M]\!] = \mathbf{1}$ (by Theorem 14.8) we may consider $M[F]$ in $\mathbf{V}^{(\mathbf{B})}$. In fact, it will turn out that $V = M[F]$ in $\mathbf{V}^{(\mathbf{B})}$. The proof of this statement is a corollary of the following theorem which shows that the method of forcing on the one hand and the method which uses the models $\mathbf{V}^{(\mathbf{B})}$ are essentially equivalent (cf. Corollary 14.23). For the remaining part of this section we assume the Axiom of Choice (in $\mathbf{V}$).

Theorem 14.22. Suppose that N is a countable standard transitive model of $ZF + AC$ such that $\mathbf{B} \in N$. Let $\mathbf{P}$ be the partial order structure associated with $\mathbf{B}$ (thus $\langle N, \mathbf{P}\rangle$ is a setting for forcing). Then for any set G_0 which is $\mathbf{P}$-generic over N we can define a mapping $h\colon (V^{(\mathbf{B})})^N \to N[G_0]$ which is onto and satisfies

(i) $h_0([\![\varphi(u_1, \ldots, u_n)]\!]) = \mathbf{1} \leftrightarrow N[G_0] \models \varphi(h(u_1), \ldots, h(u_n))$ for $u_1, \ldots, u_n \in (V^{(\mathbf{B})})^N$ and φ any formula of $\mathscr{L}_0(\{M\})$. Here $M(a)$ is interpreted in $N[G_0]$ as $a \in N$ and h_0 is the N-complete homomorphism from $|\mathbf{B}|$ into $|\mathbf{2}|$ associated with G_0. Consequently

(ii) $h_0([\![M(u)]\!]) = \mathbf{1} \leftrightarrow h(u) \in N$ for $u \in (V^{(\mathbf{B})})^N$.

Proof. Given G_0 which is $\mathbf{P}$-generic over N and $h_0\colon |\mathbf{B}| \to |\mathbf{2}|$ where h_0 is associated with G_0 in the familiar way, define $h\colon (V^{(\mathbf{B})})^N \to V$ by induction as follows:

$$h(u) = \{h(x) \mid x \in \mathscr{D}(u) \wedge h_0([\![x \in u]\!]) = \mathbf{1}\} \quad \text{for} \quad u \in (V^{(\mathbf{B})})^N.$$

Then we have for $u, v \in (V^{(\mathbf{B})})^N$,

1. $h_0([\![u = v]\!]) = \mathbf{1} \leftrightarrow h(u) = h(v)$.
2. $h_0([\![u \in v]\!]) = \mathbf{1} \leftrightarrow h(u) \in h(v)$.

1 and 2 follow from the N-completeness of h_0, since

$$h(u) = h(v) \leftrightarrow (\forall x \in \mathscr{D}(u))[h_0([\![x \in u]\!]) = \mathbf{1} \to h(x) \in h(v)] \wedge (\forall x \in \mathscr{D}(v))[h_0([\![x \in v]\!]) = \mathbf{1} \to h(x) \in h(u)]$$

and

$$h(u) \in h(v) \leftrightarrow (\exists y \in \mathscr{D}(v))[h(u) = h(y) \wedge h_0([\![y \in u]\!]) = \mathbf{1}].$$

Furthermore,

3. $(\forall k \in N)[h(\check{k}) = k]$.

This follows by induction on rank (k) using the fact that

$$\begin{aligned} h(\check{k}) &= \{h(x) \mid x \in \mathscr{D}(\check{k}) \wedge h_0([\![x \in \check{k}]\!]) = \mathbf{1}\} \\ &= \{h(x) \mid (\exists k_1 \in k)[x = \check{k}_1]\} \\ &= \{h(\check{k}_1) \mid k_1 \in k\}. \end{aligned}$$

Since

$$\begin{aligned} [\![M(u)]\!] &= \sum_{k \in N} [\![u = \check{k}]\!] \\ &= \sum_{k \in R(\alpha) \cap N} [\![u = \check{k}]\!] \end{aligned}$$

for some $\alpha \in On^N$ by Theorem 14.2.2 and the fact that N is a model of ZF. Hence, by the N-completeness of h_0,

$$\begin{aligned} h([\![M(u)]\!]) = \mathbf{1} &\leftrightarrow (\exists k \in N)[h([\![u = \check{k}]\!]) = \mathbf{1}] \\ &\leftrightarrow (\exists k \in N)[h(u) = h(\check{k})] && \text{by 1} \\ &\leftrightarrow h(u) \in N && \text{by 3.} \end{aligned}$$

Consequently

4. $h_0([\![M(u)]\!]) = \mathbf{1} \leftrightarrow h(u) \in N, u \in (V^{(\mathbf{B})})^N$.
Let F_N be $F \upharpoonright \tilde{\mathbf{B}}$, i.e., $F_N\colon |\tilde{\mathbf{B}}| \to B$ defined by

$$F_N(\check{b}) = b \quad \text{for} \quad b \in \mathbf{B}.$$

Then

$$\begin{aligned} h(F_N) &= \{h(\check{k}) \mid k \in B \wedge h_0([\![\check{k} \in F]\!]) = \mathbf{1}\} \\ &= \{h(\check{k}) \mid k \in B \wedge h_0(F(\check{k})) = \mathbf{1}\} \\ &= \{b \mid b \in B \wedge h_0(b) = \mathbf{1}\}. \end{aligned}$$

Thus

5. $G = h(F_N)$ is the N-complete ultrafilter associated with h_0 and G_0.
Next we prove that h_0 preserves all the sums
6. $\sum_{x \in (V^{(\mathbf{B})})^N} [\![\varphi(x_1, u_1, \ldots, u_n)]\!]$ where φ is a formula of $\mathscr{L}_0(\{M(\ )\})$ and $u_1, \ldots, u_n \in (V^{(\mathbf{B})})^N$.

If φ is a formula of $\mathscr{L}_0(\{M(\ )\})$ and $u_1, \ldots, u_n \in (V^{(\mathbf{B})})^N$, then the sequence $S = \langle [\![\varphi(x, u_1, \ldots, u_n)]\!] \mid x \in (V^{(\mathbf{B})})^N \rangle$ is definable in $\langle N, \in, \mathbf{B} \rangle$ and the range of S, $\mathscr{W}(S)$, is contained in $\mathbf{B}$ which is a set in N. Therefore, by the AC in N, there is a function $f \in N$ such that $\mathscr{W}(f) = \mathscr{W}(S)$ and hence 6 in N is equal to a sum over a set in N which is preserved by h_0 since h_0 is N-complete. (Note that we have used the same argument in the proof of 4.) Now let $N_1 = \{h(x) \mid x \in (V^{(\mathbf{B})})^N\}$. Then for $u_1, \ldots, u_n \in (V^{(\mathbf{B})})^N$, and φ a formula of $\mathscr{L}_0(\{M(\ )\})$,

7. $h_0([\![\varphi(u_1, \ldots, u_n)]\!]) = \mathbf{1} \leftrightarrow N_1 \vDash \varphi(h(u_1), \ldots, h(u_n))$.

This is proved by induction on the number of logical symbols in φ using 1 and 2, and, for the induction step, the fact that h_0 preserves the sums 6. Furthermore, if φ contains the symbol $M(\ )$, we understand by 7 that $M(a)$ is interpreted as $a \in N$ in N_1 in accordance with 4.

8. $\forall u \in (V^{(\mathbf{B})})^N, h(u) \in N[G] = N[G_0]$.

Let $u \in (V^{(\mathbf{B})})$. Then

$$h(u) = \{h(x) \mid x \in \mathscr{D}(u) \wedge h_0([\![x \in u]\!]) = \mathbf{1}\}$$
$$h(u) = \{h(x) \mid x \in \mathscr{D}(u) \wedge [\![x \in u]\!] \in G\}.$$

Now $[\![x \in u]\!]$ for $x, u \in (V^{(\mathbf{B})})^N$ is definable in $N[G]$ from $\mathbf{B}$, hence $h(u) \in N[G]$, since $N[G]$ is a standard transitive model of ZF. Applying 7 to φ where φ is an axiom of ZF, we see that N_1 is a standard transitive model of ZF and contains N as a subset (because of 3). Furthermore, $G = h(F_N) \in N_1$ and $N_1 \subseteq N[G]$ by 8. Since $N[G]$ is the least standard transitive model of ZF containing G as an element and N as a subset we must have $N_1 = N[G]$. Therefore h is onto and 7 is just (i).

Corollary 14.23. Suppose that $\langle N, \in, \mathbf{B} \rangle$ is elementarily equivalent to $\langle V, \in, \mathbf{B} \rangle$ where N is transitive, countable and $\mathbf{B} \in N$. Let $\mathbf{P}$ be the partial order structure related to $\mathbf{B}$. Then for every sentence φ of $\mathscr{L}_0(\{M(\ )\})$, $[\![\varphi]\!] = \mathbf{1}$ (i.e., φ holds in $V^{(\mathbf{B})}$) iff $N[G] \vDash \varphi$ for all sets G which are $\mathbf{P}$-generic over N. ($M(a)$ is interpreted in $N[G]$ as $a \in N$.)

Proof. If $[\![\varphi]\!] = \mathbf{1}$ in $\mathbf{V}^{(\mathbf{B})}$, $[\![\varphi]\!] = \mathbf{1}$ in $(\mathbf{V}^{(\mathbf{B})})^N$ and the conclusion follows from (i) of Theorem 14.22. Conversely, if $[\![\varphi]\!] \neq \mathbf{1}$ in $\mathbf{V}^{(\mathbf{B})}$, then $b = [\![\varphi]\!] \neq \mathbf{1}$ in $(\mathbf{V}^{(\mathbf{B})})^N$, hence taking some $h_0\colon |\mathbf{B}| \to |\mathbf{2}|$ which is N-complete and sends b to $\mathbf{0}$, we have, by Theorem 14.22, $N[G] \vDash \neg\varphi$ for some G which is $\mathbf{P}$-generic over N.

Remark. We give two applications of this method.

Theorem 14.24. $[\![V = M[F]]\!] = \mathbf{1}$.

Proof. Let $\mathbf{N} = \langle N, \in \rangle$ be a countable transitive model of $ZF + AC$ such that $\mathbf{B} \in N$ and

$$\langle N, \in, \mathbf{B} \rangle \quad \text{is elementarily equivalent to} \quad \langle V, \in, \mathbf{B} \rangle.$$

(The existence of such an **N** can be proved in Gödel-Bernays set theory + Mathematical Induction.)

$V = M[F]$ can be written as a formula

$$(\forall x)\varphi(M(x), F) \quad \text{where } \varphi \text{ is a formula of } \mathscr{L}_0.$$

If $[\![V = M[F]]\!] \neq \mathbf{1}$ in $\mathbf{V}^{(\mathbf{B})}$,

$$b = [\![V = M[F_N]]\!] \neq \mathbf{1} \quad \text{in} \quad (\mathbf{V}^{(\mathbf{B})})^{\mathbf{N}}$$

hence as in the proof of Corollary 14.23

1. $N[G] \models \neg(\forall x)\varphi(M(x), G)$

where $G = h(F_N)$ and $h(b) = \mathbf{0}$ for some $h\colon |\mathbf{B}| \to |\mathbf{2}|$ that is N-complete. But since $M(x)$ means $x \in N$ in $N[G]$, (i) means that $V \neq N[G]$ in $N[G]$. This is a contradiction.

Theorem 14.25. (Using the Axiom of Choice in V.) $[\![AC]\!] = \mathbf{1}$, i.e., the AC holds in $\mathbf{V}^{(\mathbf{B})}$.

Proof. Choose N and **B** as in Corollary 14.23 (again we need the system GB + Mathematical Induction). Then N satisfies the AC, and so does $N[G]$ for every G which is **P**-generic over N. Hence $[\![AC]\!] = \mathbf{1}$ by Theorem 14.24.

Remark. Later we shall give another application of Theorem 14.22.

Problem. Find a proof of Theorem 14.24 that can be carried out in ZF.

Exercise. Give a direct proof of Theorem 14.25.

15. An Elementary Embedding of $V[F_0]$ in $\mathbf{V}^{(\mathbf{B})}$

We have seen that in $\mathbf{V}^{(\mathbf{B})}$, $V = M[F]$. Since $M(u)$ expresses $u \in V$ in the Boolean sense, we might expect some relationship between the Boolean-valued structures $V[F]$ and $\mathbf{V}^{(\mathbf{B})}$. Again let $\mathbf{B}$ be a complete Boolean algebra and $F: \tilde{B} \to B$ be defined by $F(\check{b}) = b$ for $b \in B$ as in §14. Furthermore, let F_0 be the identity on B.

Definition 15.1. We define a mapping $j: V[F_0] \to V^{(\mathbf{B})}$ and a denotation operator D_j on the terms and formulas of the ramified language corresponding to $V[F_0]$ by recursion in the following way (cf. Definition 9.36):

1. $j(\underline{k}) \triangleq \check{k}, k \in V$.
2. $D_j(V(t)) \triangleq [\![M(j(t))]\!]$.
3. $D_j(F(t)) \triangleq \sum_{b \in R(\alpha) \cap B} [\![j(t) = \check{b}]\!] \cdot b$ where $\alpha = \rho(t) + 1$.
4. $D_j(t_1 \in t_2) \triangleq [\![j(t_1) \in j(t_1)]\!]$.
5. $D_j(t_1 = t_2) \triangleq [\![j(t_1) = j(t_2)]\!]$.
6. $D_j(\neg\varphi) \triangleq -D_j(\varphi)$, $D_j(\varphi_1 \wedge \varphi_2) = D_j(\varphi_1) \cdot D_j(\varphi_2)$.
7. $D_j((\forall x_n{}^\beta)\varphi(x_n{}^\beta)) \triangleq \prod_{t \in T_\beta} D_j(\varphi(t))$.
8. $j(\hat{x}_n{}^\beta \varphi(x_n{}^\beta)) \triangleq v$ where $v \in V^{(\mathbf{B})}$ is given by

$$\mathscr{D}(v) = \{j(t) \mid t \in T_\beta\}$$

and

$$v(j(t)) = D_j(\varphi(t)) \qquad \text{for } t \in T_\beta.$$

Remark. We use the notation $F(b)$ for the value of the function F at b and at the same time $F(\ )$ (e.g., in 3) is a formal symbol of the ramified language. Also $[\![\ ]\!]$ refers to $V[F_0]$ and to $\mathbf{V}^{(\mathbf{B})}$. Despite these ambiguities it is hoped that the proper meaning of F and $[\![\ ]\!]$ is always clear from the context.

Theorem 15.2. If $u, v \in V^{(\mathbf{B})}$, if $\mathscr{D}(u) \subseteq S$ and $\mathscr{D}(v) \subseteq S$ where $S \subseteq V^{(\mathbf{B})}$, then $[\![u = v]\!] = \prod_{w \in S} [\![w \in u \leftrightarrow w \in v]\!]$.

Proof. $[\![u = v]\!] \leq \prod_{w \in S} [\![w \in u \leftrightarrow w \in v]\!]$ follows from the Axioms of Equality. On the other hand,

$$\prod_{w \in S} [\![w \in u \leftrightarrow w \in v]\!] \leq \prod_{w \in \mathscr{D}(u)} [\![w \in u \rightarrow w \in v]\!] \cdot \prod_{w \in \mathscr{D}(v)} [\![w \in v \rightarrow w \in u]\!]$$

$$\leq \prod_{w \in \mathscr{D}(u)} (u(w) \Rightarrow [\![w \in v]\!]) \cdot \prod_{w \in \mathscr{D}(u)} (v(w) \Rightarrow [\![w \in u]\!])$$

by Corollary 13.4.3

$$= [\![u = v]\!].$$

Theorem 15.3. $t \in T_\alpha \rightarrow j(t) \in V_\alpha^{(\mathbf{B})}$.

Proof. (By induction on α.) If $t = \underline{k}$ and $\underline{k} \in T_\alpha$, then $k \in R(\alpha)$, i.e., $k \in R(\beta + 1)$ for some $\beta < \alpha$. Therefore

$$j(\underline{k}) = \check{k} = \{\langle \check{k}_1, \mathbf{1} \rangle \mid k_1 \in k\}$$

and

$$\mathscr{D}(\check{k}) = \{\check{k}_1 \mid k_1 \in k\}.$$

From the induction hypothesis, $\mathscr{D}(\check{k}) \subseteq V_\beta^{(\mathbf{B})}$. Hence

$$\check{k} \in V_{\beta+1}^{(\mathbf{B})} \subseteq V_\alpha^{(\mathbf{B})}.$$

If $t = \hat{x}^\beta \varphi(x^\beta)$ for some φ and $t \in T_\alpha$, then $\beta < \alpha$. For $v = j(t)$ we have, by the induction hypothesis,

$$\mathscr{D}(v) = \{j(t) \mid t \in T_\beta\} \subseteq V_\beta^{(\mathbf{B})}.$$

Therefore

$$v \in V_{\beta+1}^{(\mathbf{B})} \subseteq V_\alpha^{(\mathbf{B})}.$$

Theorem 15.4. If t_1, t_2 are constant terms and φ is a limited formula, then $[\![\varphi]\!] = D_j(\varphi)$. In particular,

1. $[\![t_1 = t_2]\!] = [\![j(t_1) = j(t_2)]\!]$.
2. $[\![t_1 \in t_2]\!] = [\![j(t_1) \in j(t_2)]\!]$.

Proof. (By induction on Ord (φ).)

1. Let $\beta = \max(\rho(t_1), \rho(t_2))$. Then

$$[\![t_1 = t_2]\!] = \prod_{t \in T_\beta} [\![t \in t_1 \leftrightarrow t \in t_2]\!] \qquad \text{by Definition 9.27.6}$$

$$= \prod_{t \in T_\beta} [\![j(t) \in j(t_1) \leftrightarrow j(t) \in j(t_2)]\!] \quad \text{by the induction hypothesis}$$

$$= [\![j(t_1) = j(t_2)]\!] \qquad \text{by Theorem 15.2}$$

since

$$\mathscr{D}(j(t_i)) \subseteq \{j(t) \mid t \in T_\beta\} \qquad \text{for } i = 1, 2.$$

2. We distinguish the following three cases:

2.1 $t_1 = \underline{k}_1 \wedge t_2 = \underline{k}_2$ for some $k_1, k_2 \in V$. Then

$$\llbracket t_1 \in t_2 \rrbracket = \llbracket \underline{k}_1 \in \underline{k}_2 \rrbracket = \llbracket \check{k}_1 \in \check{k}_2 \rrbracket = \llbracket j(\underline{k}_1) = j(\underline{k}_2) \rrbracket.$$

2.2 $t_2 = \hat{x}^\beta \varphi(x^\beta)$. Let $v = j(\hat{x}^\beta \varphi(x^\beta)) = j(t_2)$. Then

$$\llbracket t_1 \in t_2 \rrbracket = \sum_{t \in T_\beta} \llbracket t_1 = t \rrbracket \cdot \llbracket \varphi(t) \rrbracket \qquad \text{by Definition 9.27.5}$$

$$= \sum_{t \in T_\beta} \llbracket j(t_1) = j(t) \rrbracket \cdot D_j(\varphi(t)) \qquad \text{by the induction hypothesis}$$

$$= \sum_{x \in \mathscr{D}(v)} \llbracket j(t_1) = x \rrbracket \cdot v(x)$$

$$= \llbracket j(t_1) \in v \rrbracket = \llbracket j(t_1) \in j(t_2) \rrbracket.$$

2.3 $t_2 = \underline{k}_2$ for some k_2. Then

$$\llbracket t_1 \in \underline{k}_2 \rrbracket = \sum_{k \in k_2} \llbracket t_1 = \underline{k} \rrbracket \qquad \text{by Definition 9.27.4}$$

$$= \sum_{k \in k_2} \llbracket j(t_1) = \check{k} \rrbracket \qquad \text{by the induction hypothesis}$$

$$= \llbracket j(t_1) \in \check{k}_2 \rrbracket = \llbracket j(t_1) \in j(t_2) \rrbracket.$$

Among the remaining cases we need only consider the following.

3. If φ is $(\forall x^\beta)\psi(x^\beta)$, then

$$\llbracket \varphi \rrbracket = \prod_{t \in T_\beta} \llbracket \psi(t) \rrbracket \qquad \text{by Definition 9.27.8}$$

$$= \prod_{t \in T_\beta} D_j(\psi(t)) \qquad \text{by the induction hypothesis}$$

$$= D_j((\forall x^\beta)\psi(x^\beta)) = D_j(\varphi).$$

4. $$\llbracket V(t) \rrbracket = \sum_{k \in R(\alpha)} \llbracket t = \underline{k} \rrbracket \qquad \text{where } \alpha = \rho(t) + 1$$

$$= \sum_{k \in R(\alpha)} \llbracket j(t) = \check{k} \rrbracket \qquad \text{by the induction hypothesis}$$

$$= \llbracket M(j(t)) \rrbracket = D_j(V(t))$$

5. $$\llbracket F(t) \rrbracket = \sum_{b \in R(\alpha) \cap B} \llbracket t = \underline{b} \rrbracket \cdot F_0(b) \qquad \text{where } \alpha = \rho(t) + 1$$

$$= \sum_{b \in R(\alpha) \cap B} \llbracket j(t) = \check{b} \rrbracket \cdot b \qquad \text{by the induction hypothesis}$$

$$= D_j(F(t)).$$

Corollary 15.5. $(\exists \alpha) \llbracket j(\hat{x}_n{}^\alpha F(x_n{}^\alpha)) = F \rrbracket = 1$.

Proof. Choose $\alpha = \text{rank}(B)$ and use 5.

Remark. From now on we again assume that V satisfies the Axiom of Choice.

Theorem 15.6. Under the assumptions of Theorem 14.22, $h(j(t)) = D_{N[G_0]}(t)$ for each constant term t in the relative sense of N. (Cp. the relativization of $V[F_0]$ to N discussed on page 100. Here F_0 is also relativized to N i.e., restricted to $\mathbf{B}^N$ so that $F_0 \in N$. F_0 in Definition 9.36 is to be replaced by

$$\{b \in B^N \mid h_0(F_0(b)) = \mathbf{1}\} = \{b \in B^N \mid h_0(b) = \mathbf{1}\} = G.)$$

Proof. (By induction on $\rho(t)$.)

$$\begin{aligned} h(j(t_1)) \in h(j(t)) &\leftrightarrow h_0([\![j(t_1) \in j(t)]\!]) = \mathbf{1} && \text{by Theorem 14.22} \\ &\leftrightarrow h_0([\![t_1 \in t]\!]) = \mathbf{1} && \text{by Theorem 15.4} \\ &\leftrightarrow D(t_1) \in D(t) && \text{by Theorem 9.37} \end{aligned}$$

(where we have h_0 instead of h). Therefore

$$(\forall x \in N[G_0])[x \in h(j(t)) \leftrightarrow x \in D(t)]$$

by the induction hypothesis i.e., $h(j(t)) = D(t)$, where

$$D = D_{N[G]} = D_{N[G_0]}.$$

Remark. Considering j relativized to N we have mappings

$$(V[F_0])^N \xrightarrow{j} (V^{(\mathbf{B})})^N \xrightarrow{h} N[G]$$

and a denotation operator (see Definition 9.36)

$$(V[F_0])^N \xrightarrow{D_{N[G]}} N[G].$$

Theorem 15.6 shows that $D_{N[G]} = h \circ j$.

Definition 15.7. Let $\mathbf{B}$ be a complete Boolean algebra and $\mathbf{M}_1$ and $\mathbf{M}_2$ be two $\mathbf{B}$-valued structures. A mapping $i: M_1 \to M_2$ (where M_i is the universe of $\mathbf{M}_i$) is *elementary* (in the Boolean sense) iff for every formula of the language of $\mathbf{M}_1$ and every $u_1, \ldots, u_n \in M_1$,

$$[\![\varphi(u_1, \ldots, u_n)]\!]_{\mathbf{M}_1} = [\![\varphi(i(u_1), \ldots, i(u_n))]\!]_{\mathbf{M}_2}.$$

Theorem 15.8. The mapping $j: V[F_0] \to V^{(\mathbf{B})}$ (of Definition 15.1) is elementary.

Proof. Let N be a countable standard transitive model of ZF such that $\mathbf{B}^N \in N$ and $\langle N, \in, \mathbf{B}^N \rangle$ is an elementary substructure of $\langle V, \in, \mathbf{B} \rangle$. (The existence of such an N can be proved in GB + mathematical induction.) Suppose that for some formula φ and some $u_1, \ldots, u_n \in (V[F_0])^N$.

$$b_1 = [\![\varphi(u_1, \ldots, u_n)]\!] \neq [\![\varphi(j(u_1), \ldots, j(u_n)]\!] = b_2 \qquad \text{in } N.$$

Let $h_0: \mathbf{B}^N \to \mathbf{2}$ be an N-complete homomorphism such that

$$h_0(b_1) \neq h_0(b_2).$$

Let G_0 be the **P**-generic filter over N associated with h_0 where **P** is the partial order structure associated with $\mathbf{B}^N$. Define h as in Theorem 14.22. Then

$$\begin{aligned} h_0(b_1) = \mathbf{1} &\leftrightarrow D_{N[G_0]}(\varphi(u_1, \ldots, u_n)) \\ &\leftrightarrow N[G_0] \models \varphi(D(u_1), \ldots, D(u_n)). \end{aligned}$$

On the other hand, by Theorem 14.22,

$$\begin{aligned} h_0(b_2) = \mathbf{1} &\leftrightarrow N[G_0] \models \varphi(h(j(u_1), \ldots, j(u_n))) \\ &\leftrightarrow N[G_0] \models \varphi(D(u_1), \ldots, D(u_n)) \qquad \text{by Theorem 15.6.} \end{aligned}$$

Consequently $h_0(b_1) = h_0(b_2)$. This is a contradiction.

Problem. If $\mathbf{B} = \mathbf{2}$, $j\colon V[F_0] \to V^{(\mathbf{B})}$ is onto in the following sense, $(\forall v \in V^{(\mathbf{B})})(\exists x \in V[F_0])[\llbracket v = j(x) \rrbracket = \mathbf{1}]$, is j also onto in this sense for every complete Boolean algebra **B**?

16. The Maximum Principle

From now on until further notice we will assume the AC for V.

Theorem 16.1. Suppose $\{u_i \mid i \in I\} \subseteq V^{(\mathbf{B})}$ and $(\forall i \in I)[\mathscr{D}(u_i) \subseteq d]$ for some $d \subseteq V^{(\mathbf{B})}$. Then there is a family $\{u_i' \mid i \in I\}$ such that

1. $(\forall i \in I)[[\![u_i = u_i']\!] = \mathbf{1}]$,
2. $(\forall i \in I)[\mathscr{D}(u_i') = d]$.

Proof. We extend the domain of u_i to d by defining $u_i' \in V^{(\mathbf{B})}$ by $\mathscr{D}(u_i') = d$ and $(\forall x \in d)[u_i'(x) = [\![x \in u_i]\!]]$ for $i \in I$. Then for all $i \in I$,

$$
\begin{aligned}
[\![u_i = u_i']\!] &= \prod_{x \in \mathscr{D}(u_i)} (u_i(x) \Rightarrow [\![x \in u_i']\!]) \cdot \mathbf{1} \\
&\geq \prod_{x \in \mathscr{D}(u_i)} (u_i(x) \Rightarrow u_i'(x)) \\
&= \prod_{x \in \mathscr{D}(u_i)} (u_i(x) \Rightarrow [\![x \in u_i]\!]) \\
&= \mathbf{1}. \qquad \text{by Theorem 13.4.3}
\end{aligned}
$$

Theorem 16.2. (The Maximum Principle)

$$(\exists v \in V^{(\mathbf{B})})[[\![(\exists u)\varphi(u)]\!] = [\![\varphi(v)]\!]]$$

i.e., for each formula φ there is a $v \in V^{(\mathbf{B})}$ such that $[\![\varphi(v)]\!]$ maximizes the set of Boolean values $\{[\![\varphi(u)]\!] \mid u \in V^{(\mathbf{B})}\}$, i.e.,

$$[\![\varphi(v)]\!] = \sum_{u \in V^{(\mathbf{B})}} [\![\varphi(u)]\!].$$

Proof. Let $b = [\![(\exists u)\varphi(u)]\!] = \sum_{u \in V^{(\mathbf{B})}} [\![\varphi(u)]\!]$. Since $\mathbf{B}$ is a set, it follows from the AC in V, that there is a sequence $\langle u_\xi \mid \xi < \alpha \rangle$ such that $\{u_\xi \mid \xi < \alpha\} \subseteq V^{(\mathbf{B})}$ and $b = \sum_{\xi < \alpha} [\![\varphi(u_\xi)]\!]$. Define

$$b_\xi = [\![\varphi(u_\xi)]\!] \cdot \prod_{\eta < \xi} [\![\neg\varphi(u_\eta)]\!] \text{ for } \xi < \alpha.$$

Then

i) $\xi, \eta < \alpha \wedge \xi \neq \eta \rightarrow b_\xi \cdot b_\eta = \mathbf{0}$ and $b = \sum_{\xi < \alpha} b_\xi$.

Since $\langle u_\xi \mid \xi < \alpha \rangle$ is a set, there is a $d \subseteq V^{(\mathbf{B})}$ such that $(\forall \xi < \alpha)[\mathscr{D}(u_\xi) \subseteq d]$.

So by Theorem 16.1 we can assume that $(\forall \xi < \alpha)[\mathscr{D}(u_\xi) = d]$. Define $v \in V^{(\mathbf{B})}$ by

$$\mathscr{D}(v) = d \wedge (\forall x \in d)\left[v(x) = \sum_{\xi < \alpha} b_\xi \cdot u_\xi(x)\right].$$

Then by i),

ii) $\xi \in \alpha \wedge x \in d \rightarrow b_\xi \cdot v(x) = b_\xi \cdot u_\xi(x)$.

Therefore, for $\xi < \alpha$,

$$\begin{aligned}
[\![v = u_\xi]\!] &= \prod_{x \in d} (v(x) \Rightarrow [\![x \in u_\xi]\!]) \cdot \prod_{x \in d} (u_\xi(x) \Rightarrow [\![x \in v]\!]) \\
&\geq b_\xi \prod_{x \in d} (v(x) \Rightarrow [\![x \in u_\xi]\!]) \cdot \prod_{x \in d} (u_\xi(x) \Rightarrow [\![x \in v]\!]) \\
&\geq b_\xi \prod_{x \in d} (b_\xi u_\xi(x) \Rightarrow b_\xi u_\xi(x)) \cdot \prod_{x \in d} (b_\xi u_\xi(x) \Rightarrow b_\xi u_\xi(x)) \\
&= b_\xi.
\end{aligned}$$

Hence $b_\xi \leq [\![v = u_\xi]\!] \cdot [\![\varphi(u_\xi)]\!] \leq [\![\varphi(v)]\!]$ for all $\xi < \alpha$. Therefore

$$b \leq [\![\varphi(v)]\!].$$

But also $[\![\varphi(v)]\!] \leq \sum_{x \in V^{(\mathbf{B})}} [\![\varphi(x)]\!] = b$, therefore

$$b = [\![\varphi(v)]\!].$$

Remark. The last part of this proof also establishes the following corollaries.

Corollary 16.3. Suppose $\{u_\xi \mid \xi < \alpha\} \subseteq V^{(\mathbf{B})}$, $\{b_\xi \mid \xi < \alpha\} \subseteq B$, $d \subseteq V^{(\mathbf{B})}$, $(\forall \xi < \alpha)[\mathscr{D}(u_\xi) = d]$ and $(\forall \xi, \eta < \alpha)[\xi \neq \eta \rightarrow b_\xi \cdot b_\eta = \mathbf{0}]$. Then

$$(\exists u \in V^{(\mathbf{B})})[\mathscr{D}(u) = d \wedge (\forall \xi < \alpha)[b_\xi \leq [\![u = u_\xi]\!]]].$$

Corollary 16.4. $\mathbf{V}^{(\mathbf{B})}$ is complete.

Example. For $b \in B$ define $u = b \cdot \check{\alpha} + (^{-}b) \cdot \check{\beta}$. If $\alpha \neq \beta$, then

$$[\![u = \check{\alpha}]\!] = b \wedge [\![u = \check{\beta}]\!] = {}^{-}b \wedge [\![\mathrm{Ord}\,(u)]\!] = \mathbf{1}.$$

In this case (assuming $b \neq \mathbf{0}, \mathbf{1}$), u is an example of a non-standard ordinal.

Remark. We have seen that $\{u \in V^{(\mathbf{B})} \mid [\![u \in v]\!] = \mathbf{1}\}$ for $v \in V^{(\mathbf{B})}$ is an equivalence class which is a proper class. Our aim is to find representatives of v of a simple form. In §6 we defined the notion of a complete **B**-valued structure. We now call $A \subseteq V^{(\mathbf{B})}$ complete iff the **B**-valued structure $\langle A, \bar{=}, \bar{\in}\rangle$ (i.e., $\mathbf{V}^{(\mathbf{B})}$ restricted to A) is complete. Thus $A \subseteq V^{(\mathbf{B})}$ is complete iff for every partition of unity, $\{b_\xi \mid \xi < \alpha\}$, and every family $\{u_\xi \mid \xi < \alpha\} \subseteq A$, there is an element $u \in A$, denoted by $\sum_{\xi < \alpha} b_\xi \cdot u_\xi$, that has the following properties:

$$(\forall \xi < \alpha)[b_\xi \leq [\![u = u_\xi]\!]].$$

Later we shall see that for every $v \in V^{(\mathbf{B})}$ there is a $u \in V^{(\mathbf{B})}$ such that $[\![u = v]\!] = \mathbf{1}$ and $\mathscr{D}(u)$ is complete.

Definition 16.5.

$$U_0^{(\mathbf{B})} \triangleq 0$$
$$U_{\alpha+1}^{(\mathbf{B})} \triangleq B^{U_\alpha^{(\mathbf{B})}}$$
$$U_\alpha^{(\mathbf{B})} \triangleq \bigcup_{\beta<\alpha} U_\beta^{(\mathbf{B})}, \qquad \alpha \in K_{\mathrm{II}}$$
$$U^{(\mathbf{B})} \triangleq \bigcup_{\alpha \in On} U_\alpha^{(\mathbf{B})}.$$

Remark. Thus each $u \in U^{(\mathbf{B})}$ is a function from $U_\alpha^{(\mathbf{B})}$ into B for some α, whereas $u \in V^{(\mathbf{B})}$ has a domain which is in general only a subset of some $V_\alpha^{(\mathbf{B})}$. Nevertheless, $U_\alpha^{(\mathbf{B})}$ and $V_\alpha^{(\mathbf{B})}$ are essentially the same:

Theorem 16.6.

1. $U_\alpha^{(\mathbf{B})} \subseteq V_\alpha^{(\mathbf{B})}$, in particular,

$$(\forall u \in U_\alpha^{(\mathbf{B})})(\exists v \in V_\alpha^{(\mathbf{B})})[[\![u = v]\!] = \mathbf{1}].$$

2. $(\forall v \in V_\alpha^{(\mathbf{B})})(\exists u \in U_\alpha^{(\mathbf{B})})[[\![u = v]\!] = \mathbf{1}]$.
3. $U_{\alpha+1}^{(\mathbf{B})}$ is complete.

Proof. 1 and 2 are proved by induction on α. 1 is obvious.

2. Let $v \in V_{\alpha+1}^{(\mathbf{B})}$. Then $\mathscr{D}(v) \subseteq V_\alpha^{(\mathbf{B})}$.

By the induction hypothesis, there exists a function $f\colon \mathscr{D}(v) \to U_\alpha^{(\mathbf{B})}$ such that

$$(\forall x \in \mathscr{D}(v))[[\![x = f(x)]\!] = \mathbf{1}].$$

Now define $u \in U_{\alpha+1}^{(\mathbf{B})}$ by

$$u(y) = [\![y \in v]\!] \quad \text{for} \quad y \in U_\alpha^{(\mathbf{B})} \qquad \text{(cp. the proof of Theorem 16.1).}$$

If $x \in \mathscr{D}(v)$, then

$$\begin{aligned} v(x) &\leq [\![x \in v]\!] \\ &= [\![f(x) = x]\!] \cdot [\![x \in v]\!] \\ &\leq [\![f(x) \in v]\!] \\ &= u(f(x)) \\ &\leq [\![f(x) \in u]\!] \qquad \text{since } f(x) \in \mathscr{D}(u) = U_\alpha^{(\mathbf{B})} \\ &= [\![f(x) = x]\!][\![f(x) \in u]\!] \\ &\leq [\![x \in u]\!]. \end{aligned}$$

Consequently

$$\begin{aligned} [\![u = v]\!] &= \prod_{x \in \mathscr{D}(v)} (v(x) \Rightarrow [\![x \in u]\!]) \cdot \prod_{y \in U_\alpha^{(\mathbf{B})}} (u(y) \Rightarrow [\![y \in v]\!]) \\ &= \prod_{x \in \mathscr{D}(v)} (v(x) \Rightarrow [\![x \in u]\!]) \\ &= \mathbf{1}, \quad \text{since for } x \in \mathscr{D}(v),\ v(x) \leq [\![x \in u]\!]. \end{aligned}$$

3. Corollary 16.3 with $d = U_\alpha^{(\mathbf{B})}$.

Definition 16.7. 1. Let $d \subseteq V^{(\mathbf{B})}$. A function $g: d \to V^{(\mathbf{B})}$ is called *extensional* iff $(\forall x, x' \in d)[[\![x = x']\!] \leq [\![g(x) = g(x')]\!]]$.

2. Let $u \in V^{(\mathbf{B})}$. Then u is *definite* iff

$$(\forall x \in \mathscr{D}(u))[u(x) = \mathbf{1}].$$

Example. $\check{k}$ is definite.

Exercise. Let $u \in V^{(\mathbf{B})}$ and $b \in B$. Define $b \cdot u$ by

$$\mathscr{D}(b \cdot u) = \mathscr{D}(u) \wedge (\forall x \in \mathscr{D}(u))[(b \cdot u)(x) = b \cdot u(x)].$$

Then,

$$[\![v \in b \cdot u]\!] = b \cdot [\![v \in u]\!]$$

and

$$[\![b \cdot u = b \cdot v]\!] = {}^{-}b + [\![u = v]\!] \quad \text{for} \quad v \in V^{(\mathbf{B})}.$$

Remark. The importance of extensional functions rests in the fact that functions (in the sense of $V^{(\mathbf{B})}$) from definite sets into definite sets corresponding to (real) extensional functions on their domains:

Theorem 16.8. Let $u, v \in V^{(\mathbf{B})}$ be definite and $\varphi: \mathscr{D}(u) \to \mathscr{D}(v)$ be an extensional function. Then

$$(\exists f \in V^{(\mathbf{B})})[[\![f: u \to v]\!] = \mathbf{1} \wedge (\forall x \in \mathscr{D}(u))[[\![f(x) = \varphi(x)]\!] = \mathbf{1}]].$$

Proof. Define

$$f = \{\langle x, \varphi(x)\rangle^{(\mathbf{B})} \mid x \in \mathscr{D}(u)\} \times \{\mathbf{1}\}.$$

Obviously, $f \in V^{(\mathbf{B})}$. We will show that $[\![f: u \to v]\!] = \mathbf{1}$.

1. $$[\![(\forall x \in u)(\exists y \in v)[\langle x, y\rangle \in f]]\!] = \prod_{x \in \mathscr{D}(u)} \sum_{y \in \mathscr{D}(v)} [\![\langle x, y\rangle \in f]\!] \quad \text{since } u, v \text{ are definite}$$

$$\geq \prod_{x \in \mathscr{D}(u)} [\![\langle x, \varphi(x)\rangle \in f]\!] \quad \text{since } \varphi: \mathscr{D}(u) \to \mathscr{D}(v).$$

$$= \mathbf{1} \quad \text{by definition of } f.$$

Furthermore, we have to show that

2. $[\![(\forall x \in u)(\forall y)(\forall z)[\langle x, y\rangle \in f \wedge \langle x, z\rangle \in f \to y = z]]\!] = \mathbf{1}$,

i.e.,

$$\text{i) } (\forall x \in \mathscr{D}(u))(\forall y, z \in V^{(\mathbf{B})})[[\![\langle x, y\rangle \in f \wedge \langle x, z\rangle \in f]\!] \leq [\![y = z]\!]].$$

Therefore let $x \in \mathscr{D}(u)$ and $y, z \in V^{(\mathbf{B})}$. Then

$$\begin{aligned}
[\![\langle x, y\rangle \in f]\!] &= \sum_{x' \in \mathscr{D}(u)} [\![\langle x, y\rangle = \langle x', \varphi(x')\rangle]\!] \\
&= \sum_{x' \in \mathscr{D}(u)} [\![x = x']\!] \cdot [\![y = \varphi(x')]\!] \\
&\leq \sum_{x' \in \mathscr{D}(u)} [\![\varphi(x) = \varphi(x')]\!] \cdot [\![y = \varphi(x')]\!] \\
&\qquad\qquad \text{since } \varphi \text{ is extensional} \\
&\leq [\![y = \varphi(x)]\!].
\end{aligned}$$

This shows that

$$\begin{aligned}
[\![\langle x, y\rangle \in f]\!] \cdot [\![\langle x, z\rangle \in f]\!] &\leq [\![y = \varphi(x)]\!] \cdot [\![z = \varphi(x)]\!] \\
&\leq [\![y = z]\!]
\end{aligned}$$

which proves i) and hence 2. From 1 and 2 we conclude that

$$[\![f: u \to v]\!] = \mathbf{1}.$$

It remains to show that

3. $(\forall x \in \mathscr{D}(u))[[\![f(x) = \varphi(x)]\!] = \mathbf{1}]$.

Let $x \in \mathscr{D}(u)$. Since $f \in V^{(\mathbf{B})}$ and $g(x) \in V^{(\mathbf{B})}$ we interpret $f(x) = g(x)$ to mean $(\exists y)[\langle x, y\rangle \in f \wedge y = g(x)]$. Therefore

$$\begin{aligned}
[\![f(x) = \varphi(x)]\!] &= [\![(\exists y)[\langle x, y\rangle \in f \wedge y = \varphi(x)]]\!] \\
&= [\![\langle x, \varphi(x)\rangle \in f]\!] \qquad \text{since } [\![f: u \to v]\!] = \mathbf{1} \\
&= \mathbf{1}.
\end{aligned}$$

Remark. Later we shall see that Theorem 16.8 has a converse if we add an additional requirement on u and v. See Theorem 16.28.

Definition 16.9. $u \in V^{(\mathbf{B})}$ is *extensional* iff

$$(\forall x, y \in \mathscr{D}(u))[[\![x = y]\!] \cdot u(x) \leq u(y)].$$

Remark. Therefore u is extensional iff the extended structure

$$\langle V^{(\mathbf{B})}, \bar{=}, \bar{\in}, \bar{u}\rangle$$

where

$$\begin{aligned}
\bar{u}(x) &= u(x) && \text{if } x \in \mathscr{D}(u) \\
&= \mathbf{1} && \text{otherwise}
\end{aligned}$$

is still a **B**-valued structure (cf. the requirements of Definition 6.5). Another interpretation can be obtained from the following result.

Theorem 16.10. If $u \in V^{(\mathbf{B})}$.

$$u \text{ is extensional} \leftrightarrow (\forall x \in \mathscr{D}(u))[u(x) = [\![x \in u]\!]].$$

Proof. If u is extensional and $x \in \mathscr{D}(u)$, then

$$u(x) \leq [\![x \in u]\!] = \sum_{y \in \mathscr{D}(u)} u(y) \cdot [\![x = y]\!] \leq u(x),$$

Therefore

$$u(x) = [\![x \in u]\!].$$

To prove the converse, assume $(\forall x \in \mathscr{D}(u))[u(x) = [\![x \in u]\!]]$. Then, for $x, y \in \mathscr{D}(u)$

$$[\![x = y]\!] \cdot u(x) = [\![x = y]\!] \cdot [\![x \in u]\!] \leq [\![y \in u]\!] = u(y).$$

Theorem 16.11.

$$(\forall v \in V^{(\mathbf{B})})(\forall d \subseteq V^{(\mathbf{B})})[\mathscr{D}(v) \subseteq d \rightarrow$$
$$(\exists u)\ [u \text{ is extensional } \wedge\ d = \mathscr{D}(u)\ \wedge\ [\![u = v]\!] = \mathbf{1}]].$$

In particular, each $v \in V^{(\mathbf{B})}$ can be represented by an extensional set.

Proof. For $v \in V^{(\mathbf{B})}$ and $d \subseteq V^{(\mathbf{B})}$ such that $\mathscr{D}(v) \subseteq d$ define $u: d \rightarrow B$ by

$$(\forall x \in d)[u(x) = [\![x \in v]\!]].$$

Then, for $x \in d$

$$u(x) \leq [\![x \in u]\!] = \sum_{y \in d} [\![y \in v]\!] \cdot [\![x = y]\!] \leq [\![x \in v]\!] = u(x),$$

i.e., u is extensional. That $[\![u = v]\!] = \mathbf{1}$ has already been proved in Theorem 16.1.

Remark. We could restrict ourselves to extensional sets u for which $u(x)$ is simply equal to $[\![x \in u]\!]$ for $x \in \mathscr{D}(u)$. However, since one still has to evaluate $[\![x \in u]\!]$ for $x \notin \mathscr{D}(u)$, this is in general no saving, though for special cases it may be very convenient to have a particular representation of **B**-valued sets.

Given any $v \in V^{(\mathbf{B})}$, we cannot expect to have $[\![u = v]\!] = \mathbf{1}$ for some definite u. We shall prove, however, that $[\![u = b \cdot v]\!] = \mathbf{1}$ for some definite u and some $b \in B$.

Definition 16.12. 1. $u \in V^{(\mathbf{B})}$ is *uniform* iff u is extensional and $\mathscr{D}(u)$ complete.

2. $S \subseteq V^{(\mathbf{B})}$ is *complete* iff the structure $\langle S, \bar{\in}, \bar{=} \rangle$ is complete in the sense of Definition 6.8.

Remark. As a consequence of Theorem 16.11 we have the following.

Theorem 16.13. $(\forall v \in V^{(\mathbf{B})})(\exists u \in V^{(\mathbf{B})})[u \text{ is uniform } \wedge\ [\![u = v]\!] = \mathbf{1}]$.

Proof. If $v \in V^{(\mathbf{B})}$, $v \in V^{(\mathbf{B})}_{\alpha+2}$ for some α. Then, by Theorem 16.6.2, $[\![v = v_1]\!] = \mathbf{1}$ and $v_1 \in U^{(\mathbf{B})}_{\alpha+2}$ for some v_1. Since $\mathscr{D}(v_1) = U^{(\mathbf{B})}_{\alpha+2}$, is complete, by Theorem 16.11,

$$[\![v_1 = u]\!] = \mathbf{1}$$

for some $u \in V^{(\mathbf{B})}$ such that $\mathscr{D}(u) = U^{(\mathbf{B})}_{\alpha+1}$ and u is extensional. Hence

$$[\![u = v]\!] = \mathbf{1}$$

and u is uniform.

Theorem 16.14. Let u be uniform. If $\{x_i \mid i \in I\} \subseteq \mathscr{D}(u)$ and

$$\{b_i \mid i \in I\} \subseteq B$$

is a partition of unity, then

$$u\left(\sum_{i \in I} b_i x_i\right) = \sum_{i \in I} b_i u(x_i).$$

(See Remark following Theorem 6.9.)

Proof. Let $y = \sum_{i \in I} b_i x_i$. Since $\mathscr{D}(u)$ is complete, y exists, and $y \in \mathscr{D}(u)$. Since $\{x_i \mid i \in I\} \subseteq \mathscr{D}(u)$,

$$u(y) = [\![y \in u]\!] = \sum_{x \in \mathscr{D}(u)} u(x) \cdot [\![x = y]\!] \geq \sum_{i \in I} u(x_i)[\![x_i = y]\!] \geq \sum_{i \in I} u(x_i) \cdot b_i.$$

(Note that since $\{x_i \mid i \in I\} \subseteq \mathscr{D}(u)$, $[\![x_i = y]\!] \geq b_i$.)

On the other hand since $\sum_{i \in I} b_i = \mathbf{1}$,

$$b_i \cdot u(y) \leq [\![x_i = y]\!] \cdot [\![y \in u]\!] \leq [\![x_i \in u]\!] = u(x_i)$$

$$b_i u(y) \leq b_i \cdot u(x_i)$$

$$u(y) \leq \sum_{i \in I} b_i u(x_i).$$

Definition 16.15. $(\forall u \in V^{(\mathbf{B})})[\sup (u) \triangleq \sum_{x \in \mathscr{D}(u)} u(x)]$.

Theorem 16.16. $(\forall u \in V^{(\mathbf{B})})[\sup (u) = [\![(\exists x)[x \in u]]\!]]$.

Proof.

$$[\![(\exists x)[x \in u]]\!] = [\![(\exists x \in u)[x = x]]\!] = \sum_{x \in \mathscr{D}(u)} u(x) \cdot [\![x = x]\!]$$

$$= \sum_{x \in \mathscr{D}(u)} u(x) = \sup (u).$$

Corollary 16.17. $(\forall u_1, u_2 \in V^{(\mathbf{B})})[[\![u_1 = u_2]\!] = \mathbf{1} \rightarrow \sup (u_1) = \sup (u_2)]$.

Theorem 16.18. Let $u \in V^{(\mathbf{B})}$ be uniform and $b \in B$. Then

$$\{x \in \mathscr{D}(u) \mid b \leq u(x)\}$$

is complete.

Proof. Let $\{b_i \mid i \in I\}$ be a partition of unity, and $\{x_i \mid i \in I\} \subseteq \mathscr{D}(u)$ satisfying $(\forall i \in I)[b \leq u(x_i)]$. Since $\mathscr{D}(u)$ is complete, $y = \sum_{i \in I} b_i x_i$ for some $y \in \mathscr{D}(u)$.

$$u(y) = \sum_{i \in I} b_i u(x_i) \qquad \text{by Theorem 16.14}$$

$$\geq \sum_{i \in I} b_i \cdot b = b$$

i.e.,

$$y \in \{x \mid x \in \mathscr{D}(u) \wedge b \leq u(x)\}.$$

Theorem 16.19. If $u \in V^{(\mathbf{B})}$ is uniform, then $(\exists x \in \mathscr{D}(u))[u(x) = \sup(u)]$.

Proof. If we do not require $x \in \mathscr{D}(u)$, the theorem follows from the maximum principle. In fact, we can use the same argument:

Let $\langle x_\xi \mid \xi < \alpha\rangle$ be an enumeration of $\mathscr{D}(u)$, i.e.,

$$\mathscr{D}(u) = \{x_\xi \mid \xi < \alpha\}$$

and put $b_\xi = u(x_\xi)\cdot \prod_{\eta<\xi}(^-u(x_\eta))$ for $\xi < \alpha$. Then the b_ξ's are disjoint and

$$\sum_{\xi<\alpha} b_\xi = \sum_{x\in\mathscr{D}(u)} u(x) = \sup(u).$$

Therefore, adding $b_\alpha = {}^-\sup(u)$, $\langle b_\xi \mid \xi \le \alpha\rangle$ is a partition of unity. Since $\mathscr{D}(u)$ is complete, $(\forall \xi < \alpha)[b_\xi \le [\![x = x_\xi]\!]]$ for some $x \in \mathscr{D}(u)$. Hence

$$\begin{aligned} u(x) = [\![x \in u]\!] &= \sum_{\xi<\alpha} u(x_\xi)\cdot[\![x = x_\xi]\!] \\ &\ge \sum_{\xi<\alpha} b_\xi = \sup(u) \ge u(x) \qquad \text{by definition of } b_\xi \text{ and } \sup(u). \end{aligned}$$

Therefore

$$u(x) = \sup(u).$$

Theorem 16.20. Let $u \in V^{(\mathbf{B})}$ be uniform. Define $v \in V^{(\mathbf{B})}$ by $\mathscr{D}(v) = \{y \mid y \in \mathscr{D}(u) \wedge u(y) = \sup(u)\}$ and $(\forall y \in \mathscr{D}(v))[v(y) = 1]$. Then v is definite, uniform, and $[\![u = b\cdot v]\!] = 1$, where $b = \sup(u)$.

Proof. Clearly v is definite. If x and y are in $\mathscr{D}(v)$, then $[\![x = y]\!]v(x) \le v(y)$. So v is extensional. Therefore to show that v is uniform, it suffices to prove that $\mathscr{D}(v)$ is complete. For this purpose, let $\{b_\xi \mid \xi < \alpha\}$ be a partition of **1** and let $\{y_\xi \mid \xi < \alpha\} \subset \mathscr{D}(v)$. Then we have to show that there exists an $a \in \mathscr{D}(v)$ such that

1. $(\forall \xi < \alpha)[b_\xi \le [\![a = y_\xi]\!]]$.

Since $\mathscr{D}(v) \subset \mathscr{D}(u)$, $\{y_\xi \mid \xi < \alpha\} \subset \mathscr{D}(u)$. By the uniformity of u there exists an $a \in \mathscr{D}(u)$ such that 1 holds. Therefore it is enough to show that $a \in \mathscr{D}(v)$. Since

$$b_\xi \le [\![a = y_\xi]\!] \to b_\xi\cdot u(y_\xi) \le [\![a = y_\xi]\!]\cdot u(y_\xi) \le u(a)$$

it follows that

$$b = \sum_{\xi<\alpha} b_\xi\cdot b \le u(a).$$

But, since $b = \sup(u)$, we have $u(a) = b$ and hence $a \in \mathscr{D}(v)$.

Next we shall show that $[\![u = b\cdot v]\!] = \mathbf{1}$. First, if $x \in \mathscr{D}(v)$ then by Theorem 16.10, $[\![x \in u]\!] = u(x) = \mathbf{1}$; hence

$$\begin{aligned} [\![u = b\cdot v]\!] &= \prod_{x\in\mathscr{D}(u)} [u(x) \Rightarrow [\![x \in b\cdot v]\!]]\cdot \prod_{x\in\mathscr{D}(v)} [(b\cdot v)(x) \Rightarrow [\![x \in u]\!]] \\ \text{2.} \qquad &= \prod_{x\in\mathscr{D}(u)} [u(x) \Rightarrow b\cdot \sum_{t\in\mathscr{D}(v)} [\![x = t]\!]] \end{aligned}$$

since the second factor is $\mathbf{1}$. Let $x, x_0 \in \mathscr{D}(u)$ with $u(x_0) = b$. Then since $\{u(x), \, {}^-u(x)\}$ is a partition of unity and u is uniform, there exists a $z \in \mathscr{D}(u)$ such that $z = u(x)x + ({}^-u(x))x_0$. (See Remark following Theorem 6.9). By Theorem 16.14

$$\begin{aligned} u(z) &= u(x)\cdot u(x) + ({}^-u(x))\cdot u(x_0) \\ &= u(x) + ({}^-u(x))\cdot b \geq b. \end{aligned}$$

Therefore $u(z) = b$, since $b = \sup(u)$; and hence $z \in \mathscr{D}(v)$. Then

$$\begin{aligned} u(x) \Rightarrow b\cdot \sum_{t \in \mathscr{D}(v)} [\![x = t]\!] &\geq u(x) \Rightarrow b\cdot [\![x = z]\!] \\ &\geq b\cdot u(x) \Rightarrow b\cdot u(x) \quad \text{since} \quad u(x) \leq [\![z = x]\!].] \\ 3. \qquad\qquad &= \mathbf{1}. \end{aligned}$$

Since x is an arbitrary element of $\mathscr{D}(u)$, we have $[\![u = b\cdot v]\!] = \mathbf{1}$ by 2 and 3.

Corollary 16.21. $(\forall u \in V^{(\mathbf{B})})(\exists b \in B)(\exists v \in V^{(\mathbf{B})})[v$ is definite and uniform $\wedge \, [\![u = b\cdot v]\!] = \mathbf{1}]$.

Exercise.

1. Let $u \in V^{(\mathbf{B})}$ be extensional and $b \in B$. Then

$$[\![u = b\cdot u]\!] = [\sup(u) \Rightarrow b].$$

As a consequence,

$$\begin{gathered} [b \geq \sup(u)] \rightarrow [[\![u = b\cdot u]\!] = \mathbf{1}] \\ [b < \sup(u)] \rightarrow [[\![u = b\cdot u]\!] < \mathbf{1}]. \\ [\![u = \mathbf{0}]\!] = {}^-\sup(u) \end{gathered}$$

$$\begin{aligned} [\![b\cdot \check{1} \in \alpha]\!] &= [\![b\cdot \check{1} = \check{1}]\!] + [\![b\cdot \check{1} = \check{0}]\!] \qquad \text{if } \alpha > 1 \\ &= b + ({}^-b) \\ &= \mathbf{1}. \end{aligned}$$

2. Define $u \in V^{(\mathbf{B})}$ by

$$\mathscr{D}(u) = \{\check{1}, \check{2}\} \wedge u(\check{1}) = b \wedge u(\check{2}) = {}^-b.$$

Then $\sup(u) = [\![(\exists x)[x \in u]]\!] = \mathbf{1}$. Furthermore, defining $a = b\cdot \check{1} + ({}^-b)\cdot \check{2}$ (See Remark following Theorem 6.9) and $v = \{a\}^{\mathbf{B}}$ we have, $[\![u = v]\!] = \mathbf{1}$.

3. Define $u, a_1, a_2, v \in V^{(\mathbf{B})}$ by

$$\begin{aligned} \mathscr{D}(u) &= \{\check{1}, \check{2}, \check{3}\} \wedge u(\check{1}) = b \wedge u(\check{2}) = u(\check{3}) = {}^-b, \\ a_1 &= b\cdot \check{1} + ({}^-b)\cdot \check{2} \\ a_2 &= b\cdot \check{1} + ({}^-b)\cdot \check{3} \\ v &= \{a_1, a_2\}^{(\mathbf{B})}. \end{aligned}$$

Then $[\![u = v]\!] = \mathbf{1}$.

Remark. The results which we have obtained so far can be used to determine the power set (in $\mathbf{V}^{(\mathbf{B})}$) of various sets in $V^{(\mathbf{B})}$.

Theorem 16.22. Let $v \in V^{(\mathbf{B})}$ be extensional, $u \in V^{(\mathbf{B})}$, and $b = [\![v \subseteq u]\!]$. If $v' = b \cdot v$, then

1. v' is extensional,
2. $[\![v' \subseteq u]\!] = \mathbf{1}$,
3. $[\![v \subseteq u]\!] = [\![v' = v]\!]$.

Proof.

1. Is obvious.

$$2.\ [\![v' \subseteq u]\!] = \prod_{x \in \mathscr{D}(v)} (v'(x) \Rightarrow [\![x \in u]\!])$$

$$= \prod_{x \in \mathscr{D}(v)} ([\![v \subseteq u]\!] \cdot [\![x \in v]\!] \Rightarrow [\![x \in u]\!]) \qquad \text{since } v \text{ is extensional}$$

$$= \prod_{x \in \mathscr{D}(v)} [\![v \subseteq u \wedge x \in v \rightarrow x \in u]\!]$$

$$= \mathbf{1}.$$

3. $[\![v = v']\!] = [\![v = v']\!] \cdot [\![v' \subseteq u]\!] \leq [\![v \subseteq u]\!] = b$ by 2
$[\![v = v']\!] = [\![v = b \cdot v]\!] \geq b$ by Exercise 1 above.

Hence $[\![v = v']\!] = b = [\![v \subseteq u]\!]$.

Theorem 16.23. Let $u, v \in V^{(\mathbf{B})}$ be extensional and $[\![v \subseteq u]\!] = \mathbf{1}$. Define v' by the conditions

1. $\mathscr{D}(v') = \mathscr{D}(u)$.
2. $(\forall x \in \mathscr{D}(u))[v'(x) = [\![x \in v]\!]]$.

Then $[\![v = v']\!] = \mathbf{1}$.

Proof. v' is extensional and $[\![v' \subseteq v]\!] = \mathbf{1}$ by the definition of v'. It remains to show that $(\forall x \in \mathscr{D}(v))[[\![x \in v]\!] \leq [\![x \in v']\!]]$. Let $x \in \mathscr{D}(u)$. Since $[\![v \subseteq u]\!] = \mathbf{1}$,

$$[\![x \in v]\!] \leq [\![x \in u]\!]$$

$$[\![x \in v]\!] = [\![x \in v]\!] \cdot [\![x \in u]\!]$$

$$= \sum_{x' \in \mathscr{D}(u)} [\![x \in v]\!] \cdot u(x') \cdot [\![x = x']\!]$$

$$\leq \sum_{x' \in \mathscr{D}(u)} [\![x' \in v]\!] \cdot u(x')[\![x = x']\!]$$

$$= \sum_{x' \in \mathscr{D}(u)} v'(x)u(x') \cdot [\![x = x']\!]$$

$$\leq [\![x \in v']\!].$$

Remark. From Theorems 16.22, 16.23, and 16.11 we obtain the following:

Theorem 16.24.

$$(\forall u, v \in V^{(\mathbf{B})})(\exists v' \in V^{(\mathbf{B})})[\mathscr{D}(v') = \mathscr{D}(u) \wedge [\![v' \subseteq u]\!] = \mathbf{1} \wedge [\![v \subseteq u]\!] = [\![v = v']\!]].$$

Remark. This theorem is important for the treatment of power sets in $\mathbf{V}^{(\mathbf{B})}$. Namely, if $u \in V^{(\mathbf{B})}$ and we regard elements $v \in V^{(\mathbf{B})}$ which are subsets of u in the sense of $\mathbf{V}^{(\mathbf{B})}$, i.e., $[\![v \subseteq u]\!] = \mathbf{1}$, then $\mathscr{D}(v)$ may be greater than $\mathscr{D}(u)$, and, in fact, there is no set $d \subseteq V^{(\mathbf{B})}$ which includes $\mathscr{D}(v)$ for all these v. However, by Theorem 16.24, we can find some v' such that $[\![v' \subseteq u]\!] = \mathbf{1}$, $[\![v = v']\!] = \mathbf{1}$ and $\mathscr{D}(v') = \mathscr{D}(u)$. As a corollary, we have the following:

Theorem 16.25. If $u, v \in V^{(\mathbf{B})}$, and if

$$A = \{v' \in V^{(\mathbf{B})} \mid \mathscr{D}(v') = \mathscr{D}(u) \wedge [\![v' \subseteq u]\!] = \mathbf{1}\}$$

then

$$[\![v \subseteq u]\!] = \sum_{v' \in A} [\![v = v']\!],$$

Proof. If $b = \sum_{v' \in A} [\![v = v']\!]$ then from Theorem 16.24

$$[\![v \subseteq u]\!] \leq b.$$

On the other hand,

$$b \leq \sum_{v' \in A} [\![v' \subseteq u]\!] \cdot [\![v' = v]\!] \leq [\![v \subseteq u]\!].$$

Remark. As an application, we determine $R(\check{\alpha})$ in $V^{(\mathbf{B})}$.

Theorem 16.26. For $u \in V^{(\mathbf{B})}$

1. $[\![u \in R(\check{\alpha})]\!] = \sum_{v \in V_\alpha^{(\mathbf{B})}} [\![u = v]\!] = [\![u \in V_\alpha^{(\mathbf{B})} \times \{\mathbf{1}\}]\!]$,

therefore $[\![R(\check{\alpha}) = V_\alpha^{(\mathbf{B})} \times \{\mathbf{1}\}]\!] = \mathbf{1}$.

2. $\mathscr{D}(u) \subseteq V_\alpha^{(\mathbf{B})} \rightarrow [\![u \subseteq R(\check{\alpha})]\!] = \mathbf{1}$.

3. $[\![u \subseteq R(\check{\alpha})]\!] = \sum_{t \in A_\alpha} [\![u = t]\!]$

where $A_\alpha = \{t \in V^{(\mathbf{B})} \mid \mathscr{D}(t) = V_\alpha^{(\mathbf{B})}\}$.

Proof. (By induction on α.)

1. $[\![u \in R(\check{\alpha})]\!] = [\![(\exists \xi < \check{\alpha})[u \subseteq R(\xi[]\!]$

$$= \sum_{\xi < \alpha} [\![u \subseteq R(\check{\xi})]\!]$$

$$= \sum_{\xi < \alpha} \sum_{t \in A_\xi} [\![u = t]\!] \quad \text{by the induction hypothesis for 3}$$

$$= \sum_{\xi < \alpha} \sum_{t \in V_{\xi+1}^{(\mathbf{B})}} [\![u = t]\!] \quad (\text{for, if } \mathscr{D}(t) \subseteq V_\xi^{(\mathbf{B})},$$

$$(\exists t' \in V_{\xi+1}^{(\mathbf{B})})[\mathscr{D}(t') = V_\xi^{(\mathbf{B})} \wedge [\![t = t']\!] = \mathbf{1}])$$

$$= \sum_{v \in V_\alpha^{(\mathbf{B})}} [\![u = v]\!]$$

$$= \sum_{v \in V_\alpha^{(\mathbf{B})}} [\![u = v]\!](V_\alpha^{(\mathbf{B})} \times \{\mathbf{1}\})(v)$$

$$= [\![u \in V_\alpha^{(\mathbf{B})} \times \{\mathbf{1}\}]\!].$$

2. Assume $\mathscr{D}(u) \subseteq V_\alpha^{(\mathbf{B})}$. Then, since by 1 $[\![x \in R(\check{\alpha})]\!] = \mathbf{1}$ for $x \in V_\alpha^{(\mathbf{B})}$,

$$[\![u \subseteq R(\check{\alpha})]\!] = \prod_{x \in \mathscr{D}(u)} (u(x) \Rightarrow [\![x \in R(\check{\alpha})]\!]) = \mathbf{1}.$$

3. $[\![u \subseteq R(\check{\alpha})]\!] = \sum_{\substack{t \in A_\alpha \\ [\![t \subseteq R(\check{\alpha})]\!] = \mathbf{1}}} [\![u = t]\!]$ by 1 and Theorem 16.25

$$= \sum_{t \in A_\alpha} [\![u = t]\!] \quad \text{by 2}$$

Theorem 16.27. If u is definite, $[\![\mathscr{P}(u) = B^{\mathscr{D}(u)} \times \{\mathbf{1}\}]\!] = \mathbf{1}$.

Proof. Note that $B^{\mathscr{D}(u)} \subseteq V^{(\mathbf{B})}$ and $B^{\mathscr{D}(u)}$ is a set. Let $x \in V^{(\mathbf{B})}$ and $A_u = \{v \in V^{(\mathbf{B})} \mid \mathscr{D}(v) = \mathscr{D}(u)\}$. Then

$$\begin{aligned} [\![x \in B^{\mathscr{D}(u)} \times \{\mathbf{1}\}]\!] &= \sum_{v \in A_u} [\![x = v]\!] \\ &= \sum_{\substack{v \in A_u \\ [\![v \subseteq u]\!] = \mathbf{1}}} [\![x = v]\!] \quad \text{since } u \text{ is definite} \\ &= [\![x \subseteq u]\!] \quad \text{by Theorem 16.25.} \end{aligned}$$

Remark. Next we prove the converse of Theorem 16.8.

Theorem 16.28. Let $u, v \in V^{(\mathbf{B})}$ be definite and uniform. If $f \in V^{(\mathbf{B})}$ and $[\![f: u \to v]\!] = \mathbf{1}$, then there exists a (real) function $\varphi: \mathscr{D}(u) \to \mathscr{D}(v)$ such that φ is extensional and

$$(\forall x \in \mathscr{D}(u))[[\![f(x) = \varphi(x)]\!] = \mathbf{1}].$$

Proof. Since $[\![(\forall x \in u)(\exists y \in v)[f(x) = y]]\!] = \mathbf{1}$, and u and v are definite

$$(\forall x \in \mathscr{D}(u))\left[\sum_{y \in \mathscr{D}(v)} [\![f(x) = y]\!] = \mathbf{1}\right].$$

For $x \in \mathscr{D}(u)$ define $v' \in V^{(\mathbf{B})}$ by

$$\mathscr{D}(v') = \mathscr{D}(v) \wedge (\forall y \in \mathscr{D}(v))[v'(y) = [\![f(x) = y]\!]].$$

Since v is uniform, so is v'. Therefore, by Theorem 16.19, for each $x \in \mathscr{D}(v)$ we can find some $y_0 \in \mathscr{D}(v)$ such that

$$[\![f(x) = y_0]\!] = \sum_{y \in \mathscr{D}(v)} [\![f(x) = y]\!] = \mathbf{1}.$$

Using the AC, there is a function $\varphi: \mathscr{D}(u) \to \mathscr{D}(v)$ such that

$$(\forall x \in \mathscr{D}(u))[[\![f(x) = \varphi(x)]\!] = \mathbf{1}].$$

It remains to show that φ is extensional.

Let $x_1, x_2 \in \mathscr{D}(u)$. Then since $[\![f: u \to v]\!] = \mathbf{1}$

$$\begin{aligned} [\![x_1 = x_2]\!] &\leq [\![f(x_1) = f(x_2)]\!] \\ &= [\![f(x_1) = f(x_2)]\!] \cdot [\![f(x_1) = \varphi(x_1)]\!] \cdot [\![f(x_2) = \varphi(x_2)]\!] \\ &\leq [\![\varphi(x_1) = \varphi(x_2)]\!]. \end{aligned}$$

17. Cardinals in $\mathbf{V}^{(\mathbf{B})}$

The theorems of this section can be obtained from the corresponding results in the theory of forcing by translating them in the manner outlined in Corollary 14.23. However, since this translation requires the existence of elementary subsystems of V and thus cannot be carried out in ZF, we shall try to give direct proofs in $\mathbf{V}^{(\mathbf{B})}$. Corresponding to the fact that every cardinal in $M[G]$, where $\langle M, \mathbf{P}\rangle$ is a setting for forcing and G is $\mathbf{P}$-generic over M, is a cardinal in M we have the following.

Theorem 17.1. If α is not a cardinal, then $[\![\neg \text{Card}(\check{\alpha})]\!] = \mathbf{1}$.

Proof. $\neg \text{Card}(\alpha) \leftrightarrow (\exists f)(\exists \beta < \alpha)[f: \beta \to \alpha \wedge \mathscr{W}(f)^* = \alpha]$.

Therefore $\neg \text{Card}(\alpha) \leftrightarrow (\exists f)\phi(f, \alpha)$ where $\phi(f, \alpha)$ is a bounded formula. Thus, by Corollary 13.18, if α is not a cardinal,

$$[\![\phi(\check{f}, \check{\alpha})]\!] = \mathbf{1}$$

for some $f \in V$, hence

$$[\![\neg \text{Card}(\check{\alpha})]\!] = \mathbf{1}.$$

Remark. As might be expected, for finite cardinals and for ω we can prove the converse of Theorem 17.1.

Theorem 17.2. For every $\alpha \leq \omega$, $[\![\text{Card}(\check{\alpha})]\!] = \mathbf{1}$.

Proof. We have to show that

$$[\![\neg (\exists f)(\exists \beta < \check{\alpha})[f: \beta \to \check{\alpha} \wedge \mathscr{W}(f) = \check{\alpha}]]\!] = \mathbf{1}$$

i.e.,

$$(\forall f \in V^{(\mathbf{B})})(\forall \beta < \alpha)[[\![f: \check{\beta} \to \check{\alpha} \wedge \mathscr{W}(f) = \check{\alpha}]\!] = \mathbf{0}].$$

Suppose that on the contrary,

$$b = [\![f: \check{\beta} \to \check{\alpha} \wedge \mathscr{W}(f) = \check{\alpha}]\!] > \mathbf{0} \quad \text{for some} \quad f \in V^{(\mathbf{B})}, \beta < \alpha.$$

Then $b \leq [\![(\forall \eta < \check{\alpha})(\exists \xi < \check{\beta})[f(\xi) = \check{\eta}]]\!]$,

i) $b \leq \prod_{\eta < \alpha} \sum_{\xi < \beta} [\![f(\check{\xi}) = \check{\eta}]\!]$.

[Note: Let $\psi(\xi, \eta)$ be the formula that expresses $f(\xi) = \eta$. Then $[\![f(\check{\xi}) = \check{\eta}]\!]$ means $[\![\psi(\check{\xi}, \check{\eta})]\!]$.]

* $\mathscr{W}(f) = \{y \mid (\exists x)[\langle x,y\rangle \in f]\}$.

Now let us assume that $\alpha \leq \omega$. Since $\beta < \alpha$,

$$b \leq \prod_{\eta < \beta+1} \sum_{\xi < \beta} [\![f(\check{\xi}) = \check{\eta}]\!]$$
$$= \sum_{\varphi \in \beta^{(\beta+1)}} \prod_{\eta < \beta+1} [\![f(\varphi(\eta)^{\vee}) = \check{\eta}]\!]$$

by the $(\beta + 1, \beta)$-DL, (see Definition 4.1) which holds for every $\mathbf{B}$ since β is finite.

Therefore

$$\mathbf{0} < b \prod_{\eta < \beta+1} [\![f(\varphi(\eta)^{\vee}) = \check{\eta}]\!] \quad \text{for some} \quad \varphi : \beta + 1 \to \beta.$$

There must exist $n, m < \beta + 1$ such that $n \neq m \wedge \varphi(n) = \varphi(m)$. Then

$$\begin{aligned} \mathbf{0} &< b[\![f((\varphi(n))^{\vee}) = \check{n}]\!] \cdot [\![f((\varphi(m))^{\vee}) = \check{m}]\!] \\ &\leq b[\![\check{n} = \check{m}]\!] \qquad \text{since } b \leq [\![f((\varphi(n))^{\vee}) = f((\varphi(m))^{\vee})]\!] \\ &= \mathbf{0} \qquad \text{since } n \neq m. \end{aligned}$$

This is a contradiction.

Remark. It is easy to see that the same proof can be used to show: If $\mathbf{B}$ satisfies the (α, α)-DL, where α is a cardinal, then for each cardinal $\gamma \leq \alpha$, $[\![\text{Card}(\check{\gamma})]\!] = \mathbf{1}$, i.e., cardinals $\leq \alpha$ remain cardinals in $V^{(\mathbf{B})}$. (It can also be shown that we only need the $(\alpha, 2)$-DL since $(\alpha, 2)$-DL $\leftrightarrow$ (α, α)-DL.)

In general, Theorem 17.2 does not hold for all cardinals. However, corresponding to Theorem 11.8 we have a converse of Theorem 17.1. For a more general result we introduce the following definition.

Definition 17.3. Let γ be a cardinal. A Boolean algebra $\mathbf{B}$ satisfies the γ-chain condition iff

$$(\forall S \subseteq B)[(\forall x, y \in S)[x \neq y \to x \cdot y = \mathbf{0}] \to \bar{\bar{S}} \leq \gamma].$$

In particular, $\mathbf{B}$ satisfies the ω-chain condition iff $\mathbf{B}$ satisfies the c.c.c.

Theorem 17.4. Let γ be an infinite cardinal and suppose that $\mathbf{B}$ satisfies the γ-chain condition. If $\alpha > \gamma$ is a cardinal, then $[\![\text{Card}(\check{\alpha})]\!] = \mathbf{1}$.

Proof. As in the proof of Theorem 17.2, suppose that $[\![\text{Card}(\check{\alpha})]\!] \neq \mathbf{1}$ for some cardinal $\alpha > \gamma$, then defining b as before, we have for some $\beta < \alpha$, and $f \in V^{(\mathbf{B})}$,

i) $b \leq \prod_{\eta < \alpha} \sum_{\xi < \beta} [\![f(\check{\xi}) = \check{\eta}]\!]$ where $b > \mathbf{0}$.

Therefore, using the AC in V,

$$(\forall \eta < \alpha)(\exists \xi_\eta < \beta)[b \cdot [\![f(\check{\xi}_\eta) = \check{\eta}]\!] \neq \mathbf{0}].$$

For $\xi < \beta$ define

$$A_\xi = \{\eta < \alpha \mid \xi_\eta = \xi\}.$$

Then for some $\xi_* < \beta$,

ii) $\overline{\overline{A}}_{\xi_*} > \gamma$,

since otherwise $(\forall \xi_* < \beta)[\overline{\overline{A}}_{\xi_*} \leq \gamma]$. But $\alpha = \bigcup_{\xi_* < \beta} A_{\xi_*}$, so this would imply $\overline{\overline{\alpha}} \leq \overline{\overline{\beta}} \cdot \gamma < \alpha$ since $\beta, \gamma < \alpha$. This is a contradiction.

Consider

$$S = \{b \cdot [\![f(\check{\xi}_*) = \check{\eta}]\!] \mid \eta \in A_{\xi_*}\}.$$

Then for $\eta \in A_{\xi_*}$, $\xi_\eta = \xi_*$ and hence

$$b \cdot [\![f(\check{\xi}_*) = \check{\eta}]\!] = b \cdot [\![f(\check{\xi}_\eta) = \check{\eta}]\!] \neq \mathbf{0}$$

since $b \leq [\![f\colon \check{\beta} \to \check{\alpha}]\!] \wedge [\![\check{\xi}_* = \check{\xi}_\eta]\!] = \mathbf{1}$.

Therefore elements of S are $\neq \mathbf{0}$. Moreover, if $\eta_1, \eta_2 \in A_{\xi_*} \wedge \eta_1 \neq \eta_2$,

$$b \cdot [\![f(\check{\xi}_*) = \check{\eta}_1]\!] \cdot [\![f(\check{\xi}_*) = \check{\eta}_2]\!] \leq [\![\check{\eta}_1 = \check{\eta}_2]\!] = \mathbf{0}.$$

Therefore elements of S are mutually disjoint and $\overline{\overline{S}} > \gamma$, by ii). But the existence of such an S contradicts the assumption that $\mathbf{B}$ satisfies the γ-chain condition.

Corollary 17.5. If $\mathbf{B}$ satisfies the c.c.c. and α is a cardinal, then $[\![\text{Card}(\check{\alpha})]\!] = \mathbf{1}$.

Remark. This means that cardinals are absolute if $\mathbf{B}$ satisfies the c.c.c. We can express this fact also in the following way:

Corollary 17.6. If $\mathbf{B}$ satisfies the c.c.c. then $(\forall \alpha)[[\![(\omega_\alpha)^\vee = \omega_{\check{\alpha}}]\!] = \mathbf{1}]$. [Note: For the meaning of this formula see the note stated in the proof of Theorem 17.2.]

Proof. (By induction on α.)

We have already proved the case $\alpha = 0$ at the end of §13. Therefore assume $\alpha > 0$ and $(\forall \xi < \alpha)[[\![(\omega_\xi)^\vee = \omega_{\check{\xi}}]\!] = \mathbf{1}]$. Since

$$\begin{aligned} u = \omega_\alpha \leftrightarrow \text{Card}(u) \wedge (\forall \xi < \alpha)[\omega_\xi < u] \\ \wedge (\forall v)[\text{Card}(v) \wedge (\forall \xi < \alpha)[\omega_\xi < v] \to u \leq v]] \end{aligned}$$

is provable in ZF, we have

$$\begin{aligned} [\![u = \omega_{\check{\alpha}}]\!] = [\![\text{Card}(u)]\!] \cdot [\![(\forall \xi < \check{\alpha})[\omega_\xi < u]]\!] \\ \cdot [\![(\forall v)[\text{Card}(v) \wedge (\forall \xi < \check{\alpha})[\omega_\xi < v] \to u \leq v]]]\!]. \end{aligned}$$

We wish to prove that $[\![(\omega_\alpha)^\vee = \omega_{\check{\alpha}}]\!] = \mathbf{1}$. By Corollary 17.5, $[\![\text{Card}((\omega_\alpha)^\vee)]\!] = \mathbf{1}$.

$$\begin{aligned} [\![(\forall \xi < \check{\alpha})[\omega_\xi < (\omega_\alpha)^\vee]]\!] &= \prod_{\xi < \alpha} [\![\omega_{\check{\xi}} < (\omega_\alpha)^\vee]\!] \\ &= \prod_{\xi < \alpha} [\![(\omega_\xi)^\vee < (\omega_\alpha)^\vee]\!] \quad \text{by the induction hypothesis} \\ &= \mathbf{1}. \end{aligned}$$

Finally, let

$$b_0 = [\![(\forall v)[\mathrm{Card}(v) \wedge (\forall \xi < \check{\alpha})[\omega_\xi < v] \to (\omega_\alpha)^\vee \leq v]]\!]$$

$$= \prod_{\eta \in On} [([\![\mathrm{Card}(\check{\eta})]\!] \cdot \prod_{\xi < \alpha} [\![\omega_{\check{\xi}} < \check{\eta}]\!]) \Rightarrow [\![(\omega_\alpha)^\vee \leq \check{\eta}]\!]]$$

Let $\eta \in On$. If η is not a cardinal, $[\![\mathrm{Card}(\check{\eta})]\!] = \mathbf{0}$. Therefore, we need only consider the case Card (η). Then $[\![\mathrm{Card}(\check{\eta})]\!] = \mathbf{1}$ and

$$\prod_{\xi < \alpha} [\![\omega_{\check{\xi}} < \check{\eta}]\!] = \prod_{\xi < \alpha} [\![(\omega_\xi)^\vee < \check{\eta}]\!] \qquad \text{by the induction hypothesis.}$$

Hence

$$\prod_{\xi < \alpha} [\![\omega_{\check{\xi}} < \check{\eta}]\!] \neq \mathbf{0} \to (\forall \xi < \alpha)[\omega_\xi < \eta]$$
$$\to \omega_\alpha \leq \eta$$
$$\to [\![(\omega_\alpha)^\vee \leq \check{\eta}]\!] = \mathbf{1}.$$

This proves

$$[(\forall \eta \in On)[([\![\mathrm{Card}(\check{\eta})]\!] \cdot \prod_{\xi < \alpha} [\![\omega_{\check{\xi}} < \check{\eta}]\!]) \Rightarrow [\![(\omega_\alpha)^\vee \leq \check{\eta}]\!]) = \mathbf{1}].$$

Therefore $b_0 = \mathbf{1}$. Thus $[\![(\omega_\alpha)^\vee = \omega_{\check{\alpha}}]\!] = \mathbf{1}$.

Remark. Finally we mention a theorem which says that constructible sets are absolute in the same sense as ordinals, i.e., quantification of constructible sets (in the sense of $\mathbf{V}^{(\mathbf{B})}$) can be replaced by quantifications (in the Boolean sense) over the standard constructible sets. Let Const (x) be the formal predicate expressing that x is constructible in the sense of Gödel:

Definition 17.7. $\mathrm{Const}(x) \overset{\Delta}{\leftrightarrow} (\exists v)[\mathrm{Ord}(v) \wedge x = F(v)]$
where F is Gödel's constructibility function. (See Definition 15.13, Introduction to Axiomatic Set Theory.)

Theorem 17.8. $(\forall u \in V^{(\mathbf{B})})[[\![\mathrm{Const}(u)]\!] = \sum_{x \in L} [\![u = \check{x}]\!]]$.

Proof. For $u \in V^{(\mathbf{B})}$,

$$[\![\mathrm{Const}(u)]\!] = [\![(\exists v)[\mathrm{Ord}(v) \wedge u = F(v)]]\!]$$
$$= \sum_{\alpha \in On} [\![u = F(\check{\alpha})]\!] \qquad \text{by Corollary 13.23}$$

In the proof of Theorem 15.28 in Introduction to Axiomatic Set Theory we established that $x = F(\alpha)$ is equivalent to a formula $(\exists f)\phi(f, x, \alpha)$ where $\phi(f, x, \alpha)$ is a bounded formula. Then

$$x = F(\alpha) \to (\exists f)\phi(f, x, \alpha)$$
$$\to (\exists f)[\![\phi(\check{f}, \check{x}, \check{\alpha})]\!] = \mathbf{1} \qquad \text{by Corollary 13.19}$$
$$\to [\![(\exists f)\phi(f, \check{x}, \check{\alpha})]\!] = \mathbf{1}$$
$$\to [\![\check{x} = F(\check{\alpha})]\!] = \mathbf{1}.$$

Therefore, $[\![F(\alpha)^{\vee} = F(\check{\alpha})]\!] = \mathbf{1}$ and hence

$$[\![\mathrm{Const}\,(u)]\!] = \sum_{\alpha \in On} [\![u = F(\check{\alpha})]\!] \cdot [\![(F(\alpha))^{\vee} = F(\check{\alpha})]\!]$$

$$= \sum_{\alpha \in On} [\![u = (F(\alpha))^{\vee}]\!]$$

$$= \sum_{x \in L} [\![u = \check{x}]\!].$$

Remark. In the same way as we derived Corollary 13.23 from Theorem 13.22 we obtain the following corollary of Theorem 17.8.

Corollary 17.9. $[\![(\exists u)[\mathrm{Const}\,(u) \wedge \varphi(u)]]\!] = \sum_{x \in L} [\![\varphi(\check{x})]\!]$.

18. Model Theoretic Consequences of the Distributive Laws

There are several algebraic properties which are satisfied only by certain complete Boolean algebras **B** but which have important consequences for the corresponding models $\mathbf{V}^{(\mathbf{B})}$, e.g., by Theorem 17.1 and Corollary 17.5, cardinals are preserved if **B** satisfies the c.c.c. In this section we will consider certain distributive laws.

Theorem 18.1. $[\![\mathscr{P}(\omega)^{\vee} \subseteq \mathscr{P}(\check{\omega})]\!] = \mathbf{1}$

Proof.

$$(\forall s \subseteq \omega)[[\![\check{s} \subseteq \check{\omega}]\!] = \mathbf{1}]$$
$$(\forall s \subseteq \omega)[[\![\check{s} \in \mathscr{P}(\check{\omega})]\!] = \mathbf{1}]$$
$$[\![\mathscr{P}(\omega)^{\vee} \subseteq \mathscr{P}(\check{\omega})]\!] = \mathbf{1}.$$

Theorem 18.2. **B** satisfies the $(\omega, 2)$-DL iff $[\![\mathscr{P}(\omega)^{\vee} = \mathscr{P}(\check{\omega})]\!] = \mathbf{1}$.

Proof. Assume that **B** satisfies the $(\omega, 2)$-DL. We need only show that $[\![\mathscr{P}(\check{\omega}) \subseteq \mathscr{P}(\omega)^{\vee}]\!] = \mathbf{1}$. Therefore let $t \in B^{\mathscr{D}(\check{\omega})}$ such that $[\![t \subseteq \check{\omega}]\!] = \mathbf{1}$. Define $b_{n1} = [\![\check{n} \in t]\!]$, $b_{n0} = [\![\check{n} \notin t]\!]$ for $n \in \omega$. Then $(\forall n \in \omega)[b_{n1} = {}^{-}b_{n0}]$,

$$\mathbf{1} = \prod_{n<\omega} (b_{n0} + b_{n1}) = \sum_{s\in 2^{\omega}} \prod_{n\in\omega} b_{n,s(n)} \qquad \text{by the } (\omega, 2)\text{-DL}$$
$$= \sum_{s\in \mathbf{2}^{\mathscr{D}(\check{\omega})}} \prod_{n\in\omega} ([\![\check{n} \in t]\!] \Leftrightarrow s(\check{n}))$$
$$= \sum_{s\in \mathbf{2}^{\mathscr{D}(\check{\omega})}} [\![t = s]\!] \qquad \text{since } [\![t \subseteq \check{\omega}]\!] = \mathbf{1}$$
$$= [\![t \in \mathscr{P}(\omega)^{\vee}]\!] \qquad \text{since as is easily shown}$$

1. $t \in V^{(\mathbf{B})} \to \sum_{s\in \mathbf{2}^{\mathscr{D}(\check{\omega})}} [\![t = s]\!] = [\![t \in \mathscr{P}(\omega)^{\vee}]\!]$.

Hence

$$[\![t \in \mathscr{P}(\omega)^{\vee}]\!] = \mathbf{1}.$$

Therefore, by Theorem 16.25, if $A = \{t \in B^{\mathscr{D}(\check{\omega})} \mid [\![t \subseteq \check{\omega}]\!] = \mathbf{1}\}$ then

$$[\![u \in \mathscr{P}(\check{\omega})]\!] = \sum_{t\in A} [\![u = t]\!]$$
$$= \sum_{t\in A} [\![u = t]\!]\cdot[\![t \in \mathscr{P}(\omega)^{\vee}]\!]$$
$$\leq \sum_{t\in A} [\![u \in \mathscr{P}(\omega)^{\vee}]\!] = [\![u \in \mathscr{P}(\omega)^{\vee}]\!].$$

This proves that $[\![\mathscr{P}(\check{\omega}) \subseteq \mathscr{P}(\omega)^{\vee}]\!] = \mathbf{1}$. Consequently $[\![\mathscr{P}(\omega)^{\vee} = \mathscr{P}(\check{\omega})]\!] = \mathbf{1}$. To prove the converse, assume that $[\![\mathscr{P}(\check{\omega}) = \mathscr{P}(\omega)^{\vee}]\!] = \mathbf{1}$, and let

$$\{b_{ni} \mid n \in \omega \wedge i \in 2\} \subseteq B.$$

By Theorem 18.4 (below) we can assume that

$$(\forall n \in \omega)[b_{n0} = {}^{-}b_{n1}].$$

Define $u \in V^{(\mathbf{B})}$ by

$$\mathscr{D}(u) = \mathscr{D}(\check{\omega}) \wedge (\forall n \in \omega)[u(\check{n}) = b_{n1}].$$

Then

$$[\![u \subseteq \check{\omega}]\!] = \mathbf{1},$$

i.e.

$$[\![u \in \mathscr{P}(\check{\omega})]\!] = \mathbf{1}$$

and by assumption

$$[\![u \in \mathscr{P}(\omega)^{\vee}]\!] = \mathbf{1}.$$

$$\mathbf{1} = [\![u \in \mathscr{P}(\omega)^{\vee}]\!] = \sum_{s \in \mathbf{2}^{\mathscr{D}(\check{\omega})}} [\![u = s]\!] \qquad \text{by 1 above}$$

$$= \sum_{s \in \mathbf{2}^{\mathscr{D}(\check{\omega})}} \prod_{n \in \omega} ([\![\check{n} \in u]\!] \Leftrightarrow s(\check{n}))$$

$$= \sum_{s \in \mathbf{2}^{\omega}} \prod_{n \in \omega} b_{n, s(n)}.$$

Since also $\mathbf{1} = \prod_{n \in \omega} (b_{n0} + b_{n1})$, this completes the proof.

Remark.

1. In the same way as we proved Theorem 18.2 we can prove the following:

$$\mathbf{B} \text{ satisfies the } (\alpha, 2)\text{-DL} \quad \text{iff} \quad [\![\mathscr{P}(\check{\alpha}) = \mathscr{P}(\alpha)^{\vee}]\!] = \mathbf{1}.$$

2. Interpreted in the theory of forcing, Theorem 18.2 says: Let $\langle M, \mathbf{P}\rangle$ be a setting for forcing, let G be $\mathbf{P}$-generic over M and let $\mathbf{B}$ be the M-complete Boolean algebra in M associated with $\mathbf{P}$. Then $\mathbf{B}$ satisfies the $(\omega, 2)$-DL in M iff $\mathscr{P}(\omega)^M = \mathscr{P}(\omega)^{M[G]}$.

Exercise. Give a direct proof of 2.

Definition 18.3. Let α be a cardinal. $\mathbf{B}$ satisfies the *restricted* $(\alpha, 2)$-DL iff for all families $\{b_{ij} \mid i \in I \wedge j < 2\} \subseteq B$ such that $\bar{\bar{I}} = \alpha$ and

$$(\forall i \in I)[b_{i0} = {}^{-}b_{i1}]$$

we have

$$\prod_{i \in I} (b_{i0} + b_{i1}) = \sum_{f \in 2^I} \prod_{i \in I} b_{i, f(i)}.$$

Remark. Although this is a special case of the $(\alpha, 2)$-DL, the two are equivalent:

Theorem 18.4. **B** satisfies the $(\alpha, 2)$-DL iff **B** satisfies the restricted $(\alpha, 2)$-DL.

Proof. We can assume that α is an infinite cardinal. Suppose

$$\{a_{ij} \mid i \in I \wedge j < 2\} \subseteq B$$

where $\bar{\bar{I}} = \alpha$. Define $T = (I \times \{0\}) \cup (I \times \{1\})$. Then $\bar{\bar{T}} = \alpha$. For $f \in 2^T$ define $f_0, f_1 \in 2^I$ by

$$f_0(i) = f(i, 0)$$
$$f_1(i) = f(i, 1) \qquad \text{for } i \in I.$$

Because of Theorem 4.2 it suffices to show that

$$\prod_{i \in I} (a_{i0} + a_{i1}) \leq \sum_{f \in 2^I} \prod_{i \in I} a_{if(i)}.$$

Let $J = \{j \mid j = \langle i, n \rangle \text{ where } i \in I \wedge n < 2\}$. For $j = \langle i, n \rangle \in J$ define

$$b_{j0} = a_{i,n}, \qquad b_{j1} = {}^{-}a_{i,n},$$

and let

$$b = \prod_{i \in I} (a_{i0} + a_{i1}).$$

Then $(\forall j \in J)[b_{j0} = {}^{-}b_{j1}]$, and since $b \leq (a_{i0} + a_{i1})$,

i) $b \cdot b_{\langle i,0 \rangle n} \leq a_{i,n}$.
ii) $b \cdot b_{\langle i,1 \rangle n} \leq a_{i,1-n}$ for $i \in I$, $n < 2$.

Since $\prod_{j \in J} (b_{j0} + b_{j1}) = \mathbf{1}$,

$$\mathbf{1} = \sum_{f \in 2^T} \prod_{j \in T} b_{jf(j)} \qquad \text{by the restricted } (\alpha, 2)\text{-DL},$$

$$= \sum_{f_0, f_1 \in 2^I} \prod_{i \in I} b_{\langle i,0 \rangle, f_0(i)} \prod_{i \in I} b_{\langle i,1 \rangle, f_1(i)}.$$

There, using i) and ii)

$$b \leq \sum_{f_0, f_1 \in 2^I} \prod_{i \in I} a_{i, f_0(i)} \prod_{i \in I} a_{i, 1 - f_1(i)}$$

$$\leq \sum_{f \in 2^I} \prod_{i \in I} a_{i, f(i)}.$$

Remark. Other forms of the (α, β)-DL can be obtained from the following theorem.

Theorem 18.5. Let I, J be sets. Then the following conditions are equivalent:

1. For all families $\{b_{ij} \mid i \in I \wedge j \in J\} \subseteq B$,

$$\prod_{i \in I} \sum_{j \in J} b_{ij} = \sum_{f \in J^I} \prod_{i \in J} b_{if(i)}$$

(i.e. **B** satisfies the (I, J)-DL).

2. For all families $\{b_{ij} \mid i\in I \wedge j\in J\}\in B$,

$$(\forall f\in J^I)\left[\prod_{i\in I} b_{if(i)} = \mathbf{0} \rightarrow \prod_{i\in I}\sum_{j\in J} b_{ij} = \mathbf{0}\right]$$

(i.e., in order to prove the (I, J)-DL we need only show that the left-hand side is $\mathbf{0}$ if the right-hand side is $\mathbf{0}$).

3. For all families $\{b_{ij} \mid i\in I \wedge j\in J\}\subseteq B$ and any $b\in B$,

$$(\forall f\in J^I)\left[\prod_{i\in I} b_{if(i)} = \mathbf{0}\right] \wedge (\forall i\in I)\left[\sum_{j\in J} b_{ij} = b\right] \rightarrow b = \mathbf{0}.$$

Proof. We need only show that 3 implies 1. Therefore let

$$\{b_{ij} \mid i\in I \wedge j\in J\}\subseteq B.$$

Let

$$b = \prod_{i\in I}\sum_{j\in J} b_{ij} - \sum_{f\in 2^I}\prod_{i\in J} b_{if(i)},$$

$$a_{ij} = b\cdot b_{ij} \qquad \text{for } i\in I, j\in J.$$

Because of Theorem 4.2, we have to prove that $b = \mathbf{0}$.

$$\sum_{j\in J} a_{ij} = b\sum_{j\in J} b_{ij} = \sum_{j\in J} b_{ij}\left(\prod_{i'\in I}\sum_{j'\in J} b_{i'j'} - \sum_{f\in J^I}\prod_{i'\in I} b_{i'f(i')}\right)$$
$$= b$$

and for $f\in J^I$:

$$\prod_{i\in I} a_{if(i)} = b\cdot\prod_{i\in I} b_{if(i)} = \mathbf{0}.$$

Applying 3 to the family $\{a_{ij} \mid i\in I \wedge j\in J\}$, we obtain $b = \mathbf{0}$.

19. Independence Results Using the Models $\mathbf{V}^{(\mathbf{B})}$

Theorem 19.1. Let $\mathbf{B}$ be a complete Boolean algebra which does not satisfy the $(\omega, 2)$-DL. Then $[\![V \neq L]\!] = \mathbf{1}$ in $\mathbf{V}^{(\mathbf{B})}$.

Proof. (By the method of forcing.) Let M be a countable transitive model of ZF such that $\langle M, \in, \mathbf{B}^M\rangle$ is elementary equivalent to $\langle V, \in, \mathbf{B}\rangle$, let G be $\mathbf{P}$-generic over M (where $\mathbf{P}$ is the partial order structure in M associated with $\mathbf{B}^M$). Then, by Theorem 18.2

$$\mathscr{P}(\omega)^M \subsetneqq \mathscr{P}(\omega)^{M[G]}.$$

However, if $M[G]$ satisfies $V = L$, then $M[G] = M$. This is a contradiction. Therefore $M[G] \vDash V \neq L$ and hence $[\![V \neq L]\!] = \mathbf{1}$ in $\mathbf{V}^{(\mathbf{B})}$.

Exercise. Give a direct proof of Theorem 19.1 in $\mathbf{V}^{(\mathbf{B})}$.

Remark. We shall now present new proofs of the independence of certain axioms from the axioms of $ZF + AC$ by using the models $\mathbf{V}^{(\mathbf{B})}$ rather than the technique of forcing.

Theorem 19.2. If $\mathbf{B}$ is a complete Boolean algebra satisfying the c.c.c. and the cardinality of B is $\leq 2^{\aleph_0}$, then assuming the GCH in V we have that the GCH is $\mathbf{B}$-valid in $\mathbf{V}^{(\mathbf{B})}$.

Proof. Since $(\omega_\alpha)^\vee$ is definite,

$$[\![\mathscr{P}((\omega_\alpha)^\vee) = B^{\mathscr{D}((\omega_\alpha)^\vee)} \times \{\mathbf{1}\}]\!] = \mathbf{1} \qquad \text{by Theorem 16.27.}$$

By assumption, $B^{\mathscr{D}((\omega_\alpha)^\vee)}$ has cardinality $\leq (2^{\aleph_0})^{\aleph_\alpha} = \aleph_{\alpha+1}$ by the GCH in V. Therefore there exists a (real) function

$$\varphi\colon \mathscr{D}((\omega_{\alpha+1})^\vee) \xrightarrow{\text{onto}} B^{\mathscr{D}((\omega_\alpha)^\vee)}.$$

Furthermore φ is extensional:

$$\begin{aligned} u_1, u_2 \in \mathscr{D}((\omega_{\alpha+1})^\vee) &\rightarrow (\exists \delta_1, \delta_2 \in \omega_{\alpha+1})[u_1 = \check{\delta}_1 \wedge u_2 = \check{\delta}_2] \\ &\rightarrow [[\![u_1 = u_2]\!] \neq \mathbf{0} \rightarrow u_1 = u_2] \\ &\rightarrow [[\![u_1 = u_2]\!] \neq \mathbf{0} \rightarrow [\![\varphi(u_1) = \varphi(u_2)]\!] = \mathbf{1}] \\ &\rightarrow [\![u_1 = u_2]\!] \leq [\![\varphi(u_1) = \varphi(u_2)]\!]. \end{aligned}$$

By Theorem 16.8 there is an $f \in V^{(\mathbf{B})}$ such that

$$[\![f\colon (\omega_{\alpha+1})^\vee \rightarrow \mathscr{P}((\omega_\alpha)^\vee)]\!] = \mathbf{1}$$

and

$$(\forall \xi < \omega_{\alpha+1})[[\![f(\check{\xi}) = \varphi(\check{\xi})]\!] = \mathbf{1}].$$

i) $[\![(\forall \xi \in (\omega_{\alpha+1})^{\vee})[f(\xi) \in \mathscr{P}((\omega_\alpha)^{\vee})]]\!]$

$$= \prod_{\xi < \omega_{\alpha+1}} [\![f(\check{\xi}) \in \mathscr{P}((\omega_\alpha)^{\vee})]\!]$$

$$= \prod_{\xi < \omega_{\alpha+1}} [\![\varphi(\check{\xi}) \in \mathscr{P}((\omega_\alpha)^{\vee})]\!]$$

$$= \prod_{\xi < \omega_{\alpha+1}} [\![\varphi(\check{\xi}) \in B^{\mathscr{D}((\omega_\alpha)^{\vee})} \times \{\mathbf{1}\}]\!] = \mathbf{1}.$$

ii) $[\![(\forall x \in \mathscr{P}((\omega_\alpha)^{\vee}))(\exists \eta \in (\omega_{\alpha+1})^{\vee})]f(\eta) = x]\!]$

$$= \prod_{x \in B^{\mathscr{D}((\omega_\alpha)^{\vee})}} [\![(\exists \eta \in (\omega_{\alpha+1})^{\vee})[f(\eta) = x]]\!]$$

$$= \prod_{\xi < \omega_{\alpha+1}} [\![(\exists \eta \in (\omega_{\alpha+1})^{\vee})[f(\eta) = \varphi(\check{\xi})]]\!]$$

since φ is onto

$$= \mathbf{1}.$$

Therefore $[\![\mathscr{W}(f) = \mathscr{P}((\omega_\alpha)^{\vee})]\!] = \mathbf{1}$ by i) and ii). This proves that

$$[\![\overline{\overline{2^{(\omega_\alpha)^{\vee}}}} = \overline{\overline{(\omega_{\alpha+1})^{\vee}}}]\!] = \mathbf{1}$$

$$[\![\overline{\overline{2^{(\omega_\alpha)^{\vee}}}} = (\omega_{\alpha+1})^{\vee}]\!] = \mathbf{1} \qquad \text{by Corollary 17.5.}$$

$$(\forall \alpha)[\![\overline{\overline{2^{\omega_{\check{\alpha}}}}} = \omega_{\check{\alpha}+1}]\!] = \mathbf{1} \qquad \text{by Corollary 17.6}$$

$$[\![(\forall \alpha)[\overline{\overline{2^{\omega_\alpha}}} = \omega_{\alpha+1}]]\!] = \mathbf{1} \qquad \text{by Corollary 13.23, i.e.,}$$

$$[\![GCH]\!] = \mathbf{1}.$$

Theorem 19.3. Let π be an automorphism of $\mathbf{B}$ ($\pi \in \operatorname{Aut}(\mathbf{B})$). Then π can be extended to an isomorphism

$$\pi: V^{(\mathbf{B})} \to V^{(\mathbf{B})}$$

such that for every formula φ and $u_1, \ldots, u_n \in V^{(\mathbf{B})}$

$$\pi([\![\varphi(u_1, \ldots, u_n)]\!] = [\![\varphi(\pi(u_1), \ldots, \pi(u_n))]\!].$$

Proof. Note that any automorphism $\pi: |\mathbf{B}| \to |\mathbf{B}|$ is complete since $(\forall b_1, b_2 \in B)[b_1 \leq b_2 \leftrightarrow \pi(b_1) \leq \pi(b_2)]$. We define $\pi: V^{(\mathbf{B})} \to V^{(\mathbf{B})}$ by induction as follows: Let $u \in V^{(\mathbf{B})}$. Then $\pi(u) \in V^{(\mathbf{B})}$ is defined by

$$\mathscr{D}\pi(u) = \{\pi(v) \mid v \in \mathscr{D}(u)\}$$

and

$$(\forall v \in \mathscr{D}(u))[\pi(u)(\pi(v)) = \pi(u(v))].$$

It is easy to prove by transfinite induction that

$$\pi([\![u = v]\!]) = [\![\pi(u) = \pi(v)]\!]$$
$$\pi([\![u \in v]\!]) = [\![\pi(u) \in \pi(v)]\!]$$

Also $\pi: V^{(\mathbf{B})} \to V^{(\mathbf{B})}$ is onto. (Consider $\pi^{-1}: V^{(\mathbf{B})} \to V^{(\mathbf{B})}$ the extension of π^{-1}. This π^{-1} is the inverse of the extended π.) The conclusion then follows by induction on the number of logical symbols in φ.

Theorem 19.4. Let $\mathbf{P}$ be the partial order structure used in the proof of the independence of $V = L$ (Definition 11.1) and let $\mathbf{B}$ be the complete Boolean algebra of regular open sets of $\mathbf{P}$. Then $\mathbf{0}$ and $\mathbf{1}$ are the only elements of B which are invariant under all automorphisms of $\mathbf{B}$*.

Proof. Let $b \in B$ and $\mathbf{0} < b < \mathbf{1}$. Then there exist p and q such that $[p] \subseteq b$ and $[q] \subseteq {}^{-}b$. By Theorem 11.6, $\exists \pi \in \text{Aut}(\mathbf{P})$ such that $\pi(p)$ and q are compatible. Using π to also denote the automorphism on $\mathbf{B}$ induced by π we have $\pi(b) \cdot ({}^{-}b) \geq \pi([p]) \cdot [q] > \mathbf{0}$. Therefore $\pi(b) \neq b$.

Theorem 19.5. (cp. Theorem 11.7) Assuming the *GCH*, there is a Boolean algebra $\mathbf{B}$ such that the *GCH* is $\mathbf{B}$-valid in $\mathbf{V}^{(\mathbf{B})}$ but the statement "There is a definable well-ordering of $\mathscr{P}(\omega)$" is not $\mathbf{B}$-valid. (Here by "definable" we mean "definable using constants $\check{k}$ as parameters". Note that $\check{k}$ corresponds to the constant $\underline{k} \in C(M)$ in Theorem 11.7.) "There is a definable well ordering of $\mathscr{P}(\omega)$" is in fact a statement in the language of set theory.

Proof. Let $\mathbf{B}$ be as in Theorem 19.4. Then $\overline{\overline{B}} \leq 2^{\aleph_0}$ and $\mathbf{B}$ satisfies the c.c.c., hence the *GCH* is $\mathbf{B}$-valid in $V^{(\mathbf{B})}$. Suppose there is a definable well-ordering of $\mathscr{P}(\omega)$ in $V^{(\mathbf{B})}$, i.e., there is a formula φ (possibly involving constants $\check{k}$) such that φ well orders $\mathscr{P}(\omega)$, i.e.,

$$[\![\{\langle x, y\rangle \in \mathscr{P}(\omega) \times \mathscr{P}(\omega) \mid \varphi(x, y)\} \text{ is a linear ordering}]\!] = \mathbf{1}$$

and

$$[\![(\forall x \subseteq \mathscr{P}(\omega))[x \neq 0 \to (\exists!\, z \in x)(\forall y \in x)\varphi(z, y)]]\!] = \mathbf{1}.$$

Define

$$S = \{x \in B^{\mathscr{D}(\check{\omega})} \mid [\![x \in \mathscr{P}(\omega)^{\vee}]\!] = \mathbf{0}\} \times \{\mathbf{1}\}.$$

We shall prove that

i) $[\![(\exists x)[x \in S]]\!] = \mathbf{1} \wedge [\![S \subseteq \mathscr{P}(\omega)]\!] = \mathbf{1}$, but
ii) $b = [\![(\exists!\, x \in S)(\forall y \in S)\varphi(x, y)]\!] = \mathbf{0}$.

This gives a contradiction.

Define $u \in V^{(\mathbf{B})}$ by $\mathscr{D}(u) = \mathscr{D}(\check{\omega})$ and

$$(\forall n \in \omega)[u(\check{n}) = [\langle\{n\}, 0\rangle]^{-0}].$$

Claim: $[\![u \in (\mathscr{P}(\omega))^{\vee}]\!] = \mathbf{0}$.

$$[\![u \in \mathscr{P}(\omega)^{\vee}]\!] = \sum_{s \subseteq \omega} [\![u = \check{s}]\!]$$

and for $s \subseteq \omega$,

$$[\![u = \check{s}]\!] = \prod_{n \in \omega} (u(\check{n}) \Leftrightarrow \check{s}(\check{n})).$$

* Some authors then say that $\mathbf{B}$ satisfies the $\mathbf{0},\mathbf{1}$-law.

Suppose $[\![u = \check{s}]\!] \neq \mathbf{0}$ and let $\langle p_1, p_2\rangle \in \prod_{n\in\omega} (u(\check{n}) \Leftrightarrow \check{s}(\check{n}))$. Then

$$[\langle p_1, p_2\rangle] \subseteq \left(\bigcap_{n\in\omega} (u(\check{n}) \Leftrightarrow \check{s}(\check{n}))\right)$$

and hence

$$(\forall q_1 q_2)\Big[\langle q_1, q_2\rangle \in P \wedge \langle q_1, q_2\rangle \leq \langle p_1, p_2\rangle \rightarrow [\langle q_1, q_2\rangle] \cap \bigcap_{n\in\omega} (u(\check{n}) \Leftrightarrow \check{s}(\check{n})) \neq 0\Big].$$

Choose some $n \in \omega$ such that $n \notin p_1 \cup p_2$ and define $q = \langle q_1, q_2\rangle$ by

$$\begin{aligned} q_1 &= p_1 \\ q_2 &= p_2 \cup \{n\} \end{aligned} \qquad \text{if } \check{s}(\check{n}) = \mathbf{1}$$

$$\begin{aligned} q_1 &= p_1 \cup \{n\} \\ q_2 &= p_2 \end{aligned} \qquad \text{if } \check{s}(\check{n}) = \mathbf{0}$$

Since

$$\begin{aligned} (u(\check{n}) \Leftrightarrow s(\check{n})) &= u(\check{n}) &&\text{if } \check{s}(\check{n}) = \mathbf{1} \\ &= {}^{-}u(\check{n}) &&\text{if } \check{s}(\check{n}) = \mathbf{0} \end{aligned}$$

and ${}^{-}u(\check{n}) = [\langle 0, \{n\}\rangle]^{-0}$, (by the exercise following Corollary 11.4) we have $\langle q_1, q_2\rangle \leq \langle p_1, p_2\rangle$ and $[\langle q_1, q_2\rangle] \cap \bigcap_{n\in\omega} (u(\check{n}) \Leftrightarrow s(\check{n})) = 0$, a contradiction. Therefore $[\![u = \check{s}]\!] = \mathbf{0}$ for all $s \subseteq \omega$, and hence

$$[\![u \in P(\omega)^{\vee}]\!] = \mathbf{0}$$

as we claimed. Therefore $\mathscr{D}(S) \neq 0$ and hence $[\![(\exists x)[x \in S]]\!] = \mathbf{1}$. This proves i).

If $\pi \in \operatorname{Aut}(\mathbf{B})$, and $\tilde{\pi}$ is the isomorphism from $V^{(\mathbf{B})}$ into $V^{(\mathbf{B})}$ induced by π according to Theorem 19.3, then $\tilde{\pi}(\check{k}) = \check{k}$ and hence S is invariant under $\tilde{\pi}$. Therefore b is invariant under all $\pi \in \operatorname{Aut}(\mathbf{B})$. Suppose $b \neq \mathbf{0}$. Then $b = \mathbf{1}$ by Theorem 19.4. Define $v \in V^{(\mathbf{B})}$ by

$$\mathscr{D}(v) = \mathscr{D}(\check{\omega})$$

$$(\forall n \in \omega)[v(\check{n}) = [\![(\exists x \in S)(\forall y \in S)[\varphi(x, y) \wedge \check{n} \in x]]\!]].$$

Since $b = \mathbf{1}$, by the maximum principle there is a $u_0 \in V^{(\mathbf{B})}$ such that

$$[\![u_0 \in S]\!] = \mathbf{1} \wedge [\![(\forall y \in S)\varphi(u_0, y)]\!] = \mathbf{1}.$$

$$\begin{aligned} [\![u_0 = v]\!] &= \prod_{n\in\omega} [\![\check{n} \in u_0 \leftrightarrow \check{n} \in v]\!] \\ &= \prod_{n\in\omega} [\![(\exists x \in S)(\forall y \in S)[\varphi(x, y) \wedge \check{n} \in x] \leftrightarrow \check{n} \in v]\!] \qquad \text{since } b = \mathbf{1} \\ &= \mathbf{1} \qquad \text{by the definition of } v. \end{aligned}$$

Therefore $[\![v \in S]\!] = \mathbf{1}$. But for every $n \in \omega$ $v(\check{n})$ is invariant, hence $v(\check{n}) = \mathbf{0}$ or $\mathbf{1}$, i.e.,

$$v \in \mathbf{2}^{\mathscr{D}(\check{\omega})}.$$

Therefore $[\![v \in (\mathscr{P}(\omega))^{\vee}]\!] = \mathbf{1}$.

Since $v \in \mathscr{D}(S)$ this implies that

$$\mathbf{1} = [\![v \in S]\!] = \sum_{x \in \mathscr{D}(S)} [\![x = v]\!] \leq \sum_{x \in \mathscr{D}(S)} [\![x \in (\mathscr{P}(\omega))^{\vee}]\!] = \mathbf{0}.$$

This is a contradiction which proves ii).

Remark. Next we give a new proof of the independence of the Continuum Hypothesis, however this time we use a measure algebra $\mathbf{B}$. Let I be an index set of cardinality $> 2^{\aleph_0}$, let $X = 2^{\omega \times I}$ be a generalized Cantor space, $\mathscr{B}$ the σ-algebra of all Borel sets of X, $\mathbf{N}$ the σ-ideal in $\mathscr{B}$ consisting of all the null sets for the usual product measure, and finally let $\mathbf{B} = \mathscr{B}/\mathbf{N}$. Basic open sets of X are of the form

$$U(p_0) = \{p \mid p \in 2^{\omega \times I} \wedge p(j_1) = p_0(j_1) \wedge \cdots \wedge p(j_n) = p_0(j_n)\}$$

where $p_0 : \{j_1, \ldots, j_n\} \to 2$ and $j_1, \ldots, j_n \in \omega \times I$. Without proof we shall use the fact that there is a (unique) measure m for subsets of X such that

$$m(U(p_0)) = (\tfrac{1}{2})^n \quad \text{and} \quad m(X) = 1.$$

With this notation we are prepared to prove the following.

Theorem 19.6. $\mathbf{B}$ satisfies the c.c.c. and therefore $\mathbf{B}$ is complete.

Proof. Let $S \subseteq B$ be a set of mutually disjoint elements. We have to show that $\overline{\overline{S}} \leq \omega$. Therefore we can assume that $0 \notin S$. Let $S_n = \{b \mid b \in S \wedge m(b) \geq 1/n\}$ for $n \in \omega$. Since the elements of S are mutually disjoint and $m(X) = 1$, $\overline{\overline{S}}_n \leq n$ for all $n \in \omega$. Since $S = \bigcup_{n \in \omega} S_n$, this proves that $\overline{\overline{S}} \leq \omega$. Since $\mathscr{B}$ and $\mathbf{N}$ are σ-complete, so is $\mathbf{B} = \mathscr{B}/\mathbf{N}$, and therefore $\mathbf{B}$ is complete by Theorem 3.27.

Remark. For this particular Boolean algebra $\mathbf{B}$ we can prove that the negation of the Continuum Hypothesis is $\mathbf{B}$-valid in $\mathbf{V}^{(\mathbf{B})}$.

Theorem 19.7. For $\mathbf{B}$ defined as above, $[\![\neg\, CH]\!] = \mathbf{1}$ in $V^{(\mathbf{B})}$.

Proof. Define $\mathbf{B}$-valued sets $u_i \in V^{(\mathbf{B})}$ for $i \in I$ as follows:

$$\mathscr{D}(u_i) = \mathscr{D}(\check{\omega})$$

$$(\forall n \in \omega)[u_i(\check{n}) = \{p \in X \mid p(n, i) = 1\}/\mathbf{N}].$$

Obviously,

1. $(\forall i \in I)[[\![u_i \subseteq \check{\omega}]\!] = \mathbf{1}]$.

Proof. Let $i, j \in I$, $i \neq j$. Then

$$\begin{aligned}[\![u_i = u_j]\!] &= \prod_{n \in \omega} (u_i(\check{n}) \Leftrightarrow u_j(\check{n})) \\ &= \{p \in X \mid (\forall n \in \omega)[p(n, i) = p(n, j)]\}/\mathbf{N},\end{aligned}$$

since the Boolean operations in **B** are the corresponding set theoretical operations. Let

$$S_{ij} = \{p \in X \mid (\forall n \in \omega)[p(n, i) = p(n, j)]\}.$$

We have to show that $S_{ij} \in \mathbf{N}$, i.e., S_{ij} has measure 0. For arbitrary $k \in \omega$ let

$n_1, \ldots, n_k \in \omega$ be different natural numbers and
$p_1, \ldots, p_{2^k}$ be an enumeration of all functions in $2^{\{n_1, \ldots, n_k\}}$.

Define, for $1 \leq l \leq 2^k$,

$$\begin{aligned} u_l = \{p \in X \mid p(n_1, i) = p_l(n_1) \wedge \cdots \wedge p(n_k, i) = p_l(n_k) \\ \wedge\ p(n_1, j) = p_l(n_1) \wedge \cdots \wedge p(n_k, j) = p_l(n_k)\}. \end{aligned}$$

Then $S_{ij} \subseteq u_1 \cup \cdots \cup u_{2^k}$ and $m(u_l) = 1/2^{2k}$, hence $m(S_{ij}) \leq 2^k \cdot 1/2^{2k} = 1/2^k$. Since k was arbitrary, $m(S_{ij}) = 0$. Thus

2. $(\forall i, j \in I)[i \neq j \rightarrow [\![u_i = u_j]\!] = \mathbf{0}]$.

Similarly, $s \subseteq \omega \rightarrow [\![u_i = \check{s}]\!] = \mathbf{0}$.

Therefore

3. $(\forall i \in I)[[\![u_i \in (\mathscr{P}(\omega))^{\vee}]\!] = \mathbf{0}]$.

Since **B** satisfies the c.c.c., ω_1, in $\mathbf{V}^{(\mathbf{B})}$, is $\check{\omega}_1$. Therefore

$$[\![\neg\ CH]\!] = [\![\neg\ (\exists f)[f\colon \mathscr{P}(\check{\omega}) \xleftrightarrow[\text{onto}]{1-1} (\omega_1)^{\vee}]]\!]$$

If $[\![\neg\ CH]\!] < \mathbf{1}$, then by the maximum principle,

$$b = [\![f\colon \mathscr{P}(\check{\omega}) \xleftrightarrow[\text{onto}]{1-1} (\omega_1)^{\vee}]\!] > \mathbf{0} \qquad \text{for some } f \in V^{(\mathbf{B})}.$$

By 1,

$$(\forall i \in I)[[\![(\exists \xi < \check{\omega}_1)[f(u_i) = \xi]]\!] \geq b]$$

$$(\forall i \in I)\Bigg[\sum_{\xi < \omega_1} [\![f(u_i) = \check{\xi}]\!] \geq b\Bigg].$$

Now we proceed as in the proof of Theorem 17.4:

Since $\bar{\bar{I}} > 2^{\aleph_0}$, there exists an $\eta \subseteq \omega$, such that $J = \{i \in I \mid u_i = \check{\eta}\}$ is uncountable. Moreover, if $i, j \in J$ and $i \neq j$,

$$\begin{aligned} b \cdot [\![f(u_i) = \check{\eta}]\!] \cdot [\![f(u_j) = \check{\eta}]\!] &\leq b \cdot [\![f(u_i) = f(u_j)]\!] \\ &\leq [\![u_i = u_j]\!] = 0 \qquad \text{by 2.} \end{aligned}$$

Therefore

$$\{b \cdot [\![f(u_i) = \check{\eta}]\!] \mid i \in J\}$$

is an uncountable subset of B, the elements of which are pairwise disjoint. However, the existence of such a set contradicts the c.c.c. in **B**, Therefore, we must have $[\![\neg\ CH]\!] = \mathbf{1}$.

Exercise. Prove Theorem 19.7 by using the Boolean algebra of Theorem 11.10.

20. Weak Distributive Laws

Again $\mathbf{B}$ denotes a complete Boolean algebra.

Definition 20.1. $\mathbf{B}$ satisfies the (ω, ω)-*weak distributive law* ((ω, ω)-WDL) iff for every family $\{b_{nm} \mid n, m \in \omega\} \subseteq B$

$$\prod_{m<\omega} \sum_{n<\omega} b_{nm} = \sum_{f \in \omega^{\omega}} \prod_{n \in \omega} \sum_{m \leq f(n)} b_{nm}.$$

Similarly, if ω_α is not cofinal with ω, $\mathbf{B}$ satisfies the (ω, ω_α)-weak distributive law iff for every family $\{b_{n\xi} \mid n \in \omega \wedge \xi < \omega_\alpha\} \subseteq B$

$$1. \prod_{n<\omega} \sum_{\xi<\omega_\alpha} b_{n\xi} = \sum_{f \in \omega_\alpha{}^{\omega}} \prod_{n \in \omega} \sum_{\xi \leq f(n)} b_{n\xi}.$$

Remark. If $cf(\omega_\alpha) > \omega$, the right-hand side of 1 is equal to

$$\sum_{\eta<\omega_\alpha} \prod_{n<\omega} \sum_{\xi \leq \eta} b_{n\xi}.$$

Theorem 20.2. If $\mathbf{B}$ satisfies the c.c.c. and $cf(\omega_\alpha) > \omega$, then $\mathbf{B}$ satisfies the (ω, ω_α)-WDL.

Proof. Let $\{b_{n\xi} \mid n < \omega \wedge \xi < \omega_\alpha\} \subseteq B$. Then by the c.c.c., for each $n \in \omega$ there exists a countable set $C_n \subseteq B$ such that

$$\sum_{\xi<\omega_\alpha} b_{n\xi} = \sup C_n.$$

Define $\eta_0 = \sup \{\xi < \omega_\alpha \mid (\exists n \in \omega)[b_{n\xi} \in C_n]\}$. Since $cf(\omega_\alpha) > \omega$, $\eta_0 < \omega_\alpha$ and

$$(\forall n \in \omega)\left[\sum_{\xi<\omega_\alpha} b_{n\xi} = \sum_{\xi \leq \eta_0} b_{n\xi}\right],$$

hence

$$\prod_{n<\omega} \sum_{\xi<\omega_\alpha} b_{n\xi} = \sum_{\eta<\omega_\alpha} \prod_{n \in \omega} \sum_{\xi \leq \eta} b_{n\xi}.$$

Theorem 20.3. If $cf(\omega_\alpha) > \omega$, then $\mathbf{B}$ satisfies (ω, ω_α)-WDL iff

$$[\![cf(\omega_\alpha)^\vee > \check{\omega}]\!] = \mathbf{1}.$$

Proof. Assume that $\mathbf{B}$ satisfies the (ω, ω_α)-WDL. Let $f \in V^{(\mathbf{B})}$ and $b = [\![f: \check{\omega} \to (\omega_\alpha)^\vee]\!]$, i.e.,

$$\begin{aligned} b &= [\![(\forall x \in \check{\omega})(\exists y \in (\omega_\alpha)^\vee)(\forall z)[\langle x, z\rangle \in f \leftrightarrow z = y]]\!] \\ &= \prod_{n \in \omega} \sum_{\xi<\omega_\alpha} [\![(\forall z)[\langle \check{n}, z\rangle \in f \leftrightarrow z = \check{\xi}]]\!]. \end{aligned}$$

Define $b_{n\xi} = [\![f(\check{n}) = \check{\xi}]\!]$, which should be understood as

$$[\![(\forall z)[\langle \check{n}, z\rangle \in f \leftrightarrow z = \check{\xi}]]\!].$$

Then

$$\begin{aligned} b &= \prod_{n<\omega} \sum_{\xi<\omega_\alpha} b_{n\xi} \\ &= \sum_{\eta<\omega_\alpha} \prod_{n<\omega} \sum_{\xi\leq\eta} b_{n\xi} \qquad \text{by the } (\omega, \omega_\alpha)\text{-WDL} \\ &= [\![(\exists \eta < \omega_\alpha)^\vee (\forall n < \omega)[f(n) \leq \eta]]\!]. \end{aligned}$$

Since

$$[\![cf((\omega_\alpha)^\vee) > \check{\omega}]\!] = [\![(\forall f)[\text{if } f\colon \check{\omega} \to (\omega_\alpha)^\vee \text{ then } (\exists \eta < (\omega_\alpha)^\vee)(\forall n < \check{\omega})[f(n) \leq \eta]]]\!],$$

this proves $[\![cf((\omega_\alpha)^\vee) > \check{\omega}]\!] = \mathbf{1}$.

To prove the converse, let $\{b_{n\xi} \mid n < \omega \wedge \xi < \omega_\alpha\} \subseteq B$ and assume $[\![cf((\omega_\alpha)^\vee) > \check{\omega}]\!] = \mathbf{1}$. Define

$$f \in V^{(\mathbf{B})} \quad \text{by} \quad \mathscr{D}(f) = \{\langle \check{n}, \check{\xi}\rangle^{(\mathbf{B})} \mid n \in \omega \wedge \xi < \omega_\alpha\},$$
$$(\forall n \in \omega)(\forall \xi < \omega_\alpha)[f(\langle \check{n}, \check{\xi}\rangle^{(\mathbf{B})}) = b_{n\xi}].$$

Then again

i) $[\![f\colon \check{\omega} \to (\omega_\alpha)^\vee]\!] = \prod_{n<\omega} \sum_{\xi<\omega_\alpha} b_{n\xi}$

and

ii) $[\![f\colon \check{\omega} \to (\omega_\alpha)^\vee]\!] \cdot [\![(\exists \eta < (\omega_\alpha)^\vee)(\forall n < \check{\omega})[f(n) \leq \eta]]\!] = \sum_{\eta<\omega_\alpha} \prod_{n<\omega} \sum_{\xi\leq\eta} b_{n\xi}$.

But, since $[\![cf((\omega_\alpha)^\vee) > \check{\omega}]\!] = \mathbf{1}$.

$$[\![f\colon \check{\omega} \to (\omega_\alpha)^\vee]\!] \leq [\![(\exists \eta < (\omega_\alpha)^\vee)(\forall n < \check{\omega})[f(n) \leq \eta]]\!].$$

Therefore, by i) and ii),

$$\prod_{n<\omega} \sum_{\xi<\omega_\alpha} b_{n\xi} = \sum_{\eta<\omega_\alpha} \prod_{n<\omega} \sum_{\xi\leq\eta} b_{n\xi}.$$

Remark. Next we interpret the (ω, ω)-WDL:

Theorem 20.4. $\mathbf{B}$ satisfies the (ω, ω)-WDL iff

$$[\![(\forall g)[\text{if } g\colon \check{\omega} \to \check{\omega} \text{ then } (\exists f \in (\omega^\omega)^\vee)(\forall n \in \omega)[g(n) \leq f(n)]]]\!] = \mathbf{1},$$

i.e., if we define a partial ordering $\prec$ for the number theoretic functions by $f \prec g \leftrightarrow (\forall n < \omega)[f(n) \leq g(n)]$ for $f, g \in \omega^\omega$, then in $\mathbf{V}^{(\mathbf{B})}$, the standard number theoretic functions (elements of $(\omega^\omega)^\vee$) are cofinal in the set of all number theoretic functions.

Proof. Assume that $\mathbf{B}$ satisfies the (ω, ω)-WDL. Let $g \in V^{(\mathbf{B})}$ and define

$$\begin{aligned} b &= [\![g\colon \check{\omega} \to \check{\omega}]\!] \\ b_{nm} &= [\![g(\check{n}) = \check{m}]\!] \end{aligned}$$

for $n, m < \omega$ as in the previous proof.

Then

$$b = \prod_{n<\omega} \sum_{m<\omega} b_{nm}$$

$$= \sum_{f\in\omega^\omega} \prod_{n<\omega} \sum_{m\leq f(n)} b_{nm} \qquad \text{by the } (\omega, \omega)\text{-WDL}$$

$$= \sum_{f\in\omega^\omega} \prod_{n<\omega} [\![g(\check{n}) \leq \check{f}(\check{n})]\!]$$

$$b = [\![(\exists f \in (\omega^\omega)^\vee)(\forall n < \check{\omega})[g(n) \leq f(n)]]\!].$$

The converse is proved similarly.

Definition 20.5. A Boolean σ-algebra $\mathbf{B}$ is a *measure algebra* iff there exists a strictly positive σ-measure m on $\mathbf{B}$, i.e., a function from $|\mathbf{B}|$ into $[0, 1]$, the closed interval of real numbers between 0 and 1, such that

$$(\forall b \in B)[b \neq \mathbf{0} \rightarrow m(b) > 0] \wedge m(\mathbf{1}) = 1$$

and

$$(\forall b \in B^\omega)\left[(\forall i, j < \omega)[i \neq j \rightarrow b_i \cdot b_j = \mathbf{0}] \rightarrow m\left(\sum_{i<\omega} b_i\right) = \sum_{i<\omega} m(b_i)\right].$$

Remark. Note that a measure algebra always satisfies the c.c.c. and hence it is complete.

Theorem 20.6. Every measure algebra $\mathbf{B}$ satisfies the (ω, ω)-WDL.

Proof. Let $\{b_{n,k} \mid n, k < \omega\} \subseteq B$. Then for every real $\varepsilon > 0$

$$(\forall n < \omega)(\exists l \in \omega)\left[m\left(\sum_{k<\omega} b_{nk} - \sum_{k\leq l} b_{nk}\right) < \varepsilon/2^n\right].$$

Therefore

$$(\forall \varepsilon > 0)(\exists f \in \omega^\omega)(\forall n < \omega)\left[m\left(\sum_{k<\omega} b_{nk} - \sum_{k\leq f(n)} b_{nk}\right) < \varepsilon/2^n\right].$$

Since

$$\prod_{i<\omega} b_i - \prod_{i<\omega} c_i \leq \sum_{i<\omega} (b_i - c_i),$$

$$(\forall \varepsilon > 0)(\exists f \in \omega^\omega)\left[m\left(\prod_{n<\omega} \sum_{k<\omega} b_{nk} - \prod_{n<\omega} \sum_{k\leq f(n)} b_{nk}\right) < 2\varepsilon\right].$$

Therefore

$$\prod_{n<\omega} \sum_{k<\omega} b_{nk} = \sum_{f\in\omega^\omega} \prod_{n<\omega} \sum_{k\leq f(n)} b_{nk}.$$

Theorem 20.7. The Boolean algebra of all regular open sets in ω^ω does not satisfy the (ω, ω)-WDL.

Proof. Define $b_{nm} = \{p \in \omega^\omega \mid p(n) = m\}$ for $n, m < \omega$. Then b_{nm} is clopen and therefore it is regular open. Obviously,

$$\prod_{n<\omega} \sum_{m<\omega} b_{nm} = \mathbf{1},$$

but

$$\sum_{f\in\omega^\omega} \prod_{n<\omega} \sum_{m\leq f(n)} b_{mn} = \mathbf{0}.$$

for otherwise there exists some $f \in \omega^\omega$ such that

$$\mathbf{0} \neq \prod_{n<\omega} \sum_{m\leq f(n)} b_{nm} = \prod_{n<\omega} \{p \in \omega^\omega \mid p(n) \leq f(n)\}^{-0}$$

$$= \prod_{n<\omega} \{p \in \omega^\omega \mid p(n) \leq f(n)\}.$$

Then there exist $n_1, \ldots, n_i, l_1, \ldots, l_i$ such that

i) $\{p \in \omega^\omega \mid p(n_1) = l_1 \wedge \cdots \wedge p(n_i) = l_i\}$

$$\subseteq \left(\bigcap_{n<\omega} \{p \in \omega^\omega \mid p(n) \leq f(n)\}\right)^{-}.$$

Choose some $n_0 \notin \{n_1, \ldots, n_i\}$ and $l_0 > f(n_0)$. Then by i),

$$\{p \in \omega^\omega \mid p(n_0) = l_0 \wedge \cdots \wedge p(n_i) = l_i\} \cap \bigcap_{n<\omega} \{p \in \omega^\omega \mid p(n) \leq f(n)\} \neq 0,$$

Since this intersection is empty we have a contradiction.

21. A Proof of Marczewski's Theorem

Definition 21.1. A set a is *quasi-disjoint* iff

$$(\forall x, y \in a)[x \neq y \rightarrow x \cap y = \bigcap a]$$

Remark. From Definition 21.1 a set a is quasi-disjoint iff $x \notin \bigcap(a)$ implies that x is in at most one $y \in a$.

Theorem 21.2. (Erdös-Rado) Let α, γ be cardinals with $\gamma \geq \aleph_0$. Then $(\forall x)[x \in A \rightarrow \bar{\bar{x}} \leq \alpha] \wedge (\forall x \subseteq A)[x \text{ is quasi-disjoint} \rightarrow \bar{\bar{x}} \leq \gamma] \rightarrow \bar{\bar{A}} \leq \gamma^{\alpha}$.

Proof. For each $\delta < \alpha^+$ we construct a set $A_\delta \subseteq A$ such that

a. $(\forall \delta < \alpha^+)[\bar{\bar{A}}_\delta \leq \gamma^\alpha]$ and

b. $A = \bigcup_{\delta < \alpha^+} A_\delta$.

This proves the theorem since $\bar{\bar{A}} \leq 2^\alpha \cdot \gamma^\alpha = \gamma^\alpha$. We define A_δ by recursion: For convenience we start with $A_{-1} = 0$. Suppose A_δ for $\delta < \beta$ (where $\beta < \alpha^+$) has already been defined. Let

$$E_\beta = \bigcup_{\delta < \beta} \left(\bigcup A_\delta \right).$$

For each $K \subseteq E_\beta$ let

$$\bar{K} = \{x \in A \mid x \cap E_\beta = K\}$$

and let K^* be a maximal quasi-disjoint subset of $\bar{K}$. Moreover, we require that if

$$(\exists S, T \in \bar{K})[S \cap T = K]$$

then K^* contains all such S and T. In this case, $\bigcap K^* = K$, since K^* is quasi-disjoint. Finally,

$$A_\beta = \bigcup \{K^* \mid K \subseteq E_\beta\}.$$

Claim: $\{A_\delta \mid \delta < \alpha^+\}$ satisfies conditions a and b.

Clearly, $A_\delta \subseteq A$. We prove a by induction on $\beta < \alpha^+$: Suppose $\beta < \alpha^+$ and $(\forall \delta < \beta)[\bar{\bar{A}}_\delta \leq \gamma^\alpha]$. Then $\bar{\bar{E}}_\beta \leq \alpha \cdot \gamma^\alpha = \gamma^\alpha$ since $(\forall x \in A_\delta)[\bar{\bar{x}} \leq \alpha]$. For $K \subseteq E_\beta$, $\bar{\bar{K}}^* \leq \gamma$ by assumption. Since $K^* \neq 0 \rightarrow \bar{K} \neq 0 \wedge \bar{\bar{K}} \leq \alpha$,

$$A_\beta = \bigcup \{K^* \mid K \subseteq E_\beta \wedge \bar{\bar{K}} \leq \alpha\}.$$

Since E_β has at most $(\gamma^\alpha)^\alpha = \gamma^\alpha$ subsets of cardinality $\leq \alpha$, $\overline{\overline{A}}_\beta \leq \gamma^\alpha \cdot \gamma = \gamma^\alpha$

To prove b, suppose that

$$A - \bigcup_{\delta < \alpha^+} A_\delta \neq 0.$$

Let $S \in A - \bigcup_{\delta < \alpha^+} A_\delta$, and for $\delta < \alpha^+$, let $K_\delta = S \cap E_\delta$. Since $S \in \overline{K}_\delta, K_\delta^* \neq 0$.

Claim: $(\forall \delta < \alpha^+)(\exists T \in K_\delta^*)[S \cap (T - E_\delta) \neq 0]$.

Suppose $(\forall T \in K_\delta^*)[S \cap (T - E_\delta) = 0]$ for some $\delta < \alpha^+$. Then

$$(\forall T \in K_\delta^*)[S \cap T \subseteq S \cap E_\delta = K_\delta = T \cap E_\delta]$$

since $T \in \overline{K}_\delta$. Therefore

$$(\forall T \in K_\delta^*)[S \cap T = K_\delta].$$

Since $K_\delta^* \neq 0$, by our requirement on K_δ^*

$$\bigcap K_\delta^* = K_\delta.$$

Therefore $K_\delta^* \cup \{S\}$ is quasi-disjoint. Since K_δ^* is maximal, $S \in K_\delta^* \subset A_\delta$. But this contradicts our assumption that $S \in A - \bigcup_{\delta < \alpha^+} A_\delta$. Therefore we can choose x_δ, T_δ such that

$$(\forall \delta < \alpha^+)[T_\delta \in K_\delta^* \wedge x_\delta \in S \cap (T_\delta - E_\delta)].$$

$$\delta < \beta < \alpha^+ \rightarrow x_\delta \in T_\delta \subseteq \bigcup A_\delta \subseteq E_\beta \wedge x_\beta \notin E_\beta.$$

So $\{x_\delta \mid \delta < \alpha^+\} \subseteq S$ and $\{x_\delta \mid \delta < \alpha^+\}$ has cardinality α^+. This is a contradiction, since $S \in A$ and hence $\overline{\overline{S}} \leq \alpha$.

Remark. Engelking and Karłowicz used this result to prove the following theorems:

Theorem 21.3. Let α, γ be cardinals with $\gamma \geq \aleph_0$. Suppose that $A = \{A_t \mid t \in T\}$ and $\{B_t \mid t \in T\}$ satisfy the following conditions:

1. $(\forall t \in T)[\overline{\overline{A}}_t \leq \alpha \wedge \overline{\overline{B}}_t \leq \gamma]$.
2. $(\forall t, t' \in T)[t \neq t' \rightarrow A_t \cap B_{t'} \neq 0 \wedge A_t \cap B_t = 0]$.

Then $\overline{\overline{T}} \leq \gamma^\alpha$.

Proof. Note that $\overline{\overline{A}} = \overline{\overline{T}}$ by 2. We will show that A satisfies the conditions of Theorem 21.2. Let $\{A_t \mid t \in T_0\}$ be quasi-disjoint, $0 \neq T_0 \subseteq T$. Choose $t_0 \in T_0$ and define

$$C_t = A_t \cap B_{t_0} \quad \text{for} \quad t \in T_0 - \{t_0\}.$$

Then we claim that

i) $(\forall t \in T_0 - \{t_0\})[C_t \neq 0]$, and
ii) $(\forall t, t' \in T_0 - \{t_0\})[t \neq t' \rightarrow C_t \cap C_{t'} = 0]$.

Claim i) follows from 2. Claim ii) we prove in the following way.

Suppose $x \in C_t \cap C_{t'}$ for some $t, t' \in T_0 - \{t_0\}$ with $t \neq t'$. Then

$$x \in A_t \cap A_{t'} = \bigcap_{t'' \in T_0} A_{t''}$$

since $\{A_{t''} \mid t'' \in T_0\}$ is quasi-disjoint

$$x \in B_{t_0} \cap \bigcap_{t'' \in T_0} A_{t''} \subseteq B_{t_0} \cap A_{t_0} = 0.$$

This is a contradiction.

Therefore, for each $t \in T_0 - \{t_0\}$ we can pick $x_t \in C_t \subseteq B_{t_0}$ such that $t, t' \in T_0 - \{t_0\}$ and $t \neq t' \rightarrow x_t \neq x_{t'}$. Hence

$$\overline{\overline{T}}_0 \leq \overline{\overline{B}}_{t_0} \leq \gamma.$$

Corollary 21.4. Let γ be a cardinal with $\gamma \geq \aleph_0$. If

1. $(\forall t \in T)[\overline{\overline{A}}_t < \omega \wedge \overline{\overline{B}}_t \leq \gamma]$.
2. $(\forall t, t' \in T)[t \neq t' \rightarrow A_t \cap B_{t'} \neq 0 \wedge A_t \cap B_t = 0]$,

then $\overline{\overline{T}} \leq \gamma$.

Proof. Let $A^{(n)} = \{A_t \mid t \in T \wedge \overline{\overline{A}}_t = n\}$ for $n \in \omega$. By Theorem 21.3 (for $\alpha = n$), $\overline{\overline{A^{(n)}}} = \gamma^n = \gamma$.

Since $A = \{A_t \mid t \in T\} = \bigcup_{n < \omega} A^{(n)}$, $\overline{\overline{A}} = T \leq \gamma$.

Theorem 21.5. (Marczewski) Let I be a set and $\{X_i \mid i \in I\}$ be a family of topological spaces such that each X_i has a base b_i of cardinality $\leq \gamma$. Let $X = \prod_{i \in I} X_i$ be the product space. If $\{O^{(t)} \mid t \in T\}$ is a family of pairwise disjoint open sets of X, then $\overline{\overline{T}} \leq \gamma$.

Proof. We can assume that $X_i \cap X_{i'} = 0$ for $i \neq i'$ and $X_i \in b_i$ for $i, i' \in I$. Let

$$p_j\colon \prod_{i \in I} X_i \rightarrow X_j, \qquad j \in J,$$

be the canonical projection,

$$O_j^{(t)} = p_j``O^{(t)}$$

be the jth-component of $O^{(t)}$.

Since $O^{(t)}$ is open, $\{i \in I \mid O_i^{(t)} \neq X_i\}$ is finite and $O_i^{(t)}$ is open in X_i for each $i \in I$. We can assume that

$$(\forall i \in I)(\forall t \in T)[O_i^{(t)} \neq 0]$$

and

$$(\forall i \in I)(\forall t \in T)[O_i^{(t)} \in b_i],$$

since each $O_i^{(t)}$ contains a basic open set. In order to apply the previous corollary, define

$$A_t = \{O_i^{(t)} \mid i \in I \wedge O_i^{(t)} \neq X_i\}$$

$$B_t = \bigcup_{\substack{i \in I \\ O_i^{(t)} \neq X_i}} \{b \mid b \in b_i \wedge b \cap O_i^{(t)} = 0\}, \qquad t \in T.$$

Then $(\forall t \in T)[\overline{\overline{A}}_t < \omega \wedge \overline{\overline{B}}_t \leq \gamma]$ since $\{i \in I \mid O_i^{(t)} \neq X\}$ is finite and $(\forall i \in I)[\overline{\overline{b}}_i \leq \gamma]$.

1. $(\forall t \in T)[A_t \cap B_t = 0]$.

Suppose not. Then for some $t \in T$, and $i \in I$,

$$O_i^{(t)} \in A_t \cap B_t,$$

i.e., $O_i^{(t)} \in b_j$ and $O_i^{(t)} \cap O_j^{(t)} = 0$ for some $j \in I$. Since $X_i \cap X_j = 0$ for $i \neq j$, we must have $i = j$, but then

$$O_i^{(t)} \cap O_j^{(t)} = 0.$$

This is a contradiction.

2. $t, t' \in T \wedge t \neq t' \to A_t \cap B_{t'} \neq 0$.

Let $t, t' \in T$, $t \neq t'$. Since $O^{(t)} \cap O^{(t')} = 0$,

$$(\exists i \in I)[O_i^{(t)} \cap O_i^{(t')} = 0].$$

Then $O_i^{(t)} \in A_t \cap B_{t'}$ and hence we have 2. Therefore Corollary 21.4 applies, and we have $\overline{\overline{T}} \leq \gamma$.

Corollary 21.6. If $\{X_i \mid i \in I\}$ is a family of topological spaces (I a set) and each X_i has a base of cardinality $\leq \gamma$, where γ is an infinite cardinal, then the Boolean algebra $\mathbf{B}$ of all regular open sets of the product space $\prod_{i \in I} X_i$ satisfies the γ-chain condition. In particular, if each X_i is 2nd countable (i.e., X_i has a countable base) then $\mathbf{B}$ satisfies the c.c.c.

22. The Completion of a Boolean Algebra

For the following let X be a topological space, and let $S \subseteq X$ be a subspace of X with the relative topology. The topological operations $^{-}$, ^{0} and $^{-S}$, ^{0S} refer to X and S respectively.

Theorem 22.1.

1. $A \subseteq S \to A^{-S} = A^{-} \cap S$.
2. $A \subseteq S \to A^{0S} = S - (S - A)^{-}$.

Proof. 1. Let $A \subseteq S$; then obviously

$$A^{-S} \subseteq A^{-} \cap S.$$

Conversely, let $p \in A^{-} \cap S$. Then

$$p \in S \wedge (\forall N(p))[N(p) \cap A \neq 0]$$

$$(\forall N(p))[N(p) \cap S \cap A \neq 0] \qquad \text{since } A \subseteq S$$

$$(\forall N^{S}(p))[N^{S}(p) \cap A \neq 0]$$

i.e.,

$$p \in A^{-S}.$$

2. Follows from 1.

Theorem 22.2. If $A \subseteq S$, if S is dense in X, and if A is regular open in S then $A = A^{-0} \cap S$.

Proof. Let $A \subseteq S$ be regular open in S. By Theorem 22.1.1, $A^{-S} = A^{-} \cap S$. Then, since $A^{-0} \cap S$ is open in S,

$$A^{-0} \cap S \subseteq (A^{-S})^{0S} = A.$$

On the other hand, if $p \in A$, then since A is open in S, there exists a $N(p)$ such that $N(p) \cap S \subseteq A$. Since S is dense in X, $N(p) \subseteq (N(p) \cap S)^{-} \subseteq A^{-}$. Thus $p \in A^{-0} \cap S$.

Theorem 22.3. Let S be dense in X and let A, B be regular open (in X). Then $A \cap S \subseteq B \cap S \to A \subseteq B$.

Proof. $A = A \cap S^{-} \subseteq (A \cap S)^{-} \subseteq (B \cap S)^{-} \subseteq B^{-}$

$$A \subseteq B^{-0} \qquad \text{since } A \text{ is open}$$

$$A \subseteq B \qquad \text{since } B^{-0} = B.$$

Theorem 22.4. If S is dense in X and A is regular open (in X) then $A \cap S$ is regular open in S.

Proof. We need only show that $(A \cap S)^{-SOS} \subseteq A^{-0} \cap S$ for by the proof of Theorem 22.2 we know that the reverse inclusion holds. Let

$$p \in (A \cap S)^{-SOS},$$

then

$$(\exists N(p))[N(p) \cap S \subseteq (A \cap S)^{-} \cap S] \qquad \text{by Theorem 22.1}$$
$$N(p) = N(p) \cap S^{-} \subseteq (N(p) \cap S)^{-} \subseteq (A \cap S)^{-} \subseteq A^{-}.$$

Thus

$$p \in A^{-0} \cap S.$$

Remark. As a consequence of Theorems 22.2–22.4 we have the following.

Theorem 22.5. Let S be dense in X and let $\mathbf{B}$ and $\mathbf{B}_0$ be the complete Boolean algebras of all regular open sets in X and S respectively. Then $\mathbf{B}_0$ and $\mathbf{B}$ are isomorphic. An isomorphism $i\colon \mathbf{B}_0 \to \mathbf{B}$ is given by

$$b_0 = i(b_0) \cap S \quad \text{for} \quad b_0 \in B_0.$$

Proof. For $b_0 \in B_0$, define $i(b_0) = b_0{}^{-0}$. Then $i\colon B_0 \to B$ and $b_0 = i(b_0) \cap S$, by Theorem 22.2. Let $b \in B$. Since $b \cap S$ is regular open in S (by Theorem 22.4), $i(b \cap S) = b$. Therefore i is onto, is one-to-one, by Theorem 22.2, and by Theorem 22.3 it preserves $\leq$.

Definition 22.6. Let $\mathbf{B}_0$ be a Boolean algebra (which need not be complete). A completion of $\mathbf{B}_0$ is a pair $\langle \mathbf{B}, h\rangle$ such that:

1. $\mathbf{B}$ is a complete Boolean algebra,
2. $h\colon \mathrm{B}_0 \to B$ is a monomorphism (i.e., one-to-one),
3. if $\sum_{\alpha\in A} b_\alpha = b$ in $\mathbf{B}_0$, then $\sum_{\alpha\in A} h(b_\alpha) = h(b)$ in B,
4. $h``(B_0 - \{\mathbf{0}\})$ is dense in $B - \{\mathbf{0}\}$.

Remark. Our next result shows that every Boolean algebra has a completion which is unique in a certain sense.

Theorem 22.7. Let $\mathbf{B}$ be a Boolean algebra (not necessarily complete) and let $\mathbf{P} = \langle P, \leq\rangle$ be the partial order structure determined by $\mathbf{B}$, i.e., $P = B - \{\mathbf{0}\}$ and $\leq$ is $\leq$ in $\mathbf{B}$. Let $\tilde{\mathbf{B}}$ be the Boolean algebra of all regular open sets in $\mathbf{P}$ and let $j\colon B \to \tilde{B}$ be defined by

$$j(\mathbf{0}) = \mathbf{0} \wedge (\forall p \in P)[j(p) = [p]].$$

(Since $\mathbf{P}$ is fine we have by Lemma 5.22 that $[p] = [p]^{-0}$.)

Then $\langle \tilde{\mathbf{B}}, j\rangle$ is a completion of $\mathbf{B}$. Moreover, if $\langle \tilde{\mathbf{B}}_1, f\rangle$ is any completion of $\mathbf{B}$, then there exists an isomorphism

$$k\colon \tilde{\mathbf{B}} \xrightarrow{\text{onto}} \tilde{\mathbf{B}}_1$$

such that the diagram

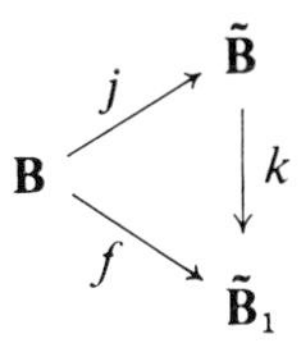

commutes, i.e., $k \circ j = f$.

Proof. $j``P$ is dense in $\tilde{\mathbf{B}}$ because $\{[p] \mid p \in P\}$ is a base for open sets in $\mathbf{P}$. Repeating the proof of Theorem 1.30 we see that $\langle \tilde{\mathbf{B}}, j\rangle$ is a completion of $\mathbf{B}$. If $\langle \tilde{\mathbf{B}}_1, f\rangle$ is any completion of $\mathbf{B}$, $S = f``(B - \{\mathbf{0}\})$ is dense in $\tilde{B}_1 - \{\mathbf{0}\}$. Thus by Theorem 22.5 the Boolean algebra of regular open sets of S is isomorphic to the complete Boolean algebra of all regular open sets in $\tilde{\mathbf{B}}_1 - \{\mathbf{0}\}$ which is isomorphic to $\tilde{\mathbf{B}}_1$ by Theorem 1.40, and hence also to $\tilde{\mathbf{B}}$. This gives an isomorphism $k: \tilde{\mathbf{B}} \to \tilde{\mathbf{B}}_1$ as required by the theorem.

Remark. By "the" completion of a Boolean algebra $\mathbf{B}$ we will mean the Boolean algebra $\tilde{\mathbf{B}}$ defined in Theorem 22.7. We will regard $\mathbf{B}$ as a subalgebra of $\tilde{\mathbf{B}}$ by identifying B and $j``B$.

Theorem 22.8. Let X, Y be topological spaces and let $f: X \to Y$ be an open continuous map onto Y. Then for $B \subseteq Y$ we have,

1. $(f^{-1})``(B^-) = ((f^{-1})``B))^-$.
2. $(f^{-1})``(B^0) = ((f^{-1})``B))^0$.

Proof. Let $x \in (f^{-1})``(B^-)$. Then $f(x) \in B^-$ and hence

$$(\forall N(f(x)))[N(f(x)) \cap B \neq 0]$$
$$(\forall N(x))[f``N(x) \cap B \neq 0] \qquad \text{since } f \text{ is open}$$
$$(\forall N(x))[N(x) \cap (f^{-1})``B \neq 0].$$

Thus

$$x \in ((f^{-1})``B)^-.$$

On the other hand, $((f^{-1})``B)^- \subseteq (f^{-1})``(B^-)$ since $(f^{-1})``(B^-)$ is closed. This proves 1.

2. Follows from 1 since f is onto.

Remark. From Theorem 22.8 the next result follows easily.

Theorem 22.9. Let X, Y be topological spaces and let $f: X \to Y$ be an open continuous map onto Y. Then f induces a complete monomorphism $i: \mathbf{B}_Y \to \mathbf{B}_X$ such that

$$(\forall b \in B_Y)[i(b) = (f^{-1})``b],$$

where $\mathbf{B}_X$ and $\mathbf{B}_Y$ are the complete Boolean algebras of all regular open sets of X and Y respectively.

Remark. Let $\mathbf{B}_0$, $\mathbf{B}_1$ be complete Boolean algebras and let $i: \mathbf{B}_0 \to \mathbf{B}_1$ be a complete monomorphism. Define

$$\mathbf{P}_0 = \langle P_0, \leq \rangle \quad \text{and} \quad \mathbf{P}_1 = \langle P_1, \leq \rangle$$

by $P_0 = B_0 - \{\mathbf{0}\}$, $P_1 = B_1 - \{\mathbf{0}\}$. We would like to define an open continuous mapping

$$f: \mathbf{P}_1 \xrightarrow{\text{onto}} \mathbf{P}_0$$

such that the associated complete monomorphism from $\mathbf{B}_0$ into $\mathbf{B}_1$ is i. For this purpose, define $\#(b_1) = \inf \{b_0 \mid b_1 \leq i(b_0)\}$ in $\mathbf{B}_0$.

(We use b_0, b_1 as variables ranging over B_0 and B_1 respectively.) Under these assumptions we can prove the following.

Theorem 22.10.

1. $b_1 \leq i(\#(b_1))$.
2. $b_1 = \mathbf{0} \to \#(b_1) = \mathbf{0}$.
3. $\#(i(b_0)) = b_0$.
4. $b_1 \leq i(b_0) \leftrightarrow \#(b_1) \leq b_0$.
5. $b_0 \cdot \#(b_1) = \#(i(b_0) \cdot b_1)$.
6. $i(b_0) \cdot b_1 = \mathbf{0} \to b_0 \cdot \#(b_1) = \mathbf{0}$.

Proof. 1–4 follow from the definition of $\#$. (Note that $\mathbf{B}_0$ and $\mathbf{B}_1$ are completed and so is i.)

5. $i(b_0) \cdot b_1 \leq i(b_0 \cdot \#(b_1))$ by 1.
$\#(i(b_0) \cdot b_1) \leq b_0 \cdot \#(b_1)$ by 4.

Suppose $\#(i(b_0) \cdot b_1) < b_0 \cdot \#(b_1)$, then

$$\begin{aligned} b_1 &= (i(b_0) \cdot b_1) + (b_1 \cdot (^- i(b_0))) \\ &\leq i(\#(i(b_0) \cdot b_1) + \#(b_1 \cdot (^- i(b_0)))) \qquad \text{by 1.} \end{aligned}$$

Since $b_1 \leq i(\#(b_1))$,

$$\begin{aligned} i((^- b_0) \cdot \#(b_1)) = i(^- b_0) \cdot i(\#(b_1)) &\geq i(^- b_0) \cdot b_1 \qquad \text{by 3,} \\ (^- b_0) \cdot \#(b_1) &\geq \#(^- i(b_0) \cdot b_1) \qquad \text{by 4,} \end{aligned}$$

hence

$$\begin{aligned} &b_1 \leq i(\#(i(b_0) \cdot b_1) + ((^- b_0) \cdot \#(b_1))) \\ &b_1 < i((b_0 \cdot \#(b_1)) + (^-(b_0) \cdot \#(b_1))) \qquad \text{by assumption} \\ &b_1 < i(\#(b_1)). \end{aligned}$$

Thus we have

$$b_1 \leq i(b_0') < i(\#(b_1))$$

where $b_0' = \#(i(b_0) \cdot b_1) + \#(b_1 \cdot (^- i(b_0)))$ which contradicts the definition of $\#$.

6. Follows from 5.

Theorem 22.11

1. $(\#^{-1})``[b_0] = [i(b_0)]$.
2. $\#$ is an open continuous map from $\mathbf{P}_1$ onto $\mathbf{P}_0$.
3. The complete monomorphism from $\mathbf{B}_0$ into $\mathbf{B}_1$ associated with $\#$ is i.

Proof.

$$
\begin{aligned}
1.\ b_1 \in (\#^{-1})``[b_0] &\leftrightarrow \#(b_1) \le b_0 \\
&\leftrightarrow b_1 \le i(b_0) \\
&\leftrightarrow b_1 \in [i(b_0)].
\end{aligned}
$$

2. $\#$ is continuous because of 1. We will show that $\#``[b_1]$ is open for every $b_1 \in B_1$.

Let $b_0 \in \#``[b_1]$. If $b_0' \le b_0$, then since $\#$ is order preserving $b_0' \le \#(b_1)$ and hence

$$
\begin{aligned}
b_0' = b_0' \cdot \#(b_1) &= \#(i(b_0') \cdot b_1) \qquad \text{by 5 of Theorem 22.10} \\
&\in \#``[b_1],
\end{aligned}
$$

hence $[b_0] \subseteq \#``[b_1]$.

$\#$ is onto by Theorem 22.10.3.

3. Obvious from 1.

Remark. Next we prove that $\#$ is uniquely determined by the properties 2 and 3 of Theorem 22.11.

Theorem 22.12. If

$$f\colon \mathbf{P}_1 \xrightarrow{\text{onto}} \mathbf{P}_0$$

is open and continuous and induces i, then

$$(\forall b_1 \in P_1)[f(b_1) = \#(b_1)],$$

i.e., $f = \#$.

Proof. There are two complete monomorphisms:

$$j\colon \mathbf{B}_{\mathbf{P}_0} \to \mathbf{B}_{\mathbf{P}_1}$$

induced by f via Theorem 22.9, and

$$i\colon \mathbf{B}_0 \to \mathbf{B}_1.$$

These monomorphisms are related to each other by

$$[i(b_0)] = j([b_0])$$

via the isomorphisms $\mathbf{B}_0 \leftrightarrow \mathbf{B}_{\mathbf{P}_0}$, $\mathbf{B}_1 \leftrightarrow \mathbf{B}_{\mathbf{P}_1}$ which are given by $b \leftrightarrow [b]$. Since f induces j in the sense of Theorem 22.9, we have, using Theorem 22.11.1

$$
\begin{aligned}
(f^{-1})``[b_0] &= j([b_0]) \\
&= [i(b_0)] = (\#^{-1})``(b_0).
\end{aligned}
$$

Therefore, for all $b_0 \in B_0$ and $b_1 \in B_1$,

$$\begin{aligned} \#(b_1) \le b_0 &\leftrightarrow b_1 \le i(b_0) \\ &\leftrightarrow f(b_1) \le b_0 \end{aligned}$$

which gives $\#(b_1) = f(b_1)$.

Theorem 22.13. Let $\mathbf{B}_0$, $\mathbf{B}_1$, $\mathbf{B}_2$ be complete Boolean algebras, let

$$\begin{aligned} i_1 &: \mathbf{B}_0 \to \mathbf{B}_1 \\ i_2 &: \mathbf{B}_1 \to \mathbf{B}_2 \end{aligned}$$

be complete monomorphisms and let $\#_j$ be the open continuous mapping associated with i_j ($j = 1, 2$), i.e.,

$$\mathbf{B}_0 \underset{\#_1}{\overset{i_1}{\rightleftarrows}} \mathbf{B}_1 \underset{\#_2}{\overset{i_2}{\rightleftarrows}} \mathbf{B}_2.$$

If $i = i_2 \circ i_1$ and $\# = \#_1 \circ \#_2$, then $\#$ induces i.

Proof. $\#: \mathbf{B}_2 \to \mathbf{B}_0$ is open, continuous and onto $\mathbf{B}_0$.

$$\#^{-1} = (\#_1 \circ \#_2)^{-1} = \#_2^{-1} \circ \#_1^{-1},$$

hence

$$(\#^{-1})``b_0 = (i_2 \circ i_1)(b_0) = i(b_0).$$

Remark. We are mostly interested in the case where $\mathbf{B}_0$ is a complete subalgebra of $\mathbf{B}_1$ and i is the identity on $\mathbf{B}_0$. Suppose that there is a big complete Boolean algebra $\mathbf{B}$ such that all the complete Boolean algebras under consideration are complete subalgebras of $\mathbf{B}$. Thus, if $\mathbf{B}_0$, $\mathbf{B}_1$ are complete subalgebras of $\mathbf{B}$ and $\mathbf{B}_0 \subseteq \mathbf{B}_1$, we denote the map $\mathbf{B}_1 \to \mathbf{B}_0$ associated with $i: \mathbf{B}_0 \to \mathbf{B}_1$, where i is the identity map on B_0, by $\#(\mathbf{B}_0, \mathbf{B}_1)$. Then by the definition of $\#$

$$b \in B_1 \to \#(\mathbf{B}_0, \mathbf{B}_1)(b) = \#(\mathbf{B}_0, \mathbf{B})(b).$$

Therefore we can simply write $\#(\mathbf{B}_0)$ for $\#(\mathbf{B}_0, \mathbf{B}_1)$.

Definition 22.14. Let κ be a cardinal or On.

$$\langle \{X^\alpha \mid \alpha < \kappa\}, \{p_{\alpha\beta} \mid \alpha \le \beta < \kappa\} \rangle$$

is called an open, continuous and onto (o.c.o.) inverse system of topological spaces iff

1. X^α is a topological space.
2. $p_{\alpha\beta}: X^\beta \to X^\alpha$ is an o.c.o. map.
3. $p_{\alpha\alpha}$ is an identity map.
4. $p_{\alpha\beta} \cdot p_{\beta\gamma} = p_{\alpha\gamma}$.

For an o.c.o inverse system, we define a topological space $X = \lim_{\alpha \to \kappa} X^\alpha$ in the following way.

Let

$$X = \{f \in \prod_{\alpha \in \kappa} X^\alpha \mid (\forall \alpha < \beta < \kappa)[p_{\alpha\beta}(f(\beta)) = f(\alpha)]\}.$$

The topology on X is defined by the following open base.

$$T = \{\tilde{G}_\alpha \mid \alpha < \kappa \wedge (G_\alpha \text{ is open in } X^\alpha)\}$$

where

$$\tilde{G}_\alpha = \{f \in X \mid f(\alpha) \in G_\alpha\}.$$

We define $p_\alpha: X \to X^\alpha$ by $p_\alpha(f) = f(\alpha)$.

Remark. Then the following result is obvious.

Theorem 22.15. $\langle X, T\rangle$ is a topological space, p_α is an o.c.o. map and

$$p_{\alpha\beta} \circ p_\beta = p_\alpha.$$

Definition 22.16. Let κ be a cardinal or On.

$$\langle\{\mathbf{B}_\alpha \mid \alpha < \kappa\}, \{i_{\alpha\beta} \mid \alpha \leq \beta < \kappa\}\rangle$$

is called a direct system of complete Boolean algebras iff $\mathbf{B}_\alpha$ $(\alpha < \kappa)$ is a complete Boolean algebra and $i_{\alpha\beta}: \mathbf{B}_\alpha \to \mathbf{B}_\beta$ is a complete isomorphism such that

1. $i_{\alpha\alpha}$ is an identity map.
2. $i_{\beta\gamma} \circ i_{\alpha\beta} = i_{\alpha\gamma}$.

Remark. We assume that $\mathbf{B}_\alpha$ is a complete subalgebra of $\mathbf{B}_\beta$ if $\alpha \leq \beta < \kappa$. Under this assumption $\bigcup_{\alpha<\kappa} B_\alpha$ becomes a Boolean algebra $\mathbf{B}'$ if we define $b_1 + b_2, b_1 \cdot b_2, {}^-b$ by $b_1 + b_2, b_1 \cdot b_2, {}^-b$ in $\mathbf{B}_\alpha$ where α is the least ordinal α such that, $b_1, b_2 \in \mathbf{B}_\alpha$ or $b \in \mathbf{B}_\alpha$ respectively. These definitions are unambiguous since $\mathbf{B}_\alpha \subseteq \mathbf{B}_\beta$ for $\alpha < \beta$. $\mathbf{B} \triangleq \lim_{\alpha\to\kappa} \mathbf{B}_\alpha$ is defined to be the completion of $\mathbf{B}'$.

Theorem 22.17. $\mathbf{B}_\alpha$ is a complete subalgebra of $\mathbf{B}$.

Proof. Let $S \subseteq B_\alpha$ and $b_0 = \prod S$ in $\mathbf{B}_\alpha$, i.e. $b_0 = \prod^{\mathbf{B}_\alpha}\{s \mid s \in S\}$. Let $b \in B$ and suppose that $(\forall x \in S)[x \geq b]$. We would like to show that $b_0 \geq b$.

Since $b \in B$ we have $b \in B_\beta$ for some β. Then either $b \in B_\alpha$, and hence $b \leq b_0$, or $\beta > \alpha$. But if $\beta > \alpha$ then $\mathbf{B}_\alpha$ is a complete subalgebra of $\mathbf{B}_\beta$ and hence $b_0 = \prod^{\mathbf{B}_\alpha} S = \prod^{\mathbf{B}_\beta} S \geq b$.

Theorem 22.18. Let κ be a cardinal or On. Let

$$\langle\{X^\alpha \mid \alpha < \kappa\}, \{p_{\alpha\beta} \mid \alpha \leq \beta < \kappa\}\rangle$$

be an o.c.o. inverse system and X be $\lim_{\alpha\to\kappa} X^\alpha$. Let $\mathbf{B}_\alpha$ be the Boolean algebra of regular open sets in X^α, let $i_{\alpha\beta}: \mathbf{B}_\alpha \to \mathbf{B}_\beta$ be the complete isomorphism induced by $p_{\alpha\beta}$ (Theorem 22.9) and let $\mathbf{B} = \lim_{\alpha\to\kappa} \mathbf{B}_\alpha$. Then $\mathbf{B}$ is isomorphic to the Boolean algebra of regular open sets B_X of X.

Proof, By Theorem 22.7, it is sufficient to show that B_X is a completion of $\bigcup_{\alpha<\kappa} B_\alpha$. For this purpose we have to show (i) each B_α is a complete subalgebra of B_X and (ii) $\bigcup_{\alpha<\kappa} B_\alpha - \{\mathbf{0}\}$ is dense in $B_X - \{\mathbf{0}\}$. Since the projection $p_\alpha: X \to X^\alpha$ is an o.c.o. map, by Theorem 22.9, p_α induces a complete monomorphism $i_\alpha: B_\alpha \to B_X$, where $i_\alpha(b) = (p_\alpha{}^{-1})``b$ for $b \in B_\alpha$. This proves (i). To prove (ii), it suffices to show that

$$(\exists\alpha)(\exists G_\alpha)[(G_\alpha \text{ is a nonempty regular open set in } X^\alpha) \wedge \tilde{G}_\alpha \subseteq G].$$

(Since $i_\alpha(G_\alpha) = \tilde{G}_\alpha$, $\tilde{G}_\alpha$ is a nonzero element of B_α which is below G and which belongs to $\bigcup_{\alpha<\kappa} B_\alpha$.) Since G is nonempty and open in X, there exists an α and a nonempty open set $\tilde{O}_\alpha$ in X^α such that $\tilde{O}_\alpha \subseteq G$. Define $G_\alpha = O_\alpha{}^{-0}$ in X^α. G_α is nonempty and regular open in X^α. It is easily seen that

$$\tilde{G}_\alpha \subseteq G^{-0} = G.$$

Theorem 22.19. Let κ be a cardinal or On. Let X_i $(i < \kappa)$ be topological spaces and define $X^\alpha = \prod_{i<\alpha} X_i$ for $\alpha < \kappa$ and $X = \prod_{i<\kappa} X_i$ with λ-product topologies. We define $p_{\alpha\beta}$ $(\alpha \le \beta < \kappa)$ to be a projection from X^β onto X^α. Then $\langle\{X^\alpha \mid \alpha < \kappa\}, \{p_{\alpha\beta} \mid a \le \beta < \kappa\}\rangle$ is an o.c.o. inverse system.

Moreover, if $cf(\kappa) \ge \lambda$, then X is homeomorphic to $\lim_{\alpha\to\kappa} X^\alpha$.

Proof. (We prove only the second part leaving the proof of the first part to the reader.) Let $\Phi: X \to X' = \lim_{\alpha\to\kappa} X^\alpha$ be defined as follows: $\Phi(f)(\alpha) = f \restriction \alpha$. Then Φ is one-to-one, onto and continuous. To show that Φ is an open map, let $G = \prod_{\nu<\kappa} O_\nu$ be an open set of X, where $O_\nu = X_\nu$ except for $< \lambda$ number of ν's. Since $cf(\kappa) \ge \lambda$, $\sup\{\nu \mid O_\nu \ne X_\nu\} < \kappa$. Let it be β and let

$$G_\beta = \prod_{\nu<\alpha} \Phi``O_\nu.$$

Then $\tilde{G}_\beta$ is a basic open set in X' and $\tilde{G}_\beta \subseteq \Phi(G)$. So Φ is open. Therefore $X \cong X'$.

Theorem 22.20. Let κ be a cardinal or On. Let

$$\langle\{X^\alpha \mid \alpha < \kappa\}, \{p_{\alpha\beta} \mid \alpha \le \beta < \kappa\}\rangle$$

be an o.c.o. inverse system and $X = \lim_{\alpha\to\kappa} X^\alpha$. Let Y be a topological space and $q_\alpha: Y \to X^\alpha$ $(\alpha < \kappa)$ satisfy the following conditions

1. q_α is an o.c.o. map.
2. $p_{\alpha\beta} \circ q_\beta = q_\alpha$.

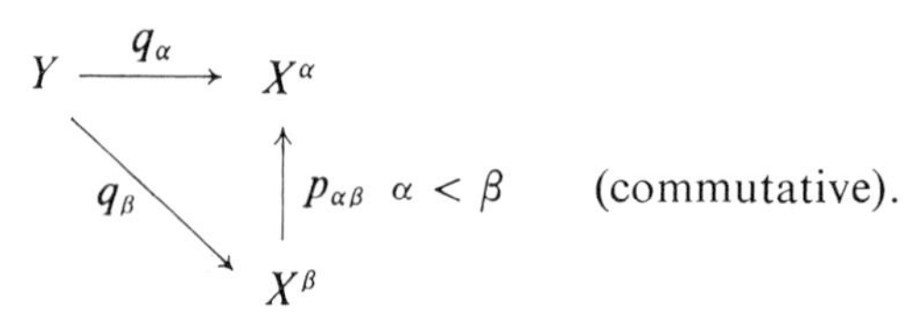

(commutative).

Then there exists a dense subset X_0 of X and an o.c.o. map $q: Y \to X_0$ such that $q_\alpha = p_\alpha \circ q$.

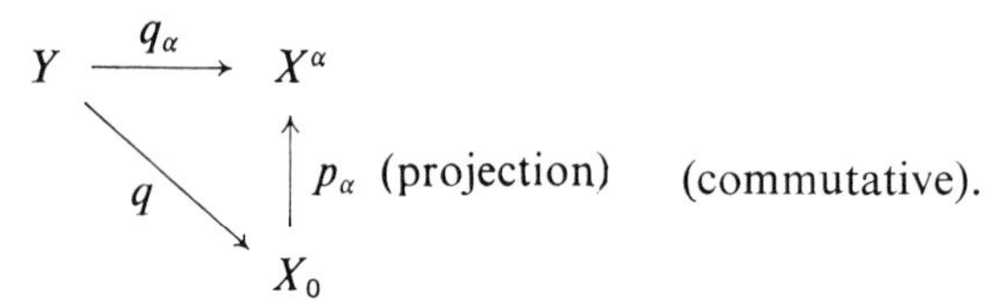

(commutative).

If, in addition, Y satisfies the following condition

$$(\forall y_1, y_2 \in Y)[y_1 \neq y_2 \rightarrow (\exists \alpha < \kappa)[q_\alpha(y_1) \neq q_\alpha(y_2)]]$$

then q is a homeomorphic map from Y to X_0.

Proof. Define q by

$$(q(y))(\alpha) = q_\alpha(y).$$

Obviously $q(y) \in X$ and $q_\alpha = p_\alpha \circ q$. Define

$$X_0 = \{q(y) \mid y \in Y\}.$$

1. X_0 is dense in X.

Let $x \in X$ and let G be an open set with $x \in G$. Then there exist $\alpha < \kappa$ and G_α such that G_α is open in X^α and

$$x \in \tilde{G}_\alpha \subseteq G.$$

Take $y \in (q_\alpha{}^{-1})``G_\alpha$ and define $x_0 = q(y)$. Then $x_0 \in X_0$ and $x_0 \in G$, i.e., $G \cap X_0 \neq 0$.

2. $q: Y \rightarrow X_0$ is o.c.o.

Obvious.

3. If Y satisfies the additional condition, then q is one-to-one.

An o.c.o. map is homeomorphic if it is one-to-one.

Remark. We cannot improve Theorem 22.20 by adding $X_0 = X$ as is easily seen from the following counterexample.

Let $\kappa = \omega$ and $X^n = 2^n$ with the discrete topology. If $p_{nm}(f) = f \upharpoonright n$, then $X = \lim_{n \to \omega} X^n$ is homeomorphic to 2^ω with the product topology. Now define $Y = \{f \in X \mid (\exists n < \omega)(\forall m)[n \leq m < \omega \rightarrow f(m) = 0]\}$ and $q_\alpha = p_\alpha \upharpoonright Y$. The desired q is uniquely determined by $(q(y))(\alpha) = p_\alpha(y)$ for $y \in Y$. Then $X_0 = q``Y \neq X$.

Definition 22.21. Let X and Y be topological spaces. A map $i: X \rightarrow Y$ is a topological embedding if $i: X \rightarrow i``X$ is a homeomorphism.

Definition 22.22. A topological space X is called *atomic* if for every $x \in X$ there exists a smallest open set G_0 such that $x \in G_0$ i.e., $(\forall G)[x \in G$ and G open $\rightarrow G_0 \subseteq G]$. This open set G_0 is denoted by $[x]$.

Remark. If $\mathbf{P} = \langle P, \leq \rangle$ is a partial order structure, then the topological space of $\mathbf{P}$ is atomic and satisfies the T_0-axiom of separation. On the other hand, if X is a T_0-space and atomic, then we define $x \leq y$ by $[x] \subseteq [y]$. This then becomes a partial order structure. Therefore we may think of the two notions partial order structure and atomic T_0-space as the same.

Theorem 22.23. Let $\mathbf{P}_1 = \langle P_1, \leq_1 \rangle$ and $\mathbf{P}_2 = \langle P_2, \leq_2 \rangle$ be partial order structures and

$$i: P_1 \xrightarrow{1-1} P_2.$$

Then "i is a topological embedding" is equivalent to

$$\text{“}(\forall p, q \in P_1)[p \leq_1 q \leftrightarrow i(p) \leq_2 i(q)].\text{”}$$

Proof. Since $\mathbf{P}_1$ and $\mathbf{P}_2$ are atomic T_0-spaces, this is clear.

Theorem 22.24. Let $\mathbf{P}_1 = \langle P_1, \leq_1 \rangle$ and $\mathbf{P}_0 = \langle P_0, \leq_0 \rangle$ be partial order structures and $p: \mathbf{P}_1 \to \mathbf{P}_0$ be o.c.o. Then

1. for every $x, y \in P_1$

$$x \leq_1 y \to p(x) \leq_0 p(y).$$

2. $p``[x]_{\mathbf{P}_1} = [p(x)]_{\mathbf{P}_0}$.

Proof.

1. Suppose $x \leq_1 y$. Then $x \in [y]_{\mathbf{P}_1} \subseteq (p^{-1})``[p(y)]_{\mathbf{P}_0}$ (since p is continuous and $[y]_{\mathbf{P}_1}$ is the smallest neighborhood of y). Therefore

$$p(x) \in [p(y)]_{\mathbf{P}_0}.$$

2. $p``[x]_{\mathbf{P}_1} \supseteq [p(x)]_{\mathbf{P}_0}$ since $p$ is an open map. Therefore $p``[x]_{\mathbf{P}_1} = [p(x)]_{\mathbf{P}_0}$.

Definition 22.25. Let κ be a cardinal or On. A system $\langle \{\mathbf{P}_a \mid \alpha < \kappa\}, \{p_{\alpha\beta} \mid \alpha \leq \beta < \kappa\} \rangle$ is called a normal limiting system of partial order structures iff

1. $\mathbf{P}_\alpha$ is a partial order structure for every $\alpha < \kappa$ and $\langle \{\mathbf{P}_\alpha \mid \alpha < \kappa\}, \{p_{\alpha\beta} \mid \alpha \leq \beta < \kappa\} \rangle$ is an o.c.o. inverse system.
2. $\mathbf{P}_0 \subseteq \mathbf{P}_1 \subseteq \cdots \subseteq \mathbf{P}_\alpha \subseteq \cdots (\alpha < \kappa)$.
3. If $x \in P_\alpha$ and $\alpha < \beta$, then

$$p_{\alpha\beta}(x) = x.$$

4. $(p_{\alpha\beta}{}^{-1})``[x]_{\mathbf{P}_\alpha} = [x]_{\mathbf{P}_\beta}$ if $x \in P_\alpha$ and $\alpha < \beta < \kappa$.

Remark. 4 is equivalent to the following:

4*. For $x \in P_\alpha$, $y \in P_\beta$, $\alpha < \beta < \kappa$,

$$p_{\alpha\beta}(y) \leq x \leftrightarrow y \leq x.$$

Example. Let $\mathbf{B}_0 \subseteq \mathbf{B}_1 \subseteq \cdots \subseteq \mathbf{B}_\alpha \subseteq \cdots (\alpha < \kappa)$ be a direct system of complete Boolean algebras. Let $\mathbf{P}_\alpha$ be the associated partial order structure for $\mathbf{B}_\alpha$ and let $p_{\alpha\beta}$ be $\#_{\alpha\beta}$ (see Theorem 22.11). It is easily seen that this is a normal limiting system.

Theorem 22.26. Let κ be a cardinal or On. Let $\mathbf{P}_0 \subseteq \mathbf{P}_1 \subseteq \cdots \subseteq \mathbf{P}_\alpha \subseteq \cdots (\alpha < \kappa)$ be a normal limiting system. If $x, y \in P_\alpha$ are compatible in $\mathbf{P}_\beta$ for $\alpha < \beta < \kappa$, then x and y are compatible in $\mathbf{P}_\alpha$.

Proof. Suppose $(\exists z \in P_\beta)[z \leq x \wedge z \leq y]$. Then $p_{\alpha\beta}(z) \leq x$ and $p_{\alpha\beta}(z) \leq y$.

Theorem 22.27. Let κ be a cardinal or On. Let $\mathbf{P}_0 \subseteq \mathbf{P}_1 \subseteq \cdots \subseteq \mathbf{P}_\alpha \subseteq \cdots$ $(\alpha < \kappa)$ be a normal limiting system.

4'. If $x \in P_\alpha$, $y \in P_\beta$, $\alpha < \beta < \kappa$, then

$$\text{Comp}(x, p_{\alpha\beta}(y)) \leftrightarrow \text{Comp}(x, y)$$

Proof. It is easily seen that the latter implies the former. Now assume x and $p_{\alpha\beta}(y)$ are compatible.

$$(\exists z \in P_\alpha)[z \le x \wedge z \in [p_{\alpha\beta}(y)]_{\mathbf{P}_\alpha} = p_{\alpha\beta}``[y]_{\mathbf{P}_\beta}] \qquad \text{by 4.}$$

Therefore

$$(\exists u \in P_\beta)[u \le y \wedge z = p_{\alpha\beta}(u)].$$
$$p_{\alpha\beta}(u) \le x \to u \le x. \qquad \text{(Use 4*.)}$$

So, x and y are compatible.

Remark. A weakly normal limiting system is obtained from a normal limiting system by replacing 4 by 4'. Actually, what we mainly use is a weakly normal limiting system. However, in many cases, the two definitions are equivalent as is seen in the following.

Theorem 22.28. Let κ be a cardinal or On. Let $\mathbf{P}_0 \subseteq \mathbf{P}_1 \subseteq \cdots \subseteq \mathbf{P}_\alpha \subseteq \cdots$ $(\alpha < \kappa)$ be a weakly normal limiting system. If $\bigcup_{\alpha<\kappa} \mathbf{P}_\alpha$ satisfies

5. $(\forall p, q \in \bigcup_{\alpha<\kappa} P_\alpha)[q \not\le p \to (\exists r \in \bigcup_{\alpha<\kappa} P_\alpha)[r \le q \wedge \neg\text{Comp}(r, p)]]$
then the system is a normal limiting system.

Proof. We have to show that 4* follows from 4' and 5. For that, let $x \in P_\alpha$, $y \in P_\beta$, and $\alpha < \beta < \kappa$. Suppose $p_{\alpha\beta}(y) \le x$ and $y \not\le x$. Then by 5 there exists a $\gamma \ge \beta$ and a $z \in P_\gamma$ such that $z \le y$ and z and x are incompatible. By 4' x and $p_{\alpha\gamma}(z)$ are incompatible. On the other hand, by Theorem 22.24, $p_{\alpha\beta}(z) \le p_{\alpha\beta}(y)$. Therefore $p_{\alpha\beta}(y) \not\le x$, a contradiction. Hence

$$p_{\alpha\beta}(y) \le x \to y \le x.$$

Conversely if $y \le x$ then, again by Theorem 22.24 $p_{\alpha\beta}(y) \le p_{\alpha\beta}(x) = x$. This proves 4*.

Theorem 22.29. Let $\kappa > \omega$ be a regular cardinal or On and

$$\langle\{\mathbf{P}_\alpha \mid \alpha < \kappa\}, \{p_{\alpha\beta} \mid \alpha \le \beta < \kappa\}\rangle$$

be a normal limiting system. If $\mathbf{P}_\alpha = \bigcup_{i<\alpha} \mathbf{P}_i$ for every $\alpha < \kappa$ with $cf(\alpha) = \omega$, and if $\overline{\overline{P}}_\alpha < \kappa$ for every $\alpha < \kappa$, then $\bigcup_{\alpha<\kappa} \mathbf{P}_\alpha$ satisfies the κ-chain condition.

Proof. For a member $x \in \bigcup_{\alpha<\kappa} P_\alpha$, define $|x|$ to be the least ordinal α such that $x \in P_\alpha$. Then we have

$$x \in \bigcup_{\alpha<\kappa} P_\alpha \to cf(|x|) \ne \omega.$$

Let A be a maximal pairwise incompatible subset of $\bigcup_{\alpha<\kappa} P_\alpha$. It suffices to show that $\overline{\overline{A}} < \kappa$.

We define the sequence of ordinals $\xi_0 < \cdots < \xi_i < \cdots < \kappa$ $(i < \omega)$ by induction on i. Define $\xi_0 = 0$. Suppose that $\xi_i < \kappa$ has been defined. Take an arbitrary element x of $P_{\xi_i} - A$. By the maximality of A, there is an element e_x of A such that e_x and x are compatible. Define

$$\xi_{i+1} = \max(\xi_i + 1, \sup\{|a_x| \mid x \in P_{\xi_i} - A\}).$$

Since $\overline{\overline{P}}_{\xi_i} < \kappa$ and κ is regular, $\xi_{i+1} < \kappa$.

Now let $\eta = \sup\{\xi_i \mid i < \omega\}$. Then $\eta < \kappa$ and $cf(\eta) = \omega$. We claim that $A \subseteq \mathbf{P}_\eta$, which implies $\overline{\overline{A}} < \kappa$. Suppose not. Let $a \in A$ and $a \notin P_\eta$. Let $a \in P_\beta$ where $\eta < \beta$. There exists an n such that $p_{\eta\beta}(a) \in P_{\xi_n}$. Then

$$(\exists b \in A \cap P_{\xi_{n+1}})[\mathrm{Comp}(b, p_{\eta\beta}(a))]$$

(Take $b = e_{p_{\eta\beta}(a)}$.) Now we have two properties.

1. a and b are incompatible.
2. $p_{\eta\beta}(a)$ and b are compatible.

This is a contradiction since by 4*, $p_{\eta\beta}(a) \leq a$.

Theorem 22.30. Under the same conditions as in the preceding theorem, let $\mathbf{B}_\alpha$ be the complete Boolean algebra of regular open sets in $\mathbf{P}_\alpha$, let $\mathbf{B} = \bigcup_{\alpha<\kappa} \mathbf{B}_\alpha$ and let $\tilde{\mathbf{B}}$ to be the completion of $\mathbf{B}$. Moreover, let $\mathbf{B}_1$ and $\mathbf{B}_2$ be the complete Boolean algebras of regular open sets in $\bigcup \mathbf{P}_\alpha$ and in $\lim_{\alpha\to\kappa} \mathbf{P}_\alpha$, respectively. Then

i) $\mathbf{B}$ satisfies the κ-chain condition,
ii) $\tilde{\mathbf{B}} = \mathbf{B}$,
iii) The three complete Boolean algebras $\tilde{\mathbf{B}}$, $\mathbf{B}_1$, and $\mathbf{B}_2$ are isomorphic.

Proof. i). Follows from ii), iii), and Theorem 22.29.

iii). Since, by Theorem 22.18 $\tilde{\mathbf{B}} \simeq \mathbf{B}_2$, it suffices to show that $\mathbf{B}_1 \simeq \mathbf{B}_2$. Clearly there exists a projection $q_\beta: \bigcup_{\alpha<\kappa} \mathbf{P}_\alpha \to \mathbf{P}_\beta$ such that

1. q_β is an o.c.o. map.
2. $p_{\alpha\beta} \cdot q_\beta = q_\alpha$ $(\alpha \leq \beta < \kappa)$.

(Let $\alpha_x = \mu_\alpha(x \in P_\alpha)$ for $x \in \bigcup_{\alpha<\kappa} P_\alpha$ and take q_β as follows:

$$\begin{aligned} q_\beta(x) &= p_{\beta\alpha_x}(x) \text{ if } \beta \leq \alpha_x, \\ &= x \text{ otherwise.}) \end{aligned}$$

Define $f_x(\beta) = q_\beta(x)$ for $x \in \bigcup_{\alpha<\kappa} P_\alpha$. Then by (2) above, $p_{\alpha\beta} \circ f(\beta) = f(\alpha)$ and hence $f_x \in \lim_{\alpha\to\kappa} \mathbf{P}_\alpha$. Moreover, if $x \neq y$ then $f_x \neq f_y$. Therefore $\bigcup_{\alpha<\kappa} \mathbf{P}_\alpha$ is densely embedded in $\lim_{\alpha\to\kappa} \mathbf{P}_\alpha$. So by Theorem 22.5, $\mathbf{B}_1 \simeq \mathbf{B}_2$.

ii). It suffices to show that $\tilde{\mathbf{B}} \subseteq \mathbf{B}$. Suppose $\tilde{b} \in \tilde{\mathbf{B}}$, $\tilde{b} \neq \mathbf{0}$. Let $S = \{b \in B \mid \mathbf{0} < b \leq \tilde{b}\}$. Take A to be a maximal incompatible subset of S. Then we have

3. $\tilde{b} = \sup A$. (For, suppose $\tilde{b} > \sup A$, and consider $\tilde{b} - \sup A$. Then we have a contradiction.)

4. $\overline{\overline{A}} < \kappa$. (This follows from $\tilde{\mathbf{B}} \simeq \mathbf{B}_1$ and Theorem 22.29.)

Therefore we have $(\exists\alpha < \kappa)[A \subseteq B_\alpha]$. Since $\mathbf{B}_\alpha$ is complete, $\tilde{b} \in B_\alpha \subseteq B$.

Corollary 22.31. Let $\kappa > \omega$ be a regular cardinal and $\mathbf{B}_0 \subseteq \mathbf{B}_1 \subseteq \cdots \subseteq \mathbf{B}_\alpha \subseteq \cdots$ $(\alpha < \kappa)$ be a direct system of complete Boolean algebras such that $\mathbf{B}_\alpha$ is a completion of $\bigcup_{\beta < \alpha} \mathbf{B}_\beta$ for every limit ordinal $\alpha < \kappa$. Define $\mathbf{B} = \bigcup_{\alpha < \kappa} \mathbf{B}_\alpha$ and $\tilde{\mathbf{B}}$ to be the completion of $\mathbf{B}$. If $(\forall \alpha < \kappa)[\overline{\overline{B}}_\alpha < \kappa]$, then

1. $\mathbf{B}$ satisfies the κ-chain condition and
2. $\tilde{\mathbf{B}} = \mathbf{B}$.

Proof. Define C_α $(\alpha \leq \kappa)$ and C as follows.

i) $C_0 = \{b \in B_0 \mid b > \mathbf{0}\}$.
ii) $C_{\alpha+1} = \{b \in B_{\alpha+1} \mid \#_\alpha(b) \in C_\alpha\}$.
iii) $C_\alpha = \bigcup_{\beta < \alpha} C_\beta$ for every limit ordinal α.
iv) $C = C_\kappa$.

Define $\mathbf{P}_\alpha = \langle C_\alpha, \leq \rangle$ and apply the theorem.

23. Boolean Algebras That Are Not Sets

When for a given axiom $(\forall\alpha)A(\alpha)$, one wishes to build a model of $ZF + (\forall\alpha)A(\alpha)$, he is often lead to the existence of complete Boolean algebras $\mathbf{B}_\beta$, for which

1. $\mathbf{V}^{\mathbf{B}_\beta}$ satisfies $ZF + (\forall\alpha < \beta)A(\alpha)$, and
2. The cardinality of $|\mathbf{B}_\beta|$ increases as β increases.

In this situation, the natural idea is to find a certain limit $\mathbf{B}$ of $\mathbf{B}_\beta$ and to prove that $\mathbf{V}^{\mathbf{B}}$ is a model of $ZF + (\forall\alpha)A(\alpha)$.

In almost all cases, however, the limit algebra $\mathbf{B}$ is not a set, it is a proper class. In general if a complete Boolean algebra is not a set, then $\mathbf{V}^{\mathbf{B}}$ may not satisfy the Axiom Schema of Replacement or the Axiom of Powers. Therefore we need a general theory about conditions we should impose upon $\mathbf{B}$ in order that $\mathbf{V}^{\mathbf{B}}$ satisfy the Axiom Schema of Replacement and the Axiom of Powers.

Another interesting problem is this: We do not have many useful ways to define limits of Boolean algebras. Therefore we tend to think in terms of limits of topological spaces or limits of partial order structures. At least one can make a partial order structure $\mathbf{P}_\beta$ which is dual to $\mathbf{B}_\beta$. Then we can take $\mathbf{P}$, a limit of $\mathbf{P}_\beta$, and define a limit of $\mathbf{B}_\beta$ as the dual Boolean algebra of $\mathbf{P}$. In our opinion, one of the most interesting problems in set theory is to investigate what effects the special kinds of limits of partial order structures or topological spaces have on the limit Boolean algebra $\mathbf{B}$ and the Boolean valued universe $\mathbf{V}^{\mathbf{B}}$. In this section we will see that a limit topological space which is simultaneously a direct limit and an inverse limit of a certain sequence of topological spaces plays an important role. We believe strongly in the importance of the investigation of many other kinds of limits of Boolean algebras.

Until now we have considered only Boolean algebras which are sets. This requirement enabled us to prove the Axiom of Powers and the Maximum Principle in $\mathbf{V}^{(\mathbf{B})}$. We shall now drop this restriction and allow B to be a class. However, in many applications B is not even a class of sets but a class of classes. Consider e.g., a partial order structure $\mathbf{P} = \langle P, \leq\rangle$ where P is a proper class. Then the complete Boolean algebra of regular open classes in $\mathbf{P}$ contains classes some of which are proper classes. In order to cope with this situation we shall consider two cases. In the following we shall use $\mathbf{B}$ for a Boolean algebra which is a class of sets and $\tilde{\mathbf{B}}$ for a Boolean algebra which is a class of classes. In the second case we obviously need a set theory which is stronger than ZF. Note also that for $\mathbf{B}$ as above the completion of $\mathbf{B}$ is of type $\tilde{\mathbf{B}}$.

If we consider $\mathbf{B}$ as above we do not require that $\mathbf{B}$ be complete but *satisfy the following weaker condition*:

$$(\forall A \subseteq B)[\mathscr{M}(A) \rightarrow (\exists x \in B)[x = \sup A]].$$

On the other hand we always require that *a Boolean algebra of type* $\tilde{\mathbf{B}}$ *be complete*, i.e., $\sup A$ *exists in* $\tilde{\mathbf{B}}$ *for every class* $A \subseteq \tilde{B}$. Under these assumptions we define $\mathbf{V}^{(\mathbf{B})}$ and $\mathbf{V}^{(\tilde{\mathbf{B}})}$ in two different ways. $V^{(\mathbf{B})}$ is defined as follows:

Definition 23.1. Let $C_\alpha = B \cap R(\alpha)$ for $\alpha \in On$. Then

1. $V_0^{(\mathbf{B})} \triangleq 0$.
2. $V_\alpha^{(\mathbf{B})} \triangleq \{u \in C_\alpha^{\mathscr{D}(u)} \mid (\exists \xi < \alpha)[\mathscr{D}(u) \subseteq V_\xi^{(\mathbf{B})}]\}$.
3. $V^{(\mathbf{B})} \triangleq \bigcup_{\alpha \in On} V_\alpha^{(\mathbf{B})}$.

Remark. This is a definition in the framework of ZF. Moreover,

$$V^{(\mathbf{B})} = \{u \mid \mathscr{D}(u) \subseteq V^{(\mathbf{B})} \wedge u\colon \mathscr{D}(u) \rightarrow B\}$$

as in the case of a set $\mathbf{B}$. Note that $V_\alpha^{(\mathbf{B})}$ is a set for each α and $V^{(\mathbf{B})}$ is a class of sets.

$V^{(\tilde{\mathbf{B}})}$ is defined as follows:

Definition 23.2.

1. $V_0^{(\tilde{\mathbf{B}})} \triangleq 0$.
2. $V_\alpha^{(\tilde{\mathbf{B}})} \triangleq \{u \in \tilde{\mathbf{B}}^{\mathscr{D}(u)} \mid (\exists \xi < \alpha)[\mathscr{D}(u) \subseteq V_\xi^{(\tilde{\mathbf{B}})} \wedge \overline{\overline{\mathscr{D}(u)}} \in On]\}$.
3. $V^{(\tilde{\mathbf{B}})} \triangleq \bigcup_{\alpha \in On} V_\alpha^{(\tilde{\mathbf{B}})}$.

Remark. Note that in general $V_\alpha^{(\tilde{\mathbf{B}})}$ is not a set for $\alpha > 1$. We tacitly assume that we have a sufficiently strong set theory to define $V^{(\tilde{\mathbf{B}})}$. In the following, $u, v, w, \ldots$ range over $V^{(\mathbf{B})}$ or $V^{(\tilde{\mathbf{B}})}$. $[\![u = v]\!]$, $[\![u \in v]\!]$, and $\check{y}$ for $y \in V$ are defined in $\mathbf{V}^{(\mathbf{B})}$ and $\mathbf{V}^{(\tilde{\mathbf{B}})}$ in the same way as in the case of a set B. The following results are obtained in the same way as in the case of Boolean algebras which are sets.

Theorem 23.3. (cf. Theorem 14.2). Let $k \in V$ and $u \in V_{\alpha+1}^{(\mathbf{B})}$. Then

1. $\alpha \leq \text{rank}(k) \rightarrow [\![\check{k} \in u]\!] = \mathbf{0}$.
2. $\alpha < \text{rank}(k) \rightarrow [\![\check{k} = u]\!] = \mathbf{0}$.

Theorem 23.4. $(\forall u \in V^{(\mathbf{B})})[[\![\text{Ord}(u)]\!] = \sum_{\alpha \in On} [\![u = \check{\alpha}]\!]]$.

Remark. This is proved in the same way as Theorem 13.21. However, since B need not be a set, we have to give a proof that for $u \in V^{(\mathbf{B})}$

$$D_u = \{\xi \mid [\![u = \check{\xi}]\!] > \mathbf{0}\}$$

is a set. But this follows directly from the preceding theorem.

We can define

1. $[\![M(u)]\!] \triangleq \sum_{k\in V} [\![u = \check{k}]\!]$

in $\mathbf{V}^{(\mathbf{B})}$ and $\mathbf{V}^{(\tilde{\mathbf{B}})}$ as before. In the case of $\mathbf{V}^{(\tilde{\mathbf{B}})}$ the existence of the supremum is assured by our assumption that $\tilde{\mathbf{B}}$ is complete. On the other hand, in the case of $\mathbf{V}^{(\mathbf{B})}$ we have from Theorem 23.3 that for $u \in V^{(\mathbf{B})}$

$$(\exists\alpha)\left[\sum_{k\in V} [\![u = \check{k}]\!] = \sum_{k\in R(\alpha)} [\![u = \check{k}]\!]\right]$$

and therefore the sum in 1 is actually only a sum over the index set $R(\alpha)$ Similar remarks apply to the definitions of $\check{B}$, $\check{+}$, $\check{\cdot}$, F below in the case of $\mathbf{V}^{(\mathbf{B})}$:

2. $[\![u \in \check{B}]\!] \triangleq \sum_{b\in B} [\![u = \check{b}]\!].$

3. $[\![\check{+}(u, v, w)]\!] \triangleq \sum_{b_1,b_2\in B} [\![u = \check{b}_1]\!]\cdot[\![v = \check{b}_2]\!]\cdot[\![w = (b_1 + b_2)^{\vee}]\!].$

4. $[\![\check{\cdot}(u, v, w)]\!] \triangleq \sum_{b_1,b_2\in B} [\![u = \check{b}_1]\!]\cdot[\![v = \check{b}_2]\!]\cdot[\![w = (b\ \cdot b_2)^{\vee}]\!].$

5. $[\![u \in F]\!] \triangleq \sum_{b\in B} [\![u = \check{b}]\!]\cdot b.$

Finally, we let

$$\mathbf{V}^{(\mathbf{B})} \triangleq \langle V^{(\mathbf{B})}, \bar{=}, \bar{\in}, M, \check{B}, \check{+}, \check{\cdot}, F\rangle,$$
$$\mathbf{V}^{(\tilde{\mathbf{B}})} \triangleq \langle V^{(\tilde{\mathbf{B}})}, \bar{=}, \bar{\in}, M, \check{B}, \check{+}, \check{\cdot}, F\rangle.$$

If we consider only $\mathbf{V}^{(\mathbf{B})}$, $\tilde{\mathbf{B}}$ always denotes the completion of $\mathbf{B}$. $[\![\varphi]\!]$ is defined as before in both cases. For $\mathbf{V}^{(\mathbf{B})}$, it is always the case that $[\![\varphi]\!] \in \tilde{B}$, but we may have $[\![\varphi]\!] \notin B$ if φ contains a quantifier; hence if $[\![\varphi]\!]$ is defined by a sum or product over a class. The definitions of $\tilde{\mathbf{B}}$ and $[\![\varphi]\!]$ are beyond ZF set theory. Nevertheless, we can manage to build a theory in ZF by using the notion of forcing, i.e., we use the relation $b \le [\![\varphi]\!]$ (where $b \in B$) instead of $[\![\varphi]\!]$. This definition can be given in ZF by recursion:

1. $b \le [\![\varphi_1 \wedge \varphi_2]\!] \leftrightarrow b \le [\![\varphi_1]\!] \wedge b \le [\![\varphi_2]\!].$
2. $b \le [\![\neg\varphi]\!] \leftrightarrow (\forall b' \in B)[b' \le [\![\varphi]\!] \to b' \le {}^{-}b]$
   (since $b \le [\![\neg\varphi]\!] \leftrightarrow [\![\varphi]\!] \le {}^{-}b$).
3. $b \le [\![(\forall x)\varphi(x)]\!] \leftrightarrow (\forall u \in V^{(\mathbf{B})})[b \le [\![\varphi(u)]\!]].$

For an atomic φ the meaning of $b \le [\![\varphi]\!]$ is obvious from 1 to 5 above. Thus, given a formula φ, $b \le [\![\varphi]\!]$ is a formula of ZF.

Theorem 23.5. $\mathbf{V}^{(\mathbf{B})}$ and $\mathbf{V}^{(\tilde{\mathbf{B}})}$ satisfy the Axioms of Extensionality, Pairing, and Infinity.

Proof. The proof is the same as in the case of a set $\mathbf{B}$.

Remark. Similarly, our former proof (using $V_\alpha^{(\mathbf{B})}$ or $V_\alpha^{(\tilde{\mathbf{B}})}$ in place of M_α) establishes the following results.

Theorem 23.6. $\mathbf{V}^{(\mathbf{B})}$ and $\mathbf{V}^{(\bar{\mathbf{B}})}$ satisfy the Axiom of Regularity.

Theorem 23.7. (cf. Theorem 13.13.)

$$[\![(\exists x \in u)\varphi(x)]\!] = \sum_{x \in \mathscr{D}(u)} (u(x)[\![\varphi(x)]\!]),$$

$$[\![(\forall x \in u)\varphi(x)]\!] = \prod_{x \in \mathscr{D}(u)} (u(x) \Rightarrow [\![\varphi(x)]\!]).$$

Theorem 23.8. $\mathbf{V}^{(\mathbf{B})}$ and $\mathbf{V}^{(\bar{\mathbf{B}})}$ satisfy the Axiom of Union.

Proof. We have to show that for a given $u \in V^{(\mathbf{B})}$ or $u \in V^{(\bar{\mathbf{B}})}$

$$[\![(\exists v)(\forall x)[x \in v \leftrightarrow (\exists y \in u)[x \in y]]]\!] = \mathbf{1}.$$

Define v by

$$\mathscr{D}(v) = \bigcup_{y \in \mathscr{D}(u)} \mathscr{D}(y)$$

and

$$(\forall x \in \mathscr{D}(v))[v(x) = [\![(\exists y \in u)[x \in y]]\!]].$$

Obviously, $v \in V^{(\bar{\mathbf{B}})}$ or $v \in V^{(\mathbf{B})}$ (use Theorem 23.7) according as $u \in V^{(\bar{\mathbf{B}})}$ or $u \in V^{(\mathbf{B})}$. Since $[\![(\forall x \in v)(\exists y \in u)[x \in y]]\!] = \mathbf{1}$ by Theorem 23.7, it remains to show that

$$[\![(\forall y \in u)(\forall x \in y)[x \in v]]\!] = \mathbf{1},$$

$$[\![(\forall y \in u)(\forall x \in y)[x \in v]]\!] = \prod_{y \in \mathscr{D}(u)} \prod_{x \in \mathscr{D}(y)} (u(y) \cdot y(x) \Rightarrow [\![x \in v]\!]).$$

Let $y \in \mathscr{D}(u)$ and $x \in \mathscr{D}(y)$. Then

$$\begin{aligned} u(y) \cdot y(x) \leq [\![x \in y]\!] \cdot [\![y \in u]\!] &\leq [\![(\exists y_0 \in u)[x \in y_0]]\!] = v(x) \\ &\leq [\![x \in v]\!]. \end{aligned}$$

Theorem 23.9. $\mathbf{V}^{(\bar{\mathbf{B}})}$ satisfies the Axiom of Subsets (Zermelo's Axiom Schema of Separation).

Proof. Let $a \in V^{(\bar{\mathbf{B}})}$. We wish to prove that

$$[\![(\exists v)(\forall y)[y \in v \leftrightarrow y \in a \wedge \varphi(y)]]\!] = \mathbf{1}.$$

Define $v \in V^{(\bar{\mathbf{B}})}$ by $\mathscr{D}(v) = \mathscr{D}(a)$ and

$$(\forall x \in \mathscr{D}(v))[v(x) = a(x) \cdot [\![\varphi(x)]\!]].$$

Then

$$\begin{aligned} [\![u \in v]\!] &= \sum_{x \in \mathscr{D}(v)} v(x) \cdot [\![u = x]\!] \\ &= \sum_{x \in \mathscr{D}(a)} a(x) \cdot [\![\varphi(x)]\!] \cdot [\![u = x]\!] \\ &= \sum_{x \in \mathscr{D}(a)} a(x) \cdot [\![u = x]\!] \cdot [\![\varphi(u)]\!] \\ &= [\![u \in a \wedge \varphi(u)]\!]. \end{aligned}$$

Remark. Note that the foregoing proof requires a Boolean algebra of type $\tilde{\mathbf{B}}$ and cannot be carried out for an algebra of type $\mathbf{B}$. The difficulty is in defining v: If $a \in V^{(\tilde{\mathbf{B}})}$ then for each $x \in \mathscr{D}(v)$ we have $a(x) \in B$, but we do not know that $[\![\varphi(x)]\!] \in B$. While $\mathbf{V}^{(\tilde{\mathbf{B}})}$ must satisfy the Axiom of Subsets it need not satisfy the Axiom of Replacement, and even if this axiom is satisfied, the Axiom of Powers need not hold in $\mathbf{V}^{(\tilde{\mathbf{B}})}$. Therefore we have to look for suitable restrictions on $\tilde{\mathbf{B}}$. An important though rather weak condition is given by the following.

Definition 23.10. $\tilde{\mathbf{B}}$ satisfies the *uniform convergence law* (UCL) iff the following condition is satisfied:

Let I be a set and for each $i \in I$, let $\{b_{\alpha i} \mid \alpha \in On\} \subseteq \tilde{B}$. If

1. $(\forall \alpha)(\forall \beta)[\alpha < \beta \rightarrow b_{\alpha i} \geq b_{\beta i}]$

and

2. $\prod_{\alpha \in On} b_{\alpha i} = \mathbf{0}$

then

$$\prod_{\alpha \in On} \sum_{i \in I} b_{\alpha i} = \mathbf{0}.$$

Remark. This means roughly that if a nonincreasing sequence $b_{\alpha i}(\alpha \in On)$ of elements of $\tilde{B}$ converges to $\mathbf{0}$ for each $i \in I$, then these sequences converge uniformly to $\mathbf{0}$.

Exercise. If we omit condition 1 from the UCL, is the corresponding law (for fixed I) equivalent to the (On, I)-DL (cf. Theorem 18.5)?

Theorem 23.11. Suppose $\{b_{\alpha i} \mid \alpha \in On \wedge i < \mathbf{2}\} \subseteq \tilde{B}$,

1. $(\forall \alpha)(\forall \beta)[\alpha < \beta \rightarrow b_{\beta 0} \leq b_{\alpha 0} \wedge b_{\beta 1} \leq b_{\alpha 1}]$

and

2. $\prod_{\alpha \in On} b_{\alpha 0} = \prod_{\alpha \in On} b_{\alpha 1} = \mathbf{0},$

then

$$\prod_{\alpha \in On} (b_{\alpha 0} + b_{\alpha 1}) = \mathbf{0}.$$

Proof. (By contradiction.) Suppose $b = \prod_\alpha (b_{\alpha 0} + b_{\alpha 1}) > \mathbf{0}$. Since $\prod_\alpha b_{\alpha 0} = \mathbf{0}$, $b_{\alpha 0} \not\geq b$ for some α. Let $p = b - b_{\alpha 0}$. Then $\mathbf{0} < p \leq b$ and $p \cdot b_{\alpha 0} = \mathbf{0}$. Similarly, since $\prod_\alpha b_{\alpha 1} = \mathbf{0}$, $b_{\beta 1} \not\geq p$ for some $\beta \geq \alpha$. With $q = p - b_{\beta 1}$ we have $\mathbf{0} < q \leq p$ and $q \cdot b_{\beta 1} = \mathbf{0}$. This gives $\mathbf{0} < q \leq b$ and $q \cdot (b_{\beta 0} + b_{\beta 1}) = \mathbf{0}$, a contradiction.

Theorem 23.12. Let $\tilde{\mathbf{B}}$ be the completion of $\mathbf{B}$. For $b \in \tilde{B}$ define

$$b_\alpha = \sup\{b' \mid b' \in B \cap R(\alpha) \wedge b' \leq b\}.$$

Then

1. $b_\alpha \leq b$.
2. $\alpha < \beta \to b_\alpha \leq b_\beta$ and $b_\alpha = \prod \{b_\beta \mid \beta \geq \alpha\}$.
3. $\sum_{\alpha \in On} b_\alpha = b$.

Proof. 1 and 2 are obvious.

3. Suppose $\sum_\alpha b_\alpha < b$. Since $\tilde{\mathbf{B}}$ is the completion of $\mathbf{B}$,

$$(\exists p \in B)\left[\mathbf{0} < p \leq b - \sum_{\alpha \in On} b_\alpha\right] \qquad \text{(by the density property).}$$

Therefore $(\exists \alpha)[p \in B \cap R(\alpha)]$ and hence $p \leq b_\alpha$. This is a contradiction.

Remark. Theorem 23.12 says that every element of $\tilde{\mathbf{B}}$ can be obtained as the limit of elements of $\mathbf{B}$ (see Theorem 23.14 below) when we define limits in the following way.

Definition 23.13. If $b \in \tilde{\mathbf{B}}$ and $\{b_\alpha \mid \alpha \in On\} \subseteq \tilde{\mathbf{B}}$ then

$$\lim_{\alpha \in On} b_\alpha = \mathbf{1} \quad \text{iff} \quad \sum_{\alpha \in On} \prod_{\alpha \leq \beta} b_\beta = \mathbf{1},$$

$$\lim_{\alpha \in On} b_\alpha = b \quad \text{iff} \quad \lim_{\alpha \in On} (b_\alpha \Leftrightarrow b) = \mathbf{1}.$$

Similarly, for $u, u_\alpha \in V^{(\mathbf{B})}$ or $V^{(\tilde{\mathbf{B}})}$

$$\lim_{\alpha \in On} u_\alpha = u \quad \text{iff} \quad \lim_{\alpha \in On} [\![u_\alpha = u]\!] = \mathbf{1}.$$

We will occasionally write $u_\alpha \to u$ for $\lim_{\alpha \in On} u_\alpha = u$.

Remark. Definition 23.13 is reminiscent of the definition of the limit of a sequence of point sets in analysis. Using this definition, we can restate Theorem 23.12 as follows:

Theorem 23.14. If $\tilde{\mathbf{B}}$ is the completion of $\mathbf{B}$ and if $b \in \tilde{\mathbf{B}}$, then there is a sequence $\{b_\alpha \mid \alpha \in On\} \subseteq B$ such that $b_\alpha \to b$.

Theorem 23.15. Suppose $\lim_{\alpha \in On} b_\alpha = b$ and $\lim_{\alpha \in On} b'_\alpha = b'$. Then

1. $\lim_{\alpha \in On} (^-b_\alpha) = {}^-b$.
2. $\lim_{\alpha \in On} (b_\alpha + b'_\alpha) = b + b'$.
3. $\lim_{\alpha \in On} (b_\alpha \cdot b'_\alpha) = b \cdot b'$.

Proof. 1. Obvious since $(b_\alpha \Leftrightarrow b) = (^-b_\alpha \Leftrightarrow {}^-b)$.

2. $(b_\alpha + b'_\alpha) \Leftrightarrow (b + b')$

$$\begin{aligned} &= {}^-(b_\alpha + b'_\alpha)\cdot(^-(b + b')) + (b_\alpha + b'_\alpha)(b + b') \\ &\geq ((^-b_\alpha)(^-b) + b_\alpha \cdot b)((^-b'_\alpha)\cdot(^-b') + (b'_\alpha \cdot b')) \\ &= (b_\alpha \Leftrightarrow b)\cdot(b'_\alpha \Leftrightarrow b). \end{aligned}$$

Since

$$\sum_{\alpha \in On} \prod_{\alpha \leq \beta} (b_\beta \Leftrightarrow b) = \sum_{\alpha \in On} \prod_{\alpha \leq \beta} (b'_\beta \Leftrightarrow b') = \mathbf{1},$$

the dual form of Theorem 23.11 gives

$$\sum_{\alpha \in On} \left(\prod_{\alpha \leq \beta} (b_\beta \Leftrightarrow b) \cdot \prod_{\alpha \leq \beta} (b'_\beta \Leftrightarrow b') \right) = \mathbf{1}$$

$$\sum_{\alpha \in On} \left(\prod_{\alpha \leq \beta} (b_\beta \Leftrightarrow b)(b'_\beta \Leftrightarrow b') \right) = \mathbf{1}$$

and hence

$$\sum_{\alpha \in On} \prod_{\alpha \leq \beta} ((b_\beta + b'_\beta) \Leftrightarrow (b + b')) = \mathbf{1}.$$

3. A consequence of 1 and 2.

Lemma. If $\tilde{\mathbf{B}}$ is the completion of $\mathbf{B}$, then $V^{(\mathbf{B})} \subseteq V^{(\tilde{\mathbf{B}})}$.

Proof. For a given $u \in V^{(\mathbf{B})}$ we can prove by induction on the least α such that $u \in V_\alpha^{(\mathbf{B})}$ that $u \in V_\alpha^{(\mathbf{B})} \rightarrow u \in V^{(\tilde{\mathbf{B}})}$.

Theorem 23.16. Let $\tilde{\mathbf{B}}$ be the completion of $\mathbf{B}$. If $\tilde{\mathbf{B}}$ satisfies the UCL, then $(\forall v \in V^{(\tilde{\mathbf{B}})})(\exists \{u_\alpha \mid \alpha \in On\} \subseteq V^{(\mathbf{B})})[\lim_{\alpha \in On} u_\alpha = v]$, i.e., every $v \in V^{(\tilde{\mathbf{B}})}$ can be obtained as the limit of elements of $V^{(\mathbf{B})}$.

Proof. (By induction on the least α such that $v \in V_\alpha^{(\tilde{\mathbf{B}})}$.) As our induction hypothesis, assume

$$(\forall x \in \mathscr{D}(v))(\exists \{y_{\alpha x} \mid \alpha \in On\} \subseteq V^{(\mathbf{B})})[y_{\alpha x} \rightarrow x].$$

By Theorem 23.14,

$$(\forall x \in \mathscr{D}(v))(\exists \{b_{\alpha x} \mid \alpha \in On\} \subseteq B)[b_{\alpha x} \rightarrow v(x)].$$

Now define $u_\alpha : \{y_{\alpha x} \mid x \in \mathscr{D}(v)\} \rightarrow B$ by

$$u_\alpha(y_{\alpha x}) = b_{\alpha x} \quad \text{for} \quad x \in \mathscr{D}(v).$$

Then $(\forall \alpha)[u_\alpha \in V^{(\mathbf{B})}]$. Moreover,

$$\begin{aligned}
[\![u_\alpha = v]\!] &= \prod_{y_{\alpha x} \in \mathscr{D}(u_\alpha)} (u_\alpha(y_{\alpha x}) \Rightarrow [\![y_{\alpha x} \in v]\!]) \cdot \prod_{x \in \mathscr{D}(v)} (v(x) \Rightarrow [\![x \in u_\alpha]\!]) \\
&\geq \prod_{x \in \mathscr{D}(v)} (b_{\alpha x} \Rightarrow v(x) \cdot [\![x = y_{\alpha x}]\!]) \cdot \prod_{x \in \mathscr{D}(v)} (v(x) \Rightarrow b_{\alpha x} \cdot [\![x = y_{\alpha x}]\!]) \\
&\qquad \text{since } (\forall x \in \mathscr{D}(w))[w(x) \cdot [\![x = w_1]\!] \leq [\![w_1 \in w]\!]] \\
&\geq \prod_{x \in \mathscr{D}(v)} (b_{\alpha x} \Leftrightarrow v(x)) \cdot [\![x = y_{\alpha x}]\!] \rightarrow \mathbf{1} \cdot \mathbf{1},
\end{aligned}$$

using the dual form of the UCL for the first factor.

Corollary 23.17. If $\tilde{\mathbf{B}}$ is the completion of $\mathbf{B}$ and $\tilde{\mathbf{B}}$ satisfies the UCL, then

$$(\forall p \in \tilde{B})[p > \mathbf{0} \rightarrow (\forall v \in V^{(\tilde{\mathbf{B}})})(\exists u \in V^{(\mathbf{B})})[p \cdot [\![u = v]\!] > \mathbf{0}]).$$

Proof. For given v and $p > \mathbf{0}$ let

$$b'_\alpha = \prod_{\beta \geq \alpha} \prod_{x \in \mathscr{D}(v)} (b_{\beta x} \Leftrightarrow v(x)).$$

Since $\sum_\alpha b'_\alpha = \mathbf{1}$, there exists an α such that $b'_\alpha \cdot p > \mathbf{0}$. Then u_α has the desired property.

Theorem 23.18. Suppose that $\tilde{\mathbf{B}}$ is the completion of $\mathbf{B}$ and $\tilde{\mathbf{B}}$ satisfies the UCL. Then $\mathbf{V}^{(\mathbf{B})}$ is a $\tilde{\mathbf{B}}$-valued elementary substructure of $\mathbf{V}^{(\tilde{\mathbf{B}})}$, i.e., for each formula φ of the language of $\mathbf{V}^{(\mathbf{B})}$ and every $u_1, \ldots, u_n \in V^{(\mathbf{B})}$,

$$[\![\varphi(u_1, \ldots, u_n)]\!]^{\mathbf{V}^{(\mathbf{B})}} = [\![\varphi(u_1, \ldots, u_n)]\!]^{\mathbf{V}^{(\tilde{\mathbf{B}})}}.$$

Proof. (By induction on the number of logical symbols in φ.) We consider only the case of a quantifier.

Let $\varphi(u_1, \ldots, u_n) = (\forall x)\psi(x, u_1, \ldots, u_n)$ where $u_1, \ldots, u_n \in V^{(\mathbf{B})}$. Let $b = [\![\varphi(u_1, \ldots, u_n)]\!]^{\mathbf{V}^{(\mathbf{B})}}$ and let v be any member of $V^{(\tilde{\mathbf{B}})}$. Claim:

$$b \leq [\![\psi(v, u_1, \ldots, u_n)]\!]^{\mathbf{V}^{(\mathbf{B})}}.$$

Suppose not: Then

$$p \cdot [\![\psi(v, u_1, \ldots, u_n)]\!]^{\mathbf{V}^{(\tilde{\mathbf{B}})}} = \mathbf{0} \qquad \text{for some } p, \mathbf{0} < p \leq b.$$

By Collorary 23.17, there exists a $u \in V^{(\mathbf{B})}$ such that $p \cdot [\![u = v]\!] > \mathbf{0}$. Then

$$\begin{aligned} \mathbf{0} &= p \cdot [\![\psi(v, u_1, \ldots, u_n)]\!]^{\mathbf{V}^{(\tilde{\mathbf{B}})}} \geq p \cdot [\![u = v]\!] \cdot [\![\psi(u, u_1, \ldots, u_n)]\!]^{\mathbf{V}^{(\tilde{\mathbf{B}})}} \\ &= p \cdot [\![u = v]\!] \cdot [\![\psi(u, u_1, \ldots, u_n)]\!]^{\mathbf{V}^{(\mathbf{B})}}, \qquad \text{by the induction hypothesis} \\ &\geq p \cdot [\![u = v]\!] \cdot b = p \cdot [\![u = v]\!] > \mathbf{0}, \qquad \text{a contradiction.} \end{aligned}$$

Therefore we have

$$b \leq \prod_{v \in \mathbf{V}^{(\tilde{\mathbf{B}})}} [\![\psi(v, u_1, \ldots, u_n)]\!]^{\mathbf{V}^{(\tilde{\mathbf{B}})}} = [\![\varphi(u_1, \ldots, u_n)]\!]^{\mathbf{V}^{(\tilde{\mathbf{B}})}}.$$

Since

$$\begin{aligned} b &= \prod_{v \in V^{(\mathbf{B})}} [\![\psi(v, u_1, \ldots, u_n)]\!]^{\mathbf{V}^{(\mathbf{B})}} = \prod_{v \in V^{(\mathbf{B})}} [\![\psi(v, u_1, \ldots, u_n)]\!]^{\mathbf{V}^{(\tilde{\mathbf{B}})}} \\ &\geq [\![\varphi(u_1, \ldots, u_n)]\!]^{\mathbf{V}^{(\tilde{\mathbf{B}})}}, \end{aligned}$$

we obtain

$$b = [\![\varphi(u_1, \ldots, u_n)]\!]^{\mathbf{V}^{(\tilde{\mathbf{B}})}}.$$

Definition 23.19. A partial order structure $\mathbf{P} = \langle P, \leq \rangle$ (where P may be a proper class) satisfies the *set-chain condition* (s.c.c.) iff every class of mutually incompatible elements of P is a set. $\mathbf{B}$ satisfies the s.c.c. iff $\mathbf{P} = \langle B - \{\mathbf{0}\}, \leq \rangle$ satisfies the s.c.c.

Remark. Thus the s.c.c. for Boolean algebras is a generalization of the $\aleph_\alpha$-chain condition.

Theorem 23.20. Suppose $\mathbf{P} = \langle P, \leq \rangle$ satisfies the s.c.c. Let $\mathbf{B}$ be the Boolean algebra of regular open classes in $\mathbf{P}$. Then $\mathbf{B}$ satisfies the s.c.c.

Remark. Under the assumption of Theorem 23.20, every regular open class in $\mathbf{P}$ can be represented by $\sum \{A``x \mid x \in a\}$ where $a$ is a set and each $A``x$ is a basic open class. These are of the form $[p]$ for some $p \in P$ and hence they are determined by sets $p \in P$. Therefore, if $\tilde{\mathbf{B}}$ is the Boolean algebra of regular open classes in $\mathbf{P}$, each element of $\tilde{\mathbf{B}}$ can be represented by a set. Thus we may assume that $\tilde{\mathbf{B}}$ is a class of sets and hence is of type $\mathbf{B}$ as discussed in the beginning of this section.

Theorem 23.21. If $\mathbf{B}$ satisfies the s.c.c. then $\mathbf{B}$ satisfies the UCL.

Proof. Suppose that there is a family of sequences $\langle b_{\alpha i} \mid \alpha \in On\rangle$ for each $i \in I$, where I is a set, such that

$$\{b_{\alpha i} \mid \alpha \in On \wedge i \in I\} \subseteq B,$$
$$(\forall i \in I)(\forall \alpha)(\forall \beta)[\alpha < \beta \rightarrow b_{\alpha i} \geq b_{\beta i}],$$

and

$$(\forall i \in I)\left[\prod_{\alpha \in On} b_{\alpha i} = \mathbf{0}\right].$$

Then, for each $i \in I$, $\langle {}^{-}b_{\alpha i} \mid \alpha \in On\rangle$ is nondecreasing and converges to $\mathbf{1}$. By the s.c.c., these sequences must eventually become constant, i.e.,

$$(\forall i \in I)(\exists \beta_i)[{}^{-}b_{\beta_i i} = \mathbf{1}],$$
$$(\forall i \in I)(\forall \alpha \geq \beta_i)[b_{\alpha i} = \mathbf{0}].$$

Let $\beta = \sup_{i \in I} \beta_i$ (note that I is a set). Then

$$(\forall \alpha > \beta)\left[\sum_{i \in I} b_{\alpha i} = \mathbf{0}\right],$$

hence

$$\prod_{\alpha \in On} \sum_{i \in I} b_{\alpha i} = \mathbf{0}.$$

In particular, $\mathbf{B}$ satisfies the UCL if $\mathbf{B}$ satisfies the $\aleph_\alpha$-chain condition for some α.

Remark. If $\tilde{\mathbf{B}}$ is the completion of $\mathbf{B}$ and $\tilde{\mathbf{B}}$ satisfies the UCL, then $\mathbf{V}^{(\mathbf{B})}$ and $\mathbf{V}^{(\tilde{\mathbf{B}})}$ satisfy the same axioms of ZF by Theorem 23.18. In Theorem 23.9 we proved the Axiom of Subsets for $\mathbf{V}^{(\tilde{\mathbf{B}})}$. Next we prove the Axiom of Replacement for $\mathbf{V}^{(\mathbf{B})}$ assuming the UCL for $\tilde{\mathbf{B}}$.

Theorem 23.22. Let $\tilde{\mathbf{B}}$ be the completion of $\mathbf{B}$. If $\tilde{\mathbf{B}}$ satisfies the UCL, then $\mathbf{V}^{(\mathbf{B})}$ satisfies the Axiom of Replacement.

Proof. As in the proof of Theorem 9.25 we have to show that

(i) $[\![(\forall x)(\exists y)\varphi'(x, y)]\!] = \mathbf{1}$

implies

(ii) $(\forall a \in V^{(\mathbf{B})})[[\![(\exists v)(\forall x \in a)(\exists y \in v)\varphi'(x, y)]\!] = \mathbf{1}]$.

Therefore assume (i) and let $a \in V^{(\mathbf{B})}$. Suppose that (ii) does not hold for a. Then for some $b \in B$,

$$\mathbf{0} < b \wedge b \cdot [\![(\exists v)(\forall x \in a)(\exists y \in v)\varphi'(x, y)]\!] = \mathbf{0}.$$

Then

$$(\forall x \in \mathscr{D}(a))\left[- \sum_{y \in V_\alpha^{(\mathbf{B})}} [\![\varphi'(x, y)]\!] = \mathbf{0} \right]$$

by (i) and Theorem 23.15, hence

$$\sum_{x \in \mathscr{D}(a)} \left(- \sum_{y \in V_\alpha^{(\mathbf{B})}} [\![\varphi'(x, y)]\!] \right) = \mathbf{0} \qquad \text{by the UCL.}$$

Therefore for some $\alpha \in On$,

$$\sum_{x \in \mathscr{D}(a)} \left(- \sum_{y \in V_\alpha^{(\mathbf{B})}} [\![\varphi'(x, y)]\!] \right) \not\geq b,$$

or by passing to complements,

$$\text{(iii)} \quad \prod_{x \in \mathscr{D}(a)} \sum_{y \in V_\alpha^{(\mathbf{B})}} [\![\varphi'(x, y)]\!] \not\leq {}^-b.$$

Let $v \in V^{(\mathbf{B})}$ be $v\colon V_\alpha^{(\mathbf{B})} \to \{\mathbf{1}\}$, i.e., v is the constant function $\mathbf{1}$ on $V_\alpha^{(\mathbf{B})}$. Then by assumption

$$\begin{aligned} \mathbf{0} &= b \cdot [\![(\forall x \in a)(\exists y \in v)\varphi'(x, y)]\!] \\ &= b \cdot \prod_{x \in \mathscr{D}(a)} \left(a(x) \Rightarrow \sum_{y \in V_\alpha^{(\mathbf{B})}} [\![\varphi'(x, y)]\!] \right) \\ &\geq b \cdot \prod_{x \in \mathscr{D}(a)} \sum_{y \in V_\alpha^{(\mathbf{B})}} [\![\varphi'(x, y)]\!] > \mathbf{0} \qquad \text{by (iii).} \end{aligned}$$

This is a contradiction.

Exercise. Prove the converse of Theorem 23.22 in the following form: If $\mathbf{V}^{(\tilde{\mathbf{B}})}$ satisfies the Axiom of Replacement, then $\tilde{\mathbf{B}}$ satisfies the UCL. Hint: (K. Gloede) Let I be a set and let $\{b_{\alpha i} \mid i \in I, \alpha \in On\} \subseteq \tilde{\mathbf{B}}$ be such that

1. $(\forall i \in I)(\forall \alpha)(\forall \beta)[\alpha < \beta \to b_{\alpha i} \leq b_{\beta i}]$.

2. $(\forall i \in I)\left[\sum_{\alpha \in On} b_{\alpha i} = \mathbf{1} \right]$.

To show that $\sum_\alpha \prod_{i \in I} b_{\alpha i} = \mathbf{1}$, we define u by

$$[\![x \in u]\!] = \sum_{\alpha \in On} \prod_{i \in I} [\![x = \langle \check{\imath}, \check{\alpha} \rangle]\!] \cdot b_{\alpha i},$$

show that $[\![(\forall x)(\exists y)[x \in \check{I} \to y \in On \wedge \langle x, y \rangle \in u]]\!] = \mathbf{1}$, then invoke the Axiom of Replacement in $\mathbf{V}^{(\tilde{\mathbf{B}})}$ together with the Axiom of Subsets and Unions in $\mathbf{V}^{(\tilde{\mathbf{B}})}$ to conclude that $[\![(\exists \alpha)(\forall i \in \check{I})(\exists \rho < \alpha)[\langle i, \rho \rangle \in u]]\!] = \mathbf{1}$.

Remark. We recall that from §16 we have assumed that V satisfies the Axiom of Choice. Thus,

Theorem 23.23. $[\![AC]\!] = \mathbf{1}$ in $\mathbf{V}^{(\mathbf{B})}$.

Proof. In most cases $\mathbf{B}$ satisfies the additional requirement that $\mathbf{V}^{(\mathbf{B})}$ is a model of ZF. In this case we can prove the AC in $\mathbf{V}^{(\mathbf{B})}$ by a forcing argument just as in Theorem 14.25 with suitable modifications as in the proof of Theorem 23.24. below. However, since we shall give an example of a Boolean algebra $\mathbf{B}$ such that $\mathbf{V}^{(\mathbf{B})}$ does not satisfy the Axiom of Powers, we indicate a direct proof of Theorem 23.23 in the general case. We take the Axiom of Choice in the following form:

$$(\forall u)(\exists v)(\forall x \in u)[(\exists y \in x)(\exists !\, x' \in u)[y \in x'] \rightarrow (\exists !\, y \in x)[y \in v]].$$

Let $\phi(x, y)$ be $y \in x \wedge (\exists !\, x' \in u)[y \in x']$, $u \in V^{(\mathbf{B})}$. Since $\bigcup_{x \in \mathscr{D}(u)} \mathscr{D}(x)$ is a set, let $\{y_\xi \mid \xi < \alpha\}$ be an enumeration of this set (using the AC in V). Then define $v \in V^{(\mathbf{B})}$ by

$$\mathscr{D}(v) = \bigcup_{x \in \mathscr{D}(u)} \mathscr{D}(x) = \{y_\xi \mid \xi < \alpha\},$$

$$(\forall y \in \mathscr{D}(v))\left[v(y) = \sum_{x \in \mathscr{D}(u)} \sum_{\xi < \alpha} ([\![y = y_\xi \wedge \phi(x, y_\xi)]\!] \cdot \prod_{\eta < \xi} [\![\neg \phi(x, y_\eta)]\!])\right].$$

Note that $v(y) \in B$, since we have only sup's and inf's over sets. Now one can show that

$$(\forall y, y' \in \mathscr{D}(v))[v(y) \cdot [\![y \in x]\!] \cdot v(y') \cdot [\![y' \in x]\!] \cdot [\![x \in u]\!] \leq [\![y = y']\!]]$$

which proves the uniqueness part and

$$[\![(\exists y)\phi(x, y)]\!] \leq [\![(\exists y \in x)[y \in v]]\!]$$

which proves the existence part of the conclusion.

Remark. Sometimes we need a stronger form of the Axiom of Choice:

$$ACH \qquad (\forall x)[x \neq 0 \rightarrow H(x) \in x]$$

where H is definable in ZF using possibly some new constants which are added to the language of ZF (e.g., H itself may be a new function constant). Thus ACH means that there is a definable well-ordering of the universe. If we assume ACH, $\mathbf{V}^{(\mathbf{B})}$ is understood to be the extended structure

$$\mathbf{V}^{(\mathbf{B})} = \langle V^{(\mathbf{B})}, \bar{=}, \bar{\in}, \overline{M}, \check{B}, \check{+}, \check{\cdot}, \check{H}, F\rangle$$

where

$$[\![u \in \check{H}]\!] = \sum_{k \in H} [\![u = \check{k}]\!].$$

Theorem 23.24. Suppose that $\mathbf{V}^{(\mathbf{B})}$ is a model of ZF and assume ACH. Then ACH is $\mathbf{B}$-valid in $\mathbf{V}^{(\mathbf{B})}$.

Proof. (The proof shows how to apply forcing arguments in the case where $\mathbf{B}$ is a class.) Let $\mathbf{M}$ be a countable transitive structure such that $\langle \mathbf{M}, \mathbf{B}^{\mathbf{M}}, H^{\mathbf{M}}\rangle$ is an elementary substructure of $\langle V, \mathbf{B}, H\rangle_*$ with respect to the language $\mathscr{L}^*$ of $\mathbf{V}^{(\mathbf{B})}$. Let $h_0\colon \mathbf{B}^{\mathbf{M}} \rightarrow \mathbf{2}$ be a homomorphism which preserves all the sums which are definable in $\mathbf{M}$. (The set of these sums is countable. Note that $\mathbf{B}^{\mathbf{M}}$ is only a class in $\mathbf{M}$, so h_0 need not be $\mathbf{M}$-complete, but h_0 preserves all the sums 6 in the proof of Theorem 14.22.) Finally, let $F = \{b \in B^M \mid h_0(b) = \mathbf{1}\}$ be the ultrafilter corresponding to h_0. As in Theorem 14.22 there is a mapping

$$h: (V^{(\mathbf{B})})^{\mathbf{M}} \xrightarrow{\text{onto}} M[F]$$

such that

$$M[F] \vDash \varphi(h(u_1), \ldots, h(u_n)) \leftrightarrow h_0([\![\varphi(u_1, \ldots, u_n)]\!]) = \mathbf{1}$$

for $u_1, \ldots, u_n \in (V^{(\mathbf{B})})^{\mathbf{M}}$, φ a formula of $\mathscr{L}^*$. We have to show that $M[F] \vDash ACH$. Using the language $\mathscr{L}^*$ we can define in $\mathbf{M}$ (a Gödelization of) the ramified language obtained from $\mathscr{L}^*$ and M (i.e., with ordinals ranging over the ordinals in M), and we can express the syntactical notion of "U is a constant term" in $\mathbf{M}$ as well as $[\![u \in v]\!]$ and $[\![u = v]\!]$. Let D be the denotation operator related to $M[F]$. $D(u)$ can be expressed as follows

$$D(u) = \{D(v) \mid \rho(v) < \rho(u) \wedge [\![v \in u]\!] \in F\}.$$

Using $H^{\mathbf{M}}$, we have an $\mathbf{M}$-definable well-ordering of M and hence an $\mathbf{M}$-definable well-ordering $\leq$ of the constant terms. For $x_1, x_2 \in M[F]$, let $C(x_1)$ be the first y (w.r.t. $\leq$) such that $x_1 = D(y)$; then

$$x_1 \leq^* x_2 \overset{\Delta}{\leftrightarrow} C(x_1) \leq C(x_2).$$

Then $\leq^*$ is a well-ordering of $M[F]$ which is definable in $M[F]$ using the language $\mathscr{L}^*$. Therefore ACH holds in $M[F]$.

Remark. Next we construct two counterexamples to show that $\mathbf{V}^{(\mathbf{B})}$ need not satisfy the Axiom of Replacement, and even if it does, it need not satisfy the Axiom of Powers.

Theorem 23.25. There is a Boolean algebra $\mathbf{B}$ such that $\mathbf{V}^{(\mathbf{B})}$ does not satisfy the Axiom of Replacement.

Proof. Define a partial order structure $\mathbf{P} = \langle P, \leq \rangle$ as follows:

$$P \overset{\Delta}{=} \{p \mid \bar{\bar{p}} < \omega \wedge (\forall p' \in p)(\exists i \in \omega)(\exists \alpha \in On)[p' = \langle i, \alpha \rangle] \\ \wedge (\forall i \in \omega)(\forall \alpha)(\forall \beta)[\langle i, \alpha \rangle \in p \wedge \langle i, \beta \rangle \in p \rightarrow \alpha = \beta]\},$$

i.e., elements of P are functions from a into On for some $a \subseteq \omega$, with $\bar{\bar{a}} < \omega$.

$$p_1 \leq p_2 \overset{\Delta}{\leftrightarrow} p_1 \supseteq p_2 \quad \text{for} \quad p_1, p_2 \in P.$$

If we replace On in P by some α_0, then the resulting $\mathbf{P}_{\alpha_0}$ is a set. If $\mathbf{B}_{\alpha_0}$ is the Boolean algebra of regular open sets in $\mathbf{P}_{\alpha_0}$, then α_0 becomes countable in $\mathbf{V}^{(\mathbf{B}_{\alpha_0})}$. Similarly, we shall now obtain a function from ω onto On contradicting the Axiom of Replacement. Let $\tilde{\mathbf{B}}$ be the complete Boolean algebra of all regular open classes in $\mathbf{P}$ and let $\mathbf{B}$ be the Boolean algebra of all regular open classes which are of the form

$$\sum \{A``x \mid x \in a\}$$

where a is a set and for each x in a, $A``x$ is a basic open class. Then $\tilde{\mathbf{B}}$ is the completion of $\mathbf{B}$ and, referring to the remark preceding Theorem 23.21, $\mathbf{B}$ can be regarded as a class of sets. We define $\check{P}$, $\check{\leq}$, G by

$$[\![u \in \check{P}]\!] \triangleq \sum_{p \in P} [\![u = \check{p}]\!]$$

$$[\![\check{\leq}(u, v)]\!] \triangleq \sum_{\substack{p_1, p_2 \in P \\ p_1 \leq p_2}} [\![u = \check{p}_1]\!] \cdot [\![u = \check{p}_2]\!]$$

$$[\![u \in G]\!] \triangleq \sum_{p \in P} [\![u = \check{p}]\!] \cdot [p]^{-0}.$$

Now consider

$$\mathbf{V}^{(\mathbf{B})} \triangleq \langle V^{(\mathbf{B})}, \bar{=}, \bar{\in}, M, \check{P}, \check{\leq}, G \rangle.$$

First we prove that

1. $[\![(\forall i \in \omega)(\exists \alpha)(\exists p \in \check{P})[\langle i, \alpha \rangle \in p \wedge p \in G]]\!] = \mathbf{1}.$

By Theorem 23.4,

$$[\![(\forall i \in \omega)(\exists \alpha)(\exists p \in \check{P})[\langle i, \alpha \rangle \in p \wedge p \in G]]\!] = \prod_{i < \omega} \sum_{\alpha \in On} \sum_{\substack{p \in P \\ \langle i, \alpha \rangle \in p}} [\![\check{p} \in G]\!].$$

But for each $i \in \omega$

$$\sum_{\alpha \in On} \sum_{\substack{p \in P \\ \langle i, \alpha \rangle \in p}} [\![\check{p} \in G]\!] \geq \sum_{\alpha \in On} [\{\langle i, \alpha \rangle\}]^{-0} = \mathbf{1}.$$

This proves 1.

We next prove that

2. $[\![(\forall i \in \omega)(\forall \alpha)(\forall \beta)[(\exists p \in \check{P})[\langle i, \alpha \rangle \in p \wedge p \in G]$
$\wedge (\exists g \in \check{P})[\langle i, \beta \rangle \in g \wedge g \in G] \rightarrow \alpha = \beta]]\!] = \mathbf{1}.$

Let $i \in \omega$ and $b_\alpha = [\![(\exists p \in \check{P})(\langle \check{i}, \check{\alpha} \rangle \in p \wedge p \in G)]\!]$. We have to show that $b_\alpha \cdot b_\beta \leq [\![\check{\alpha} = \check{\beta}]\!]$. Therefore we can assume $\alpha \neq \beta$. Then

$$b_\alpha = \sum_{\substack{p \in P \\ \langle i, \alpha \rangle \in p}} [p]^{-0} = [\{\langle i, \alpha \rangle\}]^{-0},$$

hence

$$b_\alpha \cdot b_\beta = [\{\langle i, \alpha \rangle\}]^{-0} \cdot [\{\langle i, \beta \rangle\}]^{-0} = \mathbf{0}.$$

Finally, we will prove that

3. $[\![(\forall \alpha)(\exists i < \omega)(\exists p \in \check{P})[\langle i, \alpha \rangle \in p \wedge p \in G]]\!] = \mathbf{1}.$

$$[\![(\exists i < \omega)(\exists p \in \check{P})[\langle i, \check{\alpha} \rangle \in p \wedge p \in G]]\!]$$
$$= \sum_{i < \omega} \sum_{\substack{p \in P \\ \langle i, \alpha \rangle \in p}} [p]^{-0} = \sum_{i < \omega} [\{\langle i, \alpha \rangle\}]^{-0} = \mathbf{1}$$

for each α.

Now, in order to show that the Axiom of Replacement does not hold in $\mathbf{V}^{(\mathbf{B})}$, let $\varphi(i, \alpha)$ be $(\exists p \in \check{P})[\langle i, \alpha\rangle \in p \wedge p \in G]$. Then, by 1–3,

$$[\![(\forall i \in \omega)(\exists!\, \alpha)\varphi(i, \alpha) \wedge (\forall \alpha)(\exists i \in \omega)\varphi(i, \alpha)]\!] = \mathbf{1}$$

i.e., φ determines a function from ω onto On in $V^{(\mathbf{B})}$, but

$$[\![(\exists v)(\forall i < \omega)(\exists x \in v)\varphi(i, x)]\!] = \mathbf{0},$$

therefore the Axiom of Replacement does not hold in $\mathbf{V}^{(\mathbf{B})}$.

Theorem 23.26. There is a Boolean algebra $\tilde{\mathbf{B}}$ which satisfies the UCL (even more, $\tilde{\mathbf{B}}$ satisfies the c.c.c.), but $\mathbf{V}^{(\tilde{\mathbf{B}})}$ does not satisfy the Axiom of Powers.

Proof. To prove Theorem 11.11 we used the partial order structure $\mathbf{P}_\alpha = \langle P_\alpha, \leq\rangle$ where

$$P_\alpha = \{p \mid (\exists d)[d \subseteq \alpha \times \omega \wedge \bar{\bar{d}} < \omega \wedge p: d \to 2]\}$$
$$p_1 \leq p_2 \leftrightarrow p_1 \supseteq p_2 \quad \text{for} \quad p_1, p_2 \in P_\alpha,$$

to add α-many subsets of ω to M, so that $\overline{\overline{\mathscr{P}(\omega)}} \geq \alpha$ in $M[G]$ (assuming that α is a cardinal in M). In terms of Boolean-valued models this means

$$[\![\overline{\overline{\mathscr{P}(\omega)}} \geq \check{\alpha}]\!] = \mathbf{1}$$

in $\mathbf{V}^{(\mathbf{B}_\alpha)}$ for each cardinal α. (Note that P_α satisfies the c.c.c., so cardinals are absolute.)

Now we take $\tilde{\mathbf{B}}$ as the Boolean algebra of all regular open classes of 2^{On}. Then $\tilde{\mathbf{B}}$ satisfies the c.c.c. as can be seen from Theorem 11.10 which can be proved for $\tilde{\mathbf{B}}$ with suitable modifications. Hence $\tilde{\mathbf{B}}$ satisfies the UCL and may be considered as a class of sets. Since $On \times \omega \simeq On$, we obtain, in the same way as we proved this result for $\mathbf{V}^{(\mathbf{B}_\alpha)}$,

$$[\![\overline{\overline{\mathscr{P}(\omega)}} \geq \check{\alpha}]\!] = \mathbf{1}$$

in $\mathbf{V}^{(\tilde{\mathbf{B}})}$ for every cardinal α.

Therefore

$$[\![(\forall \alpha)[\overline{\overline{\mathscr{P}(\omega)}} \geq \check{\alpha}]]\!] = \mathbf{1},$$

and hence $[\![(\exists v)(\forall u)[u \in v \leftrightarrow u \leq \omega]]\!] = \mathbf{0}$, i.e., the Axiom of Powers fails in $\mathbf{V}^{(\tilde{\mathbf{B}})}$. (Note that $\mathbf{V}^{(\mathbf{B})}$ satisfies the AC by Theorem 23.23.)

Definition 23.27. $\langle b', \{b_i \mid i \in I\}\rangle$ is called an I-sieve iff $b' > \mathbf{0}$. An element $b > \mathbf{0}$ is sifted by this sieve iff

$$b \leq b' \wedge (\forall i \in I)[b \cdot b_i = \mathbf{0} \vee b \leq b_i].$$

Let $\mathbf{B}_0$ be a complete subalgebra and $\#$ be $\#(B_0, B)$. An element $b > \mathbf{0}$ is *$\#$-sifted* by this sieve iff

$$b \leq b' \wedge (\forall i \in I)[\#(b \cdot b_i) \cdot b \leq b_i].$$

Let I be a set. $\mathbf{B}$ satisfies the I-sieve law (I-SL) iff there exists a complete subalgebra $\mathbf{B}_0$ of $\mathbf{B}$ for which

1. $\mathbf{B}_0$ is a set,
2. for every sieve $\langle b', \{b_i \mid i \in I\}\rangle$ there exists a $\#$-sifted element b.

$\mathbf{B}$ satisfies the sieve law (SL) iff $\mathbf{B}$ satisfies the I-SL for *every* set I.

Definition 23.28. Let $\mathbf{B}_0$ be a complete subalgebra of $\mathbf{B}$ and $\#$ be $\#(\mathbf{B}_0, \mathbf{B})$. Then $\Delta(\mathbf{B}_0) \triangleq \{b \in B \mid \#(b) = \mathbf{1}\}$.

Exercises.

1. $b \in \Delta(\mathbf{B}_0) \to b > \mathbf{0}$.
2. $b' \in \mathbf{B}_0, b \in \Delta(\mathbf{B}_0) \to \#(b' \cdot b) = b'$.
3. $0 < b' \in \mathbf{B}_0, b \in \Delta(\mathbf{B}_0) \to b' \cdot b > \mathbf{0}$.
4. $b + (-\#(b)) \in \Delta(\mathbf{B}_0)$.
5. $b = \#(b) \cdot (b + (-\#(b)))$.

Definition 23.29. Let I be a set. $\mathbf{B}$ satisfies the I-Δ sieve law (I-ΔSL) iff there exists a complete subalgebra $\mathbf{B}_0$ of $\mathbf{B}$ for which

1. $\mathbf{B}_0$ is a set,
2. for every sieve $\langle b', \{b_i \mid i \in I\}\rangle$ defined in $\Delta(\mathbf{B}_0)$ there exists a sifted element in $\Delta(\mathbf{B}_0)$.

Exercise. If $\mathbf{B}$ satisfied the I-ΔSL, then $\mathbf{B}$ satisfies the I-SL.

Theorem 23.30. Suppose that $\tilde{\mathbf{B}}$ satisfies the UCL and $\mathbf{B}$ satisfies the SL, where $\tilde{\mathbf{B}}$ is the completion of $\mathbf{B}$. Then $\mathbf{V}^{(\mathbf{B})}$ and $\mathbf{V}^{(\tilde{\mathbf{B}})}$ satisfy the Axiom of Powers and hence both are Boolean-valued models of $ZF + AC$.

Proof. Again let $\mathbf{M}$ be a countable transitive structure such that $\langle \mathbf{M}, \mathbf{B}^{\mathbf{M}}\rangle$ is an elementary substructure of $\langle V, \mathbf{B}\rangle$ with respect to the language $\mathscr{L}^*$ of $\mathbf{V}^{(\mathbf{B})}$, let $h_0: \mathbf{B}^{\mathbf{M}} \to \mathbf{2}$ be a homomorphism which preserves all $\mathbf{M}$-definable sums and let $F = \{b \in B^{\mathbf{M}} \mid h_0(b) = \mathbf{1}\}$ be the corresponding ultrafilter. $M[F]$ is defined by ramified type theory, so it need not be a model of ZF; however, as in Theorem 9.38, $M[F]$ satisfies all the axioms of ZF except possibly the Axiom of Powers since these axioms are $\mathbf{B}$-valid in $\mathbf{V}^{(\mathbf{B})}$. So we have to prove that $M[F]$ satisfies the Axiom of Powers.

Let u be a constant term, i.e., $D(u) \in M[F]$ where D is the denotation operator related to $M[F]$, and let I be the set of all constant terms with rank $<\rho(u)$. Since "t is a constant term" is definable in $\mathbf{M}$ using the language $\mathscr{L}^*$ (cf. the proof of Theorem 23.22), $I \in M$. Since the SL holds in $\langle \mathbf{M}, \mathbf{B}^{\mathbf{M}}\rangle$, for this particular I there is an $\mathbf{M}$-complete subalgebra $\mathbf{B}_0$ of $\mathbf{B}^{\mathbf{M}}$ satisfying the condition 2 in the definition of the I-SL in $\langle \mathbf{M}, \mathbf{B}^{\mathbf{M}}\rangle$, and $\mathbf{B}_0 \in M$. For the remaining part of the proof, let $p, r, \bar{r}, \bar{r}', \ldots$ range over $\mathbf{B}^{\mathbf{M}} - \{\mathbf{0}\}$. We will reserve $p, q, \bar{q}, q'$, etc., to denote elements of $\mathbf{P}$. For $S \subseteq B_0 \times I$ and $S \in M$ we define $K(S)$ as follows:

$K(S) = D(w)$ if there exists an $\bar{r}$ and a w such that w is a constant term and

(i) $\bar{r} \in F \wedge \bar{r} \leq [\![w \subseteq u]\!]$.
(ii) $(\forall v \in I)[\#(\bar{r} \cdot [\![v \in w]\!]) \cdot \bar{r} \leq [\![v \in w]\!]]$.
(iii) $S = \{\langle p, v\rangle \mid \mathbf{0} < p \wedge p \in B_0 \wedge v \in I \wedge \mathbf{0} < p \cdot \bar{r} \leq [\![v \in w]\!]\}$,

$K(S) = 0$ otherwise.

We claim that

(iv) $K: \mathscr{P}^{\mathbf{M}}(B_0 \times I) \xrightarrow{\text{onto}} \mathscr{P}^{M[F]}(D(u))$.

This proves the theorem. For $B_0 \times I \in M$, so $\mathscr{P}^{\mathbf{M}}(B_0 \times I) \in M \subseteq M[F]$, since M satisfies the Axiom of Powers. Moreover, K is definable in $M[F]$ and $M[F]$ satisfies the Axiom of Replacement, so $\mathscr{P}^{M[F]}(D(u)) \in M[F]$ by (iv).

We prove (iv) in the following way.

1. K is a function.

Let $S \subseteq B_0 \times I$ and $S \in M$. We have to show that $K(S)$ is uniquely determined by the above requirements, i.e., assuming that conditions (i)–(iii) are satisfied by $w_0, \bar{r}_0$ and by $w_1, \bar{r}_1$ for the same S, we have to show that $D(w_0) = D(w_1)$. By symmetry, it suffices to show that

$$D(v) \in D(w_0) \rightarrow D(v) \in D(w_1).$$

Therefore let $D(v) \in D(w_0)$. We can assume $v \in I$. Then $[\![v \in w_0]\!] \in F$. (See the proof of Theorem 23.24.) Let $r' = \bar{r}_0 \cdot [\![v \in w_0]\!]$. Then $r' \in F$ by (i), and

$$\#(r') \cdot \bar{r}_0 \leq [\![v \in w_0]\!] \qquad \text{by (ii).}$$

Thus, by (iii),

$$\langle \#(r'), v\rangle \in S.$$

But also

$$\#(r') \cdot \bar{r}_1 \leq [\![v \in w_1]\!]$$

since by our assumption both $w_0, \bar{r}_0$ and $w_1, \bar{r}_1$ satisfy (iii) for the same S. Therefore, since $r' \in F \wedge r' \leq \#(r')$, $\#(r') \cdot \bar{r}_1 \in F$, and hence $[\![v \in w_1]\!] \in F$. Therefore $D(v) \in D(w_1)$. (See the proof of Theorem 23.24.)

2. The range of K is $\mathscr{P}^{M[F]}(D(u))$.

Let $D(w) \subseteq D(u)$ for some constant term w. Then $[\![w \subseteq u]\!] \in F$. For $v \in I$ define $b_v = [\![v \in w]\!]$. Then by condition 2 of the I-SL,

$$(\forall r)[r \leq [\![w \subseteq u]\!] \rightarrow (\exists \bar{r})[\bar{r} \leq r \wedge (\forall v \in I)[\#(\bar{r} \cdot b_v) \cdot \bar{r} \leq b_v]]].$$

Obviously, such $\bar{r}$'s are dense beneath $[\![w \subseteq u]\!] \in F$. Therefore

$$(\exists \bar{r})[\bar{r} \leq [\![w \subseteq u]\!] \wedge \bar{r} \in F \wedge (\forall v \in I)[\#(\bar{r} \cdot b_v) \cdot \bar{r} \subseteq b_v]].$$

Define

$$S = \{\langle p, v\rangle \mid \mathbf{0} < p \wedge p \in B_0 \wedge v \in I \wedge \mathbf{0} < p \cdot \bar{r} \leq [\![v \in w]\!]\}.$$

Then $K(S) = D(w)$ and $S \in \mathscr{P}^{\mathbf{M}}(B_0 \times I)$.

Definition 23.31. $\mathbf{B}$ is $(\aleph_\alpha, \aleph_\beta)$-splitable if the following condition is satisfied. Suppose $\mathbf{0} < q \leq b$, $q, b \in B$, $\{b_{ij} \mid i \in I \wedge j \in \aleph_\beta\} \subseteq B$, $\bar{\bar{I}} \leq \aleph_\alpha$, and $(\forall i \in I)[\sum_{j < \aleph_\beta} b_{ij} = b]$. Then

$$(\exists \bar{q} \in B)(\exists \Lambda \subseteq B)[\mathbf{0} < \bar{q} \leq q \wedge \bar{\bar{\Lambda}} \leq \aleph_\alpha \wedge (\forall q' \in B)$$
$$[\mathbf{0} < q' \leq \bar{q} \rightarrow (\forall i \in I)(\exists p \in \Lambda)(\exists \gamma < \aleph_\beta)[p \cdot q' > \mathbf{0} \wedge p \cdot \bar{q} \leq \sum_{j<\gamma} b_{ij}\Big]\Big]\Big].$$

Theorem 23.32. If $\mathbf{B}$ satisfies the c.c.c. and $cf(\aleph_\beta) > \aleph_0$, then $\mathbf{B}$ is $(\aleph_\alpha, \aleph_\beta)$-splitable.

Proof. If $\mathbf{B}$ satisfies the c.c.c. and $cf(\aleph_\beta) > \aleph_0$, then

$$b = \sum_{j < \aleph_\beta} b_{ij} = \sum_{j<\gamma} b_{ij} \quad \text{for some} \quad \gamma < \aleph_\beta.$$

Therefore we can simply take $\bar{q} = q$ and $\Lambda = \{b\}$.

Theorem 23.33. Suppose $cf(\aleph_\beta) > \aleph_\alpha$, $\mathbf{B}$ is $(\aleph_\alpha, \aleph_\beta)$-splitable and $\mathbf{V}^{(\mathbf{B})}$ satisfies the axioms of $ZF + AC$. Then $[\![cf((\aleph_\beta)^\vee) > (\aleph_\alpha)^\vee]\!] = \mathbf{1}$.

Proof. Let $\mathbf{M}$ be a countable transitive structure such that $\langle \mathbf{M}, \mathbf{B}^{\mathbf{M}}\rangle$ is an elementary subsystem of $\langle V, \mathbf{B}\rangle$. We will prove that in $(\mathbf{V}^{(\mathbf{B})})^{\mathbf{M}}$, $[\![cf((\aleph_\beta)^\vee) > (\aleph_\alpha)^\vee]\!] = \mathbf{1}$. Let $f \in (V^{(\mathbf{B})})^{\mathbf{M}}$ and

$$b = [\![f\colon (\aleph_\alpha^{\mathbf{M}})^\vee \rightarrow (\aleph_\beta^{\mathbf{M}})^\vee]\!].$$

We have to show that

$$b \leq [\![(\exists y < (\aleph_\beta^{\mathbf{M}})^\vee)(\forall \xi < (\aleph_\alpha^{\mathbf{M}})^\vee)[f(\xi) < y]]\!].$$

Therefore we can assume $b > \mathbf{0}$. Let $h_0\colon \mathbf{B}^{\mathbf{M}} \rightarrow \mathbf{2}$ be a homomorphism preserving all sums which are definable in $\mathbf{M}$ and such that $h_0(b) = \mathbf{1}$. For the remaining part we work in $\langle \mathbf{M}, \mathbf{B}^{\mathbf{M}}\rangle$. In order to avoid cumbersome notations, we will write $\aleph_\alpha, \aleph_\beta, \mathbf{B}, \ldots$ instead of $\aleph_\alpha^{\mathbf{M}}, \aleph_\beta^{\mathbf{M}}, \mathbf{B}^{\mathbf{M}}, \ldots$. Also $\bar{\bar{a}}$ means $\bar{\bar{a}}^{(\mathbf{M})}$. Let $I = \aleph_\alpha$ and for $\xi < \aleph_\alpha$, $\eta < \aleph_\beta$

$$b_{\xi\eta} = b \cdot [\![f(\xi) = \check{\eta}]\!].$$

Then

$$(\forall \xi < \aleph_\alpha)\left[\sum_{\eta < \aleph_\beta} b_{\xi\eta} = b \cdot [\![(\exists \eta < (\aleph_\beta)^\vee)[f(\xi) = \eta]]\!]\right] = b.$$

Since $\mathbf{B}$ is $(\aleph_\alpha, \aleph_\beta)$-splitable, for each $r \leq b$ there exist $\bar{r} \in B$ and $\Lambda \subseteq B$ such that $\mathbf{0} < \bar{r} \leq r$ and

$$\bar{\bar{\Lambda}} \leq \aleph_\alpha \wedge (\forall \xi < \aleph_\alpha)(\forall r' \in B)$$
$$\left[\mathbf{0} < r' \leq r \rightarrow (\exists \eta < \aleph_\beta)(\exists p \in \Lambda)\left[p \cdot r' \wedge p \cdot \bar{r} \leq \sum_{\eta' < \eta} b_{\xi\eta'}\right]\right].$$

Since such $\bar{r}$'s are dense beneath b and $h_0(b) = \mathbf{1}$, we can find an $\bar{r} \leq b$ such that $h_0(\bar{r}) = \mathbf{1}$ and $\bar{r}$ has the above properties. In particular,

$$(\forall r' \in B)[\mathbf{0} < r' \leq \bar{r} \rightarrow (\exists p \in \Lambda)[p \cdot r' > \mathbf{0}]].$$

Choose $B' \subseteq \{r' \mid r' \in B \wedge \mathbf{0} < r' \leq \bar{r}\}$ such that

$$(\forall r', r'' \in B')[(\exists p \in \Lambda)[p \cdot r' > \mathbf{0} \wedge p \cdot r'' > \mathbf{0}) \rightarrow r' = r'']$$

and

$$(\forall r' \in B)[\mathbf{0} < r' \leq \bar{r} \rightarrow (\exists r'' \in B')(\exists p \in \Lambda)[p \cdot r' > \mathbf{0} \wedge p \cdot r'' > \mathbf{0}]],$$

i.e., we identify those $r' \cdot r'' \in B$ for which $\mathbf{0} < r' \leq \bar{r}$, $\mathbf{0} < r'' \leq \bar{r}$, and $pr' > \mathbf{0} \wedge pr'' > \mathbf{0}$ for the same $p \in \Lambda$. Since $\bar{\bar{\Lambda}} \leq \aleph_\alpha$, $\bar{\bar{B'}} \leq \aleph_\alpha$. Since

$$(\forall \xi < \aleph_\alpha)(\forall r' \in B')(\exists \eta < \aleph_\beta)(\exists p \in \Lambda)\left[p \cdot r' > \mathbf{0} \wedge p \cdot \bar{r} \leq \sum_{\eta' \in \eta} b_{\xi\eta'}\right]$$

and $cf(\aleph_\beta) > \aleph_\alpha$, we can obtain a single $\eta > \aleph_\beta$ such that

$$(\forall \xi < \aleph_\alpha)(\forall r' \in B')(\exists p \in \Lambda)\left[p \cdot r' > \mathbf{0} \wedge p \cdot \bar{r} \leq \sum_{\eta' < \eta} b_{\xi\eta'}\right].$$

Because of the choice of B', it can again be replaced by $\{r' \mid \mathbf{0} < r' \leq \bar{r}\}$,

(i) $(\forall \xi < \aleph_\alpha)(\forall r' \in B)$

$$\left[\mathbf{0} < r' \leq \bar{r} \rightarrow (\exists p \in \Lambda)\left[p \cdot r' > \mathbf{0} \wedge p\bar{r} \leq \sum_{\eta' < \eta} b_{\xi\eta'}\right]\right].$$

Define F and

$$h \colon (V^{(\mathbf{B})})^{\mathbf{M}} \xrightarrow{\text{onto}} M[F]$$

as in the proof of Theorem 23.24. Suppose $h(f)(\xi) = \eta'$ for $\xi < \aleph_\alpha$ and $\eta' < \aleph_\alpha$. Then $h_0([\![f(\check{\xi}) = \check{\eta}']\!]) = \mathbf{1}$, and since $h_0(\bar{r}) = \mathbf{1}$,

$$[\![f(\check{\xi}) = \check{\eta}']\!] \cdot \bar{r} > \mathbf{0}.$$

Therefore, by (i),

$$(\exists p \in \Lambda)[p \cdot [\![f(\check{\xi}) = \check{\eta}']\!]\bar{r} > \mathbf{0} \wedge p \cdot \bar{r} \leq [\![f(\check{\xi}) < \check{\eta}]\!]].$$
$$\mathbf{0} < p \cdot [\![f(\check{\xi}) = \check{\eta}']\!] \cdot \bar{r} \cdot [\![f(\check{\xi}) < \check{\eta}]\!] \leq [\![\check{\eta}' < \check{\eta}]\!],$$

hence $\eta' < \eta$. Therefore η is a bound for $h(f)$. This proves the theorem.

Corollary 23.34. If $\mathbf{V}^{(\mathbf{B})}$ satisfies the axioms of $ZF + AC$ and $\mathbf{B}$ is $(\aleph_\alpha, \aleph_\beta)$-splitable for all $\aleph_\alpha$, $\aleph_\beta$ such that $cf(\aleph_\beta) > \aleph_\alpha$ (e.g., if $\mathbf{B}$ satisfies the c.c.c.), then cardinals are absolute, i.e., $(\forall \alpha)[[\![\text{Card}((\aleph_\alpha)^\vee)]\!] = \mathbf{1}]$.

Proof. One can easily prove that if for all α and β

$$cf(\aleph_\beta) > \aleph_\alpha \rightarrow [\![cf(\aleph_\beta)^\vee > (\aleph_\alpha)^\vee]\!] = \mathbf{1}$$

then $(\forall \alpha)[[\![\text{Card}(\aleph_\alpha)^\vee]\!] = \mathbf{1}]$.

Definition 23.35. **B** satisfies the $(\omega_\alpha, \omega_\beta)$-WDL iff for every family

$$\{b_{\xi\eta} \mid \xi < \omega_\alpha \wedge \eta < \omega_\beta\} \subseteq B,$$

$$\prod_{\xi < \omega_\alpha} \sum_{\eta < \omega_\beta} b_{\xi\eta} = \sum_{f \in \omega_\beta{}^{\omega_\alpha}} \prod_{\zeta < \omega_\alpha} \sum_{\eta < f(\zeta)} b_{\zeta\eta}.$$

Remark. This is a natural generalization of the (ω, ω_α)-WDL (see Definition 20.1).

Theorem 23.36. If **B** satisfies the $(\omega_\alpha, \omega_\beta)$-WDL, then **B** is $(\omega_\alpha, \omega_\beta)$-splitable.

Proof. Let $\mathbf{0} < r \leq b \wedge (\forall \xi < \omega_\alpha)[b = \sum_{\eta < \omega_\beta} b_{\xi\eta}]$. Then, by the $(\omega_\alpha, \omega_\beta)$-WDL,

$$b = \prod_{\xi < \omega_\alpha} \sum_{\eta < \omega_\beta} b_{\xi\eta} = \sum_{f \in \omega_\beta{}^{\omega_\alpha}} \prod_{\xi < \omega_\alpha} \sum_{\eta < f(\xi)} b_{\xi\eta}.$$

Therefore, for some $f \in \omega_\beta{}^{\omega_\alpha}$,

$$\bar{r} = r \cdot \prod_{\xi < \omega_\alpha} \sum_{\eta < f(\xi)} b_{\xi\eta} > \mathbf{0}.$$

Let $\Lambda = \{\sum_{\eta < f(\xi)} b_{\xi\eta} \mid \xi < \omega_\alpha\}$. Then $\overline{\overline{\Lambda}} \leq \aleph_\alpha$. Thus it remains to show that for $r' \in B$

$$\mathbf{0} < r' \leq \bar{r} \rightarrow (\forall \xi < \omega_\alpha)(\exists \eta < \omega_\beta)(\exists p \in \Lambda)\left[(p \cdot r' > \mathbf{0}) \wedge p\bar{r} \leq \sum_{\eta' < \eta} b_{\xi\eta'}\right].$$

Therefore let $\mathbf{0} < r' \leq \bar{r}$ and $\xi < \omega_\alpha$. Then

$$r' \leq \sum_{\eta < f(\xi)} b_{\xi\eta},$$

and we can simply take

$$p = \sum_{\eta < f(\xi)} b_{\xi\eta} \in \Lambda$$

$$p \cdot r' = r' > \mathbf{0} \wedge p \cdot \bar{r} \leq p = \sum_{\eta < f(\xi)} b_{\xi\eta}.$$

Corollary 23.37. Suppose $cf(\aleph_\beta) > \aleph_\alpha$ and $\mathbf{V}^{(\mathbf{B})}$ satisfies the axioms of $ZF + AC$. Then the following conditions are equivalent.

(i) **B** satisfies the $(\omega_\alpha, \omega_\beta)$-WDL.
(ii) **B** is $(\omega_\alpha, \omega_\beta)$-splitable.
(iii) $[\![cf((\aleph_\beta)^\vee) > (\aleph_\alpha)^\vee]\!] = \mathbf{1}$ in $\mathbf{V}^{(\mathbf{B})}$.

Proof. We can prove (i) $\leftrightarrow$ (iii) in the same way we proved Theorem 20.3. By Theorem 23.36, (i) $\rightarrow$ (ii); and by Theorem 23.32, (ii) $\rightarrow$ (iii).

Definition 23.38. Let $\mathbf{P} = \langle P, \leq, 1\rangle$ be a partial order structure with a largest member 1. An element $p \in P$ is said to be *coatomic* if

$$1 \neq p \wedge (\forall x)[x \geq p \rightarrow x = 1 \vee x = p].$$

For any $p \in P$, define

$$CA(p) = \{q \mid q \text{ is coatomic } \wedge q \geq p\}.$$

P is said to be coatomic if

$$(\forall p, q \in P)[CA(q) \subseteq CA(p) \leftrightarrow p \leq q].$$

A coatomic partial order structure **P** is said to be *strongly coatomic* if

$$(\forall p \in P)(\forall S \subseteq CA(p))(\exists q \in P)[S = CA(q)].$$

Remark. If **P** is coatomic, then

$$(\forall p \in P)[p \neq 1 \rightarrow CA(p) \neq 0]$$

and

$$(\forall p, q \in P)[p = q \leftrightarrow CA(p) = CA(q)).$$

Definition 23.39. Let $\mathbf{P} = \langle P, \leq, 1\rangle$ be strongly coatomic. **P** is said to be $\aleph_\alpha$*-bounded* iff

1. $\overline{\overline{CA(p)}} < \aleph_\alpha$.
2. Define $IC(p) = \{q \mid q \text{ is coatomic } \wedge p \text{ and } q \text{ are incompatible}\}$. Then

$$\overline{\overline{IC(p)}} \leq \aleph_\alpha.$$

3. If p and q are incompatible, then there exists $q_0 \in IC(p)$ such that $q \leq q_0$.
4. Let A be a set of coatomic elements with $\overline{\overline{A}} \leq \aleph_\alpha$. Then

$$\overline{\overline{\{p \mid CA(p) \subseteq A\}}} \leq \aleph_\alpha.$$

Remark. Condition 3 implies that the set of all q's that are incompatible with p is

$$\bigcup_{q \in IC(p)} [q].$$

Definition 23.40. Let $\mathbf{P}_0 = \langle \Gamma, \leq_0, 1_0\rangle$ and $\mathbf{P}_1 = \langle \Delta, \leq_1, 1_1\rangle$ be two partial order structures. We say that $\mathbf{P}_0$ and $\mathbf{P}_1$ form an $\aleph_\alpha$*-Easton pair* iff

1. $\mathbf{P}_0$ is a set and an $\aleph_\alpha$-bounded strongly coatomic partial order structure,
2. For every $\beta \leq \aleph_\alpha$ and for every

$$q_0 \geq_1 q_1 \geq_1 \cdots \geq_1 q_\gamma \geq_1 \cdots \qquad (\gamma < \beta)$$

there exists a $q \in \Delta$ such that

$$(\forall \gamma < \beta)[q \leq_1 q_\gamma].$$

(Next condition is dispensable. We add this in order to simplify the argument.)

3.1. $(\forall p_1, p_2 \in \Gamma)[p_1 \not\leq_0 p \rightarrow (\exists p \in \Gamma)[p \leq_0 p_1 \wedge \neg\text{Comp}\,(p_1 p_2)]]$.

3.2. $(\forall q_1, q \in \Delta)[q_1 \not\leq_1 q \rightarrow (\exists q \in \Delta)[q \leq_1 q_1 \wedge \neg\text{Comp}\,(q_1 q_2)]]$.

That is, Γ and Δ are fine in the sense of Definition 5.21.

Remark. Let $\mathbf{P}_0 = \langle \Gamma, \leq_0, 1_0 \rangle$ and $\mathbf{P}_1 = \langle \Delta, \leq_1, 1_1 \rangle$ form an $\aleph_\alpha$-Easton pair. Define $\mathbf{P} = \langle P, \leq, 1 \rangle$ as $\mathbf{P}_0 \times \mathbf{P}_1$. We use an abbreviated notation such that $p \in \Gamma$ also denotes $\langle p, 1_1 \rangle$ and $q \in \Delta$ also denotes $\langle 1_0, q \rangle$. With this abbreviation, every member of P can be denoted by $p \cdot q$ where $p \in \Gamma$ and $q \in \Delta$ and $1 = 1_0 = 1_1$.

Let $\mathbf{B}$ be the Boolean algebra of all regular open sets in $\mathbf{P}$.

$$\tilde{P} = \{b \in B \mid b \neq \mathbf{0}\}.$$

Let $p_1 \cdot q_1$ and $p_2 \cdot q_2$ be two members of P. $p_1 \cdot q_1$ and $p_2 \cdot q_2$ are compatible iff p_1 and p_2 are compatible and q_1 and q_2 are compatible. Then P is fine, hence

$$[p \cdot q]^{-0} = [p \cdot q]$$

and we may assume that P is a dense subset of $\tilde{P}$, where $1 = \mathbf{1}$. For the member $p \cdot q$ of P, we shall intentionally confuse $p \cdot q \in P$ and $[p \cdot q] \in \tilde{P}$, i.e., we sometimes use $p \cdot q$ in the place of $[p \cdot q]$ and vice versa. Therefore we sometimes express "$p_1 \cdot q_1$ and $p_2 \cdot q_2$ are compatible" by $p_1 \cdot q_1 \cdot p_2 \cdot q_2 > 0$. The former is considered in P and the latter is considered in $\tilde{P}$.

In what follows, we assume that an $\aleph_\alpha$-Easton pair $\mathbf{P}_0$, $\mathbf{P}_1$ is given as above.

Theorem 23.41. If $p, p_0 \in \Gamma$ and $q, q_0 \in \Delta$ then

1. $p_0 \cdot q_0 \leq p \leftrightarrow p_0 \leq p$.
2. $p_0 \cdot q_0 \leq q \leftrightarrow q_0 \leq q$.
3. $p_0 \cdot q_0 \leq p \cdot q \leftrightarrow p_0 \leq p \wedge q_0 \leq q$.
4. $p_0 \cdot q_0 = p \cdot q \leftrightarrow p_0 = p \wedge q_0 = q$.

Theorem 23.42. Suppose $b \in B$, $\{b_j \mid j \in J\} \subseteq B$, $b = \sum_{j \in J} b_j$, where J may be a proper class, and $b' \in \tilde{P}$. Then

$$(\exists p \in \Gamma)(\exists q \in \Delta)[p \cdot q \leq b' \wedge [p \cdot q \leq {}^{-}b \vee (\exists j \in J)[p \cdot q \leq b_j]]].$$

Proof. Case 1: $b' \cdot b > \mathbf{0}$. Then $b' \cdot b_j > \mathbf{0}$ for some $j \in J$. Hence

$$(\exists p \in \Gamma)(\exists q \in \Delta)[p \cdot q \leq b' \cdot b_j]$$

since P is dense in $\tilde{P}$.

Case 2: $b' \cdot b = \mathbf{0}$. Then $b' \leq {}^{-}b$. For the same reason

$$(\exists p \in \Gamma)(\exists q \in \Delta)[p \cdot q \leq b' \leq {}^{-}b].$$

Lemma 23.43. (Easton's main lemma.) Suppose $\aleph_\alpha$ is regular, $q \in \Delta$ and $b = \sum_{j \in J} b_j$, where J may be a proper class. Then

$$\begin{aligned}&(\exists \bar{q} \in \Delta)(\exists \Lambda \subseteq \Gamma)[\bar{q} \leq q \wedge \overline{\overline{\Lambda}} \leq \aleph_\alpha \\ &\quad \wedge (\forall p \in \Lambda)(\exists j \in J)[p \cdot \bar{q} \leq b_j \vee p \cdot \bar{q} \leq {}^{-}b] \wedge (\forall r \in \tilde{P})(\exists p \in \Lambda)[r \cdot p > 0]].\end{aligned}$$

Proof. We construct, in $\aleph_\alpha$ stages, $p_\mu \in \Gamma$ and $q_\mu \in \Delta$ for $\mu < \aleph_\alpha \cdot \aleph_\alpha$ (the ordinal product) satisfying

$$(\forall \mu_1, \mu_2 < \aleph_\alpha \cdot \aleph_\alpha)[\mu_1 < \mu_2 \rightarrow q_{\mu_2} \leq_1 q_{\mu_1}].$$

Stage 0. We pick $p_0 \in \Gamma$ and $q_0 \in \Delta$ such that

$$p_0 \cdot q_0 \leq q \wedge (\exists j \in J)[p_0 \cdot q_0 \leq q \cdot b_j \vee p_0 \cdot q_0 \leq q \cdot (^{-}b)].$$

(See the proof of the preceding theorem.) For all $\nu < \aleph_\alpha$ set $p_\nu = p_0$ and $q_\nu = q_0$.

Stage μ (where $0 < \mu < \aleph_\alpha$). Define $S_\mu = \bigcup_{\nu < \aleph_\alpha \cdot \mu} IC(p_\nu)$. Then by 2 of Definition 23.39, $\overline{\overline{IC(p_\nu)}} < \aleph_\alpha$ for $\nu < \aleph_\alpha \cdot \mu$ and hence $\overline{\overline{S_\mu}} = \aleph_\alpha$. Therefore by 4 of Definition 23.39

$$\overline{\overline{\{p \mid CA(p) \subseteq S_\mu\}}} \leq \aleph_\alpha.$$

So we can enumerate the set $\{p \mid CA(p) \subseteq S_\mu\}$ as follows:

$$\{p \mid CA(p) \subseteq S_\mu\} = \{p_\nu{}^0 \mid \aleph_\alpha \cdot \mu \leq \nu < \aleph_\alpha \cdot (\mu + 1)\}.$$

For $\aleph_\alpha \cdot \mu \leq \nu < \aleph_\alpha \cdot (\mu + 1)$ we pick $p_\nu \in \Gamma$, q'_ν, $q_\nu \in \Delta$ such that

1. $q'_\nu \leq_1 q_\lambda$ for every λ such that $\aleph_\alpha \cdot \mu \leq \lambda < \nu$.
2. $p_\nu \cdot q_\nu \leq p_\nu{}^0 \cdot q'_\nu \wedge [p_\nu \cdot q_\nu \leq {}^{-}b \vee (\exists j \in J)[p_\nu \cdot q_\nu \leq b_j]]$ where the existence of q'_ν in 1 follows from the property of $\mathbf{P}_1$ (2 of Definition 23.40) and the existence of p_ν, q_ν in 2 follows from Theorem 23.42. It is easily seen that

$$\aleph_\alpha \cdot \mu \leq \lambda < \nu \rightarrow q_\nu \leq_1 q_\lambda.$$

Finally, we pick $\bar{q}$ such that $(\forall \mu < \aleph_\alpha \cdot \aleph_\alpha)[\bar{q} \leq_1 q_\mu]$ and let

$$\Lambda = \{p_\mu \mid \mu < \aleph_\alpha \cdot \aleph_\alpha\}.$$

Obviously,

$$\bar{q} \leq q \wedge \Lambda \subseteq \Gamma \wedge \overline{\overline{\Lambda}} \leq \aleph_\alpha \wedge (\forall p \in \Lambda)(\exists j \in J)[p \cdot \bar{q} \leq b_j \vee p \cdot \bar{q} \leq {}^{-}b].$$

Thus it remains to show that

$$(\forall r \in \tilde{P})(\exists p \in \Lambda)[r \cdot p > \mathbf{0}].$$

Let $r = p' \cdot q'$. It suffices to show that

$$(\exists p \in \Lambda)[\mathrm{Comp}\,(p, p')],$$

since $r \cdot p = p' \cdot q' \cdot p \cdot 1_1$. Define $f: CA(p') \rightarrow On$ by the following condition: If $p^* \in CA(p')$ then

$$\begin{aligned} f(p^*) &= \mu_\beta(\beta < \aleph_\alpha \wedge p^* \in S_\beta) \text{ if there is such an } S_\beta \\ f(p^*) &= 0 \text{ otherwise.} \end{aligned}$$

Since $CA(p') < \aleph_\alpha$ and $\aleph_\alpha$ is regular.

$$(\exists \bar{\mu} < \aleph_\alpha)[f\text{``}CA(p') \subseteq \bar{\mu}].$$

Define $IC(\Lambda) = \bigcup_{p \in \Lambda} IC(p)$. It is easily seen that $IC(\Lambda) \subseteq \Lambda$ and

$$IC(\Lambda) \cap CA(p') \subseteq S_{\bar{\mu}}.$$

Therefore there exists a ν such that

$$\aleph_\alpha \cdot \bar{\mu} \le \nu < \aleph_\alpha \cdot (\bar{\mu} + 1),$$

and

$$CA(p_\nu{}^0) = IC(\Lambda) \cap CA(p')$$

by the definition of the notion of being strongly coatomic. Therefore

$$CA(p_\nu) \supseteq IC(\Lambda) \cap CA(p').$$

Suppose p_ν and p' are incompatible. Then

$$CA(p') \cap IC(p_\nu) \neq 0.$$

Consequently,

$$IC(p_\nu) \cap CA(p_\nu) \neq 0.$$

This is a contradiction.

Remark. Next we generalize the lemma for the case of $\aleph_\alpha$-many sequences $\{b_{ij} \mid j \in J_i\}$. However, the conclusion is somewhat weaker than in the main lemma.

Theorem 23.44. Let $\aleph_\alpha$ be given. Suppose

$$r \in P, \bar{\bar{I}} \le \aleph_\alpha, \{b_i \mid i \in I\} \subseteq B, \{b_{ij} \mid i \in I \wedge j \in J_i\} \subseteq B$$

and

$$(\forall i \in I)\left[b_i = \sum_{j \in J_i} b_{ij}\right].$$

Then for some $\bar{r} \in P$ and some $\Lambda \subseteq \Gamma$

1. $\bar{r} \le r$
2. $\bar{\bar{\Gamma}} \le \aleph_\alpha$ and
3. for each $r' \le \bar{r}$ and each $i \in I$

$$(\exists p \in \Lambda)(\exists j \in J_i)\,[p \cdot r' > \mathbf{0} \wedge [p \cdot \bar{r} \le b_{ij} \vee p \cdot \bar{r} \le {}^-b_i]].$$

Proof. Without loss of generality, we may take $I = \aleph_\alpha$. First of all, we assume that $\aleph_\alpha$ is regular and $r \in \Delta$. Taking $q = r$ we can then apply Easton's main lemma $\aleph_\alpha$-many times to define q_μ, Λ_μ for $\mu < \aleph_\alpha$ in the following way:

At stage $\mu < \aleph_\alpha$, we pick q'_μ, q_μ and $\Lambda_\mu \subseteq \Gamma$ such that

(i) $(\forall \nu < \mu)(q'_\mu \le q_\nu)$,
(ii) $q_\mu \le q \cdot q'_\mu$,
(iii) $\bar{\bar{\Lambda}}_\mu \le \aleph_\alpha$,
(iv) $p \in \Lambda_\mu \rightarrow p \cdot q_\mu \le {}^-b_\mu \vee (\exists j \in J_\mu)[p \cdot q_\mu \le b_{\mu j}]$,
(v) $(\forall r' \in P)(\exists p \in \Lambda_\mu)[r' \cdot p > \mathbf{0}]$.

Then $\mu' < \mu < \aleph_\alpha \to q_\mu \leq q'_\mu$. Therefore by 2 of Definition 23.40 we can pick a $\bar{q} \in \Delta$ such that $(\forall \mu < \aleph_\alpha)[\bar{q} \leq q_\mu]$. Define $\Lambda = \bigcup_{\mu < \aleph_\alpha} \Lambda_\mu$ and take $\bar{r} = \bar{q}$. Then $\bar{r}$ and Λ satisfy 2 and 3.

Next we consider an arbitrary $r \in P$ but still assume that $\aleph_\alpha$ is regular. Let $r = p \cdot q$ and for this q let $\bar{q}$ and Λ be constructed as above. Define $\bar{r} = p \cdot \bar{q}$. Then $\bar{r}$ and Λ satisfy 2 and 3.

Finally, we consider the case where $\aleph_\alpha$ is singular. Then $\aleph_\alpha = \sup_{\mu < \alpha'} \aleph_{\alpha_\mu + 1}$ for some sequence $\langle \alpha_\mu \mid \mu < \alpha' \rangle$ where $\alpha' = cf(\aleph_\alpha) < \aleph_\alpha$. Since $\aleph_{\alpha_\mu + 1}$ is regular we can apply the preceding proof for each $\aleph_{\alpha_\mu + 1}$ (inductively on μ). Thus for each $\mu < \alpha'$ we can pick $r_\mu \in P$ and $\Lambda_\mu \subseteq \Gamma$ in the following way. We can assume that for previously defined r_ν's $(\nu < \mu)$ $r_\nu \leq r_{\nu'}$ holds if $\nu' < \nu$. There is an $r \in P$ such that $r \leq r_\nu$ for all $\nu < \mu$. Then $r_\mu \leq r$, $\bar{\bar{\Lambda}}_\mu \leq \aleph_{\alpha_\mu + 1}$, and for each $r' \leq r_\mu$ and each $\xi < \aleph_{\alpha_\mu + 1}$

$$(\exists p \in \Lambda_\mu)[p \cdot r' > 0 \wedge [p \cdot r_\mu \leq {}^-b_\xi \vee (\exists j \in J_\xi)[p \cdot r_\mu \leq b_{\xi_j}]]].$$

We pick $\bar{r} \in P$ such that $\bar{r} \leq r_\mu$ for all $\mu < \alpha'$ and set $\Lambda = \bigcup_{\mu < \alpha'} \Lambda_\mu$. Then $\bar{r}$ and Λ satisfy the desired conditions.

Theorem 23.45. Let $\bar{\bar{I}} \leq \aleph_\alpha$ and $\aleph_\alpha$ be regular. Then $\mathbf{B}$ (the Boolean algebra of regular open sets in $\mathbf{P}$) satisfies the I-SL.

Proof. We claim that $\mathbf{B}_0$ (the Boolean algebra of regular open sets in $\mathbf{P}_0$) satisfies the condition for $\mathbf{B}_0$ in the I-SL. Let $\{b_i \mid i \in I\} \subseteq B$ and $r > \mathbf{0}$. By Theorem 23.44, there exists $\bar{r} \leq r$ and $\Lambda \subseteq \Gamma$ such that

1. $(\forall r' \leq \bar{r})(\forall i \in I)(\exists p \in \Lambda)[p \cdot r' > \mathbf{0} \wedge [p \cdot \bar{r} \leq b_i \vee p \cdot \bar{r} \leq {}^-b_i]]$.

Let $\#$ be $\#(\mathbf{B}_0, \mathbf{B})$. We have to prove that

$$(\forall i \in I)[\#(\bar{r} \cdot b_i) \cdot \bar{r} \leq b_i].$$

Suppose $\#(\bar{r} \cdot b_i) \cdot \bar{r} \nleq b_i$ for some $i \in I$. Then

$$(\exists r' \leq r)[r' \leq \#(\bar{r} \cdot b_i) \cdot \bar{r} \wedge r' \leq {}^-b_i],$$

and hence by 1:

2. $(\exists p \in \Lambda)[p \cdot r' > \mathbf{0} \wedge [p \cdot \bar{r} \leq b_i \vee p \cdot \bar{r} \leq {}^-b_i]]$.

Since $r' \leq {}^-b_i \wedge \mathbf{0} < p \cdot r' \leq p \cdot \bar{r}$, we cannot have $p \cdot \bar{r} \leq b_i$. Therefore $p \cdot \bar{r} \leq b_i$, by 2. Then $p \cdot \bar{r} \cdot b_i = \mathbf{0}$. Therefore

$$p \cdot \#(\bar{r} \cdot b_i) = \mathbf{0}$$

by Theorem 22.10.6, since $p \in \mathbf{B}_0$. Thus $\mathbf{0} < p \cdot r' \leq p \cdot \#(\bar{r} \cdot b_i)\bar{r} = \mathbf{0}$. This is a contradiction.

Theorem 23.46. $\mathbf{B}$ is $(\aleph_\alpha, \aleph_\beta)$-splitable.

Proof. Let $r \leq b$, $\{b_{ij} \mid i \in I \wedge j < \aleph_\beta\} \subseteq B$, $\bar{\bar{I}} \leq \aleph_\alpha$, and

$$(\forall i \in I)\left[b = \sum_{j < \aleph_\beta} b_{ij}\right].$$

Then by Theorem 23.44, there exists an $\bar{r} \in P$ and a $\Lambda \subseteq \Gamma$ such that $\bar{r} \leq r$, $\bar{\bar{\Lambda}} \leq \aleph_\alpha$ and

$$(\forall r' \leq \bar{r})(\forall i \in I)(\exists p \in \Lambda)(\exists j < \aleph_\beta)\ [p \cdot r' > \mathbf{0} \wedge [p \cdot \bar{r} \leq b_{ij} \vee p \cdot \bar{r} \leq {}^{-}b].$$

Since $\mathbf{0} < p \cdot \bar{r} \leq b$, we cannot have $p \cdot \bar{r} \leq {}^{-}b$.

Therefore there exists an $\bar{r} \in P$ and a $\Lambda \subseteq \Gamma$ such that $\bar{r} \leq b$, $\bar{\bar{\Lambda}} \leq \aleph_\alpha$, and

$$(\forall r' \leq \bar{r})(\forall i \in I)(\exists p \in \Lambda)(\exists j < \aleph_\beta)[p \cdot r' > \mathbf{0} \wedge p \cdot \bar{r} \leq b_{ij}].$$

Hence **B** is $(\aleph_\alpha, \aleph_\beta)$-splitable.

24. Easton's Model

In this section we will consider the question of alternatives to the *GCH*. If $G: On \to On$ we wish to know for what choices of G

$$GCH_G \qquad (\forall\alpha)[\overline{\overline{2^{\aleph_\alpha}}} = \aleph_{G(\alpha)}]$$

will be consistent with the axioms of $ZF + AC$.

There are two results, provable in $ZF + AC$, that restrict the choice of G:

$$\alpha \leq \beta \to \overline{\overline{2^{\aleph_\alpha}}} \leq \overline{\overline{2^{\aleph_\beta}}}$$

and

$$(\forall\alpha)[cf(\overline{\overline{2^{\aleph_\alpha}}}) > \aleph_\alpha] \qquad \text{(König's Theorem)}$$

From these results we see that it is necessary that G have the following properties.

1. $(\forall\alpha)(\forall\beta)[\alpha \leq \beta \to G(\alpha) \leq G(\beta)]$
2. $(\forall\alpha)[cf(\aleph_{G(\alpha)}) > \aleph_\alpha]$

Solovay conjectured that 1 and 2 are also sufficient. Solovay's conjecture is at this time still an open question. Strong supporting evidence for the conjecture was established in 1964 by Easton who, using forcing techniques, proved that for any G satisfying 1 and 2 there is a model of $ZF + AC$ in which the GCH_G holds for regular cardinals.

In this section we will prove the existence of Easton's models by showing that for each G satisfying 1 and 2 there is a Boolean algebra $\mathbf{B}$ such that $\mathbf{V}^{(\mathbf{B})}$ is a $\mathbf{B}$-valued model of $ZF + AC$ and in $\mathbf{V}^{(\mathbf{B})}$

$$[\![(\forall\alpha)[\alpha \in \text{Reg} \to \overline{\overline{2^{\aleph_\alpha}}} = \aleph_{G(\alpha)}]]\!] = \mathbf{1}$$

Throughout this section we assume that V satisfies the *GCH* and the Strong Axiom of Choice.

As our first step in the construction of the Easton model that satisfies the GCH_G we define a special partial order structure.

Definition 24.1

1. $q \in P$ iff there exists a sequence $\langle q^\alpha \mid \alpha \in On\rangle$ such that

   (i) $q^\alpha \subseteq \{\langle i, \gamma, \alpha, \eta\rangle \mid i < 2 \wedge \gamma < \aleph_\alpha \wedge \eta < \aleph_{G(\alpha)}\}$ for $\alpha \in \text{Reg}$
   $= 0$ otherwise,

(ii) $q = \bigcup_{\alpha \in On} q^\alpha$

(iii) $\alpha \in \text{Reg} \rightarrow \overline{\bigcup_{\beta \le \alpha} q^\beta} < \aleph_\alpha$

(iv) $(\forall \gamma)(\forall \alpha)(\forall \eta) \neg [\langle 0, \gamma, \alpha, \eta\rangle \in q \wedge \langle 1, \gamma, \alpha, \eta\rangle \in q]$

2. $p \le q \overset{\Delta}{\leftrightarrow} q \subseteq p$ for $p, q \in P$
3. $\mathbf{P} \overset{\Delta}{=} \langle P, \le\rangle$

Remark. Intuitively the conditions defining P may be understood as follows. If $\langle \mathbf{M}, \mathbf{P}^{\mathbf{M}}\rangle$ *is an elementary subsystem of* $\langle V, \mathbf{P}\rangle$, *then for each* $\alpha \in \text{Reg}^{\mathbf{M}}$, $\mathbf{P}^{\mathbf{M}}$ *adds* $\aleph_{G(\alpha)}$*-many subsets of* $\aleph_\alpha$ *to* $\mathbf{M}$ (cf. the $\mathbf{P}$ used in the proof of Theorem 11.10). The additional requirements, in particular 1.iii, are necessary to assure that sets added at the αth level do not affect the cardinalities at higher levels. Obviously, $\mathbf{P}$ is a proper class, and no $\mathbf{P}$ which is a set will suffice for our problem. Consequently the success of our efforts depends upon results of the preceding section and certain theorems that we must now prove.

From Definition 24.1 it is clear that each $q \in P$ uniquely determines its decomposition sequence $\langle q^\alpha \mid \alpha \in On\rangle$. So for each $q \in P$ we will use q^α to denote the αth element of this decomposition sequence.

Our first result is simply a list of elementary properties of two families of subclasses of P:

$$\Gamma_\alpha \overset{\Delta}{=} \{q \in P \mid (\forall \beta > \alpha)[q^\beta = 0]\},$$
$$\Delta_\alpha \overset{\Delta}{=} \{q \in P \mid (\forall \beta \le \alpha)[q^\beta = 0]\}.$$

Theorem 24.2.

1. Γ_α is a set, but Δ_α is a proper class.
2. $p \in \Gamma_\alpha \wedge q \in \Delta_\alpha \rightarrow p \cap q = 0 \wedge p \cup q \in P$.
3. $p_0 \in P \rightarrow (\exists! p \in \Gamma_\alpha)(\exists! q \in \Delta_\alpha)[p_0 = p \cup q]$.
4. $\alpha \in \text{Reg} \wedge p \in \Gamma_\alpha \rightarrow \bar{\bar{p}} < \aleph_\alpha$.
5. $\{q_\beta \mid \beta < \aleph_\alpha\} \subseteq \Delta_\alpha \wedge (\forall \beta)(\forall \delta)[\beta \le \delta < \aleph_\alpha \rightarrow q_\delta \le q_\beta]$
$$\rightarrow q = \bigcup_{\beta < \aleph_\alpha} q_\beta \in \Delta_\alpha.$$

Proof. 1–4 are obvious from the definitions.

5. We need only prove that $q = \bigcup_{\beta < \aleph_\alpha} q_\beta \in P$. Let $q^\gamma = \bigcup_{\beta < \aleph_\alpha} q_\beta{}^\gamma$. Then 1.i–1.iii of Definition 24.1 are satisfied. To check 1.iv of Definition 24.1, let $\gamma \in \text{Reg}$. We want to show that $\overline{\bigcup_{\gamma' \le \gamma} q^{\gamma'}} < \aleph_\gamma$. Since $(\forall \gamma \le \alpha)[q^\gamma = 0]$, we can assume $\gamma > \alpha$. Then

$$\bigcup_{\gamma' \le \gamma} q^{\gamma'} = \bigcup_{\gamma' \le \gamma} \bigcup_{\beta < \aleph_\alpha} q_\beta{}^{\gamma'} = \bigcup_{\beta < \aleph_\alpha} \bigcup_{\gamma' \le \gamma} q_\beta{}^{\gamma'}.$$

Since $\bigcup_{\gamma' \le \gamma} q_\beta{}^{\gamma'}$ has cardinality $< \aleph_\gamma$ for each $\beta < \aleph_\alpha$, since $\aleph_\alpha < \aleph_\gamma$, and since $\aleph_\gamma$ is regular, $\bigcup_{\beta < \aleph_\alpha} \bigcup_{\gamma' < \gamma} q_\beta{}^{\gamma'}$ has cardinality $< \aleph_\gamma$. Finally since

$$(\forall \beta)(\forall \delta)[\beta \le \delta < \aleph_\alpha \rightarrow q_\beta \subseteq q_\delta],$$

condition 1.v of Definition 24.1 is satisfied for q.

Definition 24.3. $\mathbf{P}_\alpha \triangleq \langle \Gamma_\alpha, \leq \rangle$.

$\mathbf{B}_\alpha$ is the Boolean algebra of all regular open sets in $\mathbf{P}_\alpha$. For $\alpha < \beta$ define $j: \Gamma_\beta \to \Gamma_\alpha$ by

$$j(p) = \bigcup_{\gamma \leq \alpha} p^\gamma \quad \text{for} \quad p \in \Gamma_\beta.$$

Remark. It is easy to see that j is an open continuous map onto Γ_α. Therefore j induces a complete monomorphism $i: \mathbf{B}_\alpha \to \mathbf{B}_\beta$ (Theorem 22.9), and we may regard $\mathbf{B}_\alpha$ as a complete subalgebra of $\mathbf{B}_\beta$ for $\alpha < \beta$. Note that each $\mathbf{B}_\alpha$ is a set.

Theorem 24.4. The map $j: \Gamma_\beta \to \Gamma_\alpha$ has the following elementary properties:

1. $q \leq p \to j(q) \leq j(p)$.
2. $j``[p]_{\mathbf{P}_\beta} = [j(p)]_{\mathbf{P}_\alpha}$.

Proof. 1. Obvious.

2. $j``[p]_{\mathbf{P}_\beta} \subseteq [j(p)]_{\mathbf{P}_\alpha}$ is obvious from 1. Now take $q \in [j(p)]_{\mathbf{P}_\alpha}$. Then $q \cup p \in \Gamma_\beta$. It is easy to see that $(q \cup p) \in [p]_{\mathbf{P}_\beta}$ and $j(q \cup p) = q$.

Theorem 24.5. Let $p \in \Gamma_\beta$ and $\beta > \alpha$. Then $\#([p]_{\mathbf{P}_\beta}) = [j(p)]_{\mathbf{P}_\alpha}$.

Proof. By the definition of $\#$. (See Remark following Theorem 22.9.)

$$\begin{aligned} \#([p]_{\mathbf{P}_\beta}) &= \inf\{b \in \mathbf{B}_\alpha \mid [p]_{\mathbf{P}_\beta} \leq i(b)\} \\ &= \inf\{b \in \mathbf{B}_\alpha \mid [p]_{\mathbf{P}_\beta} \leq (j^{-1})``b\} \\ &= \inf\{b \in \mathbf{B}_\alpha \mid j``[p]_{\mathbf{P}_\beta} \leq b\} \\ &= \inf\{b \in \mathbf{B}_\alpha \mid [j(p)]_{\mathbf{P}_\alpha} \leq b\} \\ &= [j(p)]_{\mathbf{P}_\alpha}. \end{aligned}$$

Remark. Each $\mathbf{P}_\alpha$ is fine (Definition 5.21) and hence by Lemma 5.22, $[q]_{\mathbf{P}_\alpha}{}^{-0} = [q]_{\mathbf{P}_\alpha}$ for $q \in P_\alpha$. So $[j(p)]_{\mathbf{P}_\alpha} \in B_\alpha$. Let

$$\mathbf{B} = \bigcup_{\alpha \in On} \mathbf{B}_\alpha, \quad \text{i.e.,} \quad B = \bigcup_{\alpha \in On} B_\alpha.$$

For the operations in $\mathbf{B}$, see Remark following Definition 22.16. Moreover, if $A \subseteq B$ and A is a set, then $A \subseteq B_\alpha$ for some α. Since $\sup A$ exists in $\mathbf{B}_\alpha$, $\sup A$ exists in $\mathbf{B}$. Therefore $\mathbf{B}$ is of the type considered in §23. Let $\tilde{\mathbf{B}}$ be the completion of $\mathbf{B}$.

Obviously, $\langle \{\mathbf{P}_\alpha \mid \alpha < On\}, \{j_{\alpha\beta} \mid \alpha \leq \beta < On\} \rangle$ is a normal limiting system of partial order structures. Then, by Theorem 22.30,

1. $\mathbf{B}$ satisfies the s.c.c.
2. $\mathbf{B} = \tilde{\mathbf{B}}$.
3. $\mathbf{B}$ is isomorphic to the Boolean algebra of all regular open subsets of P, since $\mathbf{P} \cong \lim_{\alpha \to On} \mathbf{P}_\alpha$.

By 1 above and Theorem 23.21, $\mathbf{B}$ satisfies the UCL. But by 2 $\mathbf{B} = \tilde{\mathbf{B}}$. Therefore $\tilde{\mathbf{B}}$ satisfies the UCL and hence, by Theorem 23.22, $V^{(\mathbf{B})}$ satisfies the Axiom of Replacement.

It is clear that $\mathbf{P}_\alpha$ and $\mathbf{P}'_\alpha \triangleq \langle \Delta_\alpha, \leq \rangle$ form an $\aleph_\alpha$-Easton pair and $\mathbf{P} \simeq \mathbf{P}_\alpha \times \mathbf{P}'_\alpha$. Therefore by Theorem 23.45 $\mathbf{B}$ satisfies the SL. Consequently, by Theorem 23.30 $\mathbf{V}^{(\mathbf{B})}$ satisfies $ZF + AC$. Furthermore, by Theorem 23.46 $\mathbf{B}$ is $(\aleph_\alpha, \aleph_\beta)$-splitable. Hence by Corollary 23.34 cardinals are absolute, i.e.,

$$(\forall \alpha)[[\![\mathrm{Card}\,(\aleph_\alpha)^\vee]\!] = \mathbf{1}].$$

Using only the properties 1 and 2 of G stated at the beginning of this section one can prove

$$[\![\overline{\overline{2^{(\aleph_\alpha)^\vee}}} = (\aleph_{G(\alpha)})^\vee]\!] = \mathbf{1} \quad \text{for} \quad \alpha \in \mathrm{Reg}.$$

In order to determine $\overline{\overline{2^{\aleph_\alpha}}}$ even in the case $\alpha \notin \mathrm{Reg}$, we require that in this case $\aleph_{G(\alpha)}$ be the least cardinal which is not cofinal with $\aleph_\alpha$ and which is greater than or equal to $\aleph_{G(\beta)}$ for each $\beta < \alpha$. Then it will turn out that in $\mathbf{V}^{(\mathbf{B})}$, $\overline{\overline{2^{\aleph_\alpha}}} = \aleph_{G(\alpha)}$, i.e., $\overline{\overline{2^{\aleph_\alpha}}}$ is the least cardinal allowed by König's Theorem. Thus we obtain in general

$$(\forall \alpha)[\![\overline{\overline{2^{(\aleph_\alpha)^\vee}}} = (\aleph_{G(\alpha)})^\vee]\!] = \mathbf{1} \quad \text{in} \quad \mathbf{V}^{(\mathbf{B})}.$$

We shall prove this statement by a forcing argument. Thus, let $\mathbf{M}$ be a transitive countable structure such that $\langle \mathbf{M}, \mathbf{B}^{\mathbf{M}} \rangle$ is an elementary subsystem of $\langle V, \mathbf{B} \rangle$. Let $h_0 \colon \mathbf{B}^{\mathbf{M}} \to \mathbf{2}$ be a homomorphism preserving all the sums which are $\mathbf{M}$-definable in the language of $\mathbf{V}^{(\mathbf{B})}$, and let

$$h \colon (V^{(\mathbf{B})})^{\mathbf{M}} \xrightarrow{\text{onto}} M[h_0]$$

be defined from h_0 as before. (See the proof of Theorem 23.24.) For the remaining part of this section we are working in $\langle \mathbf{M}, \mathbf{B}^{\mathbf{M}} \rangle$, i.e., ordinals α are ordinals in $\mathbf{M}$, $\alpha \in On^{\mathbf{M}}$, $P, \Gamma_\alpha, \Delta_\alpha, \mathrm{Reg}, \ldots$ stand for $P^{\mathbf{M}}, \Gamma_\alpha{}^{\mathbf{M}}, \Delta_\alpha{}^{\mathbf{M}}, \mathrm{Reg}^{\mathbf{M}}, \ldots$ and also $\aleph_\alpha$ always means $\aleph_\alpha{}^{\mathbf{M}}$ (which is equal to $\aleph_\alpha{}^{M[h_0]}$ since cardinals are absolute). Similarly, $p, q, \bar{q}, \ldots$ now range over $P^{\mathbf{M}}$ (or $B^{\mathbf{M}} - \{\mathbf{0}\}$).

Lemma 24.6. Assume $h_0([\![u \subseteq (\aleph_\alpha)^\vee]\!]) = \mathbf{1}$ for some $u \in (V^{(\mathbf{B})})^{\mathbf{M}}$. Then there exists a $p \in P$ and a $\Lambda \subseteq \Gamma_\alpha$ such that $h_0(p) = \mathbf{1}$, and in $\langle \mathbf{M}, \mathbf{B}^{\mathbf{M}} \rangle$:

1. $p \leq [\![u \subseteq (\aleph_\alpha)^\vee]\!]$
2. $\overline{\overline{\Lambda}} \leq \aleph_\alpha$ and
3. for each $q' \leq p$ and each $\gamma < \aleph_\alpha$

$$(\exists p' \in \Lambda)[p' \cdot q' > \mathbf{0} \wedge [p \cdot p' \leq [\![\check{\gamma} \in u]\!] \vee p \cdot p' \leq -[\![\check{\gamma} \in u]\!]]].$$

Proof. Applying Theorem 23.44 in $\langle \mathbf{M}, \mathbf{B}^{\mathbf{M}} \rangle$ to any $q \leq [\![u \subseteq (\aleph_\alpha)^\vee]\!]$, with $r = q$, $I = \aleph_\alpha$, $b_\gamma = [\![\check{\gamma} \in u]\!]$ for $\gamma < \aleph_\alpha$, and $b_{\gamma j} = b_\gamma$ for $j \in J$, we establish the existence of a $p \in P$ and a $\Lambda \subseteq \Gamma_\alpha$ satisfying 1–3 and such that $p \leq q$. Therefore the set of p's for which there exists a $\Lambda \subseteq \Gamma_\alpha$ satisfying 1–3 is dense beneath $[\![u \subseteq (\aleph_\alpha)^\vee]\!]$, so we can find a $p \leq [\![u \subseteq (\aleph_\alpha)^\vee]\!]$ which satisfies the additional requirement $h_0(p) = \mathbf{1}$ (Theorem 10.11).

Theorem 24.7. $\overline{\overline{2^{\aleph_\alpha}}} = \aleph_{G(\alpha)}$ in $M[h_0]$.

Proof.

1. $\overline{\overline{2^{\aleph_\alpha}}} \geq \aleph_{G(\alpha)}$ in $M[h_0]$:

Consider first the case $\alpha \in \text{Reg}$. Let

$$a_\eta = \{\gamma \in \aleph_\alpha \mid (\exists p \in P)[\langle 1, \gamma, \alpha, \eta\rangle \in p \wedge h_0(p) = \mathbf{1}]\}.$$

Then, as in the proof of Theorem 11.10,

$$(\forall \eta < \aleph_{G(\alpha)})[a_\eta \subseteq \aleph_\alpha]$$

and

$$(\forall \eta, \eta' < \aleph_{G(\alpha)})[\eta \neq \eta \rightarrow a_\eta \neq a_{\eta'}].$$

Since $(\forall \eta < \aleph_{G(\alpha)})[a_\eta \in M[h_0]]$, this proves 1.

Now suppose $\alpha \notin \text{Reg}$. Then $\overline{\overline{2^{\aleph_\alpha}}} \geq \overline{\overline{2^{\aleph_\beta}}} \geq \aleph_{G(\beta)}$ for each $\beta < \alpha$, $\beta \in \text{Reg}$, hence $\overline{\overline{2^{\aleph_\alpha}}} \geq \aleph_{G(\alpha)}$ by our additional requirement on G that $\aleph_{G(\alpha)}$ is the smallest cardinal greater than or equal to $\aleph_{G(\beta)}$ for all $\beta < \alpha$ that are not cofinal with $\aleph_\alpha$.

2. $\overline{\overline{2^{\aleph_\alpha}}} \leq \aleph_{G(\alpha)}$ in $M[h_0]$.

Define $\sum$ in **M** as follows:

$$\sum = \{\langle \Lambda, S\rangle \mid \Lambda \subseteq \Gamma_\alpha \wedge \Lambda \in M \wedge S \subseteq \Lambda \times \aleph_\alpha \wedge S \in M\}.$$

We first show that $\overline{\overline{\sum}} = \aleph_{G(\alpha)}$ in **M**. Clearly $\overline{\overline{\sum}} \geq \aleph_{G(\alpha)}$. Furthermore in **M**

$$\overline{\overline{\Gamma}}_\alpha \leq \aleph_\alpha \cdot \aleph_\alpha \cdot \aleph_{G(\alpha)} = \aleph_{G(\alpha)}.$$

Since the *GCH* holds in **M**,

$$\overline{\overline{\{\Lambda \mid \Lambda \subseteq \Gamma_\alpha \wedge \overline{\overline{\Lambda}} \leq \aleph_\alpha\}}} \leq \overline{\overline{\aleph_{G(\alpha)}^{\aleph_\alpha}}} = \aleph_{G(\alpha)}.$$

Therefore $\overline{\overline{\sum}} \leq \aleph_{\alpha+1} \cdot \aleph_{G(\alpha)} = \aleph_{G(\alpha)}$ and hence $\overline{\overline{\sum}} = \aleph_{G(\alpha)}$ in **M**.

Thus, to prove 2 it suffices to find a function in $M[h_0]$ which maps $\sum$ onto $\mathscr{P}(\aleph_\alpha)$ in $M[h_0]$. For $\langle \Lambda, S\rangle \in \sum$ let

$K(\langle \Lambda, S\rangle) = y$ if there are $w, \bar{q}$ such that

(i) $h_0(\bar{q}) = \mathbf{1} \wedge \bar{q} \leq [\![w \subseteq (\aleph_\alpha)^{\vee}]\!]$

(ii) for each $q' \leq \bar{q}$ and each $\gamma < \aleph_\alpha$

$$(\exists p \in \Lambda)[p \cdot q' > \mathbf{0} \wedge [p \cdot \bar{q} \leq [\![\check{\gamma} \in w]\!] \vee p \cdot \bar{q} \leq {}^{-}[\![\check{\gamma} \in w]\!]]]$$

(iii) $S = \{\langle p, \gamma\rangle \mid p \in \Lambda \wedge \gamma < \aleph_\alpha \wedge p \cdot \bar{q} \leq [\![\check{\gamma} \in w]\!]\}$

(iv) $y = h(w)$.

$K(\langle \Lambda, S\rangle) = 0$ otherwise.

The proof that

$$K: \sum \xrightarrow{\text{onto}} \mathscr{P}^{M[h_0]}(\aleph_\alpha)$$

is similar to the corresponding proof in Theorem 23.30. To prove that K is a function suppose that conditions (i)–(iv) are satisfied by $w_0, \bar{q}_0$ and $w_1, \bar{q}_1$

for the same S. We must then show that $h(w_0) = h(w_1)$. Note that $h(w_0), h(w_1) \subseteq \aleph_\alpha$. Suppose

$$\gamma \in h(w_0) \wedge \gamma \notin h(w_1) \quad \text{for some} \quad \gamma \in \aleph_\alpha.$$

Then since this is true in $M[h_0]$, we have from the definition of h (see the proof of Theorem 23.24)

$$h_0(\bar{q}_0 \cdot \bar{q}_1 \cdot [\![\check{\gamma} \in w_0]\!] \cdot [\![\check{\gamma} \notin w_1]\!]) = \mathbf{1}$$

and by (ii):

$$(\exists p \in \Lambda)[p \cdot \bar{q}_0 \cdot \bar{q}_1 \cdot [\![\check{\gamma} \in w_0]\!] \cdot [\![\check{\gamma} \notin w_1]\!] > \mathbf{0} \wedge [p \cdot \bar{q}_0 \leq [\![\check{\gamma} \in w_0]\!] \vee p \cdot \bar{q}_0 \leq [\![\check{\gamma} \notin w_0]\!]]].$$

We must have $p \cdot \bar{q}_0 \leq [\![\check{\gamma} \in w_0]\!]$, therefore by (iii)

$$\langle p, \gamma \rangle \in S.$$

But by hypothesis, $w_0, \bar{q}_0$ and $w_1, \bar{q}_1$ satisfy (i)–(iv) for the same S. Hence

$$p \cdot \bar{q}_1 \leq [\![\check{\gamma} \in w_1]\!].$$

This is a contradiction.

To prove that K is onto suppose $h(u) \subseteq \aleph_\alpha$. Then $h_0([\![u \subseteq (\aleph_\alpha)\check{}\,]\!]) = \mathbf{1}$, so by Lemma 24.6 there are $p, \Lambda \in M$ such that $h_0(p) = \mathbf{1}$ and, in $\langle \mathbf{M}, \mathbf{B}^{\mathbf{M}} \rangle$, $p \leq [\![u \subseteq (\aleph_\alpha)\check{}\,]\!]$, $\Lambda \subseteq \Gamma_\alpha$, $\bar{\bar{\Lambda}} \leq \aleph_\alpha$, and for each $q' \leq p$ and each $\gamma < \aleph_\alpha$

$$(\exists p' \in \Lambda)[p' \cdot q' > 0 \wedge [p \cdot p' \leq [\![\check{\gamma} \in u]\!] \vee p \cdot p' \leq -[\![\check{\gamma} \in u]\!]]].$$

If $S = \{\langle p', \gamma \rangle \mid p' \in \Lambda \wedge \gamma < \aleph_\alpha \wedge p' \cdot p \leq [\![\check{\gamma} \in u]\!]\}$, then $\langle \Lambda, S \rangle \in \sum$ and

$$K(\langle \Lambda, S \rangle) = h(u)$$

i.e., K is onto. Consequently, in $M[h_0]$

$$\overline{\overline{2^{\aleph_\alpha}}} = \aleph_{G(\alpha)}.$$

Bibliography

Bar-Hillel, Y. (Ed.): *Essays on the foundations of mathematics.* Jerusalem: Magnes Press 1966.

Benacerraf, P., Putnam, H.: *Philosophy of mathematics selected readings.* Englewood Cliffs: Prentice-Hall Inc. 1964.

Bernays, P., Fraenkel, A. A.: *Axiomatic set theory.* Amsterdam: North-Holland 1958.

Birkhoff, G., MacLane, S.: *A survey of modern algebra.* Macmillan 1965.

Church, A.: *Introduction to mathematical logic.* Princeton: Princeton University Press 1956.

Cohen, P.: The independency of the continuum hypothesis. *Proc. Nat. Acad. Sci. U.S.* **50**, 1143–1148 (1963).

——— *Set theory and the continuum hypothesis.* New York: W. A. Benjamin Inc. 1966.

Fraenkel, A. A., Bar-Hillel, Y.: *Foundations of set theory.* Amsterdam: North-Holland 1958.

Gödel, K.: *The consistency of the continuum hypothesis.* Princeton: Princeton University Press 1940.

——— What is Cantor's continuum problem? *Am. Math. Monthly* **54**, 515–525 (1947).

Halmos, Paul R.: *Lectures on Boolean algebras.* D. van Nostrand Co. Inc. 1963.

van Heijenoort, J.: *From Frege to Godel.* Cambridge: Harvard University Press 1967.

Jeck, Thomas: *Lectures in set theory.* New York: Springer-Verlag 1971.

Moore, T. O.: *Elementary general topology.* Prentice-Hall 1964.

Quine, W. V.: *Set theory and its logic.* Cambridge: Belknap Press 1969.

Rosser, J. Barkley: *Boolean valued models of set theory.* Academic Press 1969.

Rubin, J. E.: *Set theory for the mathematician.* San Francisco: Holden-Day 1967.

Schoenfield, J. R.: *Mathematical logic.* Reading: Addison Wesley 1967.

Sikorski, Roman: *Boolean algebras.* Springer-Verlag 1964.

Simmons, G. F.: *Introduction to topology and modern analysis.* McGraw-Hill 1963.

Takeuti, G., Zaring, W. M.: *Introduction to axiomatic set theory.* New York: Springer-Verlag 1971.

Problem List

by Paul E. Cohen

Section 1

1. Prove, for a Boolean algebra $\langle B, +, \cdot, ^-, \mathbf{0}, \mathbf{1}\rangle$, that

   i) $(\forall a)(\forall b)[a + b = a \rightarrow b - a = \mathbf{0}]$, and

   ii) $(\forall a)(\forall b)[ab = a \rightarrow a - b = \mathbf{0}]$.

2. If $\langle B, +, \cdot, ^-, \mathbf{0}, \mathbf{1}\rangle$ is a Boolean algebra, prove that $\langle B, \cdot, +, ^-, \mathbf{1}, \mathbf{0}\rangle$ is a Boolean algebra.

3. Let $\langle B, +, \cdot, ^-, \mathbf{0}, \mathbf{1}\rangle$ be a Boolean algebra and let $\leq$ be the natural order on B. Show that $+, \cdot, ^-, \mathbf{0}, \mathbf{1}$ can be defined in $\langle B, \leq\rangle$.

4. Give a partial order structure in which there is a p such that $[p]$ is not regular open.

5. Which sets are regular open in a linear order structure?

Section 2

6. Show that in Definition 2.2, condition 3 may be replaced by

$$3'.\ S \in A \wedge S \subseteq P \wedge S^- = P \wedge S \text{ is open} \rightarrow G \cap S \neq 0.$$

7. If $\langle P, \leq\rangle$ is a partial order structure, if M is a model of ZF, if Q is a dense subset of P, if $Q \in M$, and if G is $\langle P, \leq\rangle$-generic over M, then $Q \cap G$ is $\langle Q, \leq\rangle$-generic over M.

8. Let $\mathbf{P} = \langle P, \leq\rangle$ be a partial order structure with

$$P = \{p \mid (\exists d \subseteq \omega)[d \text{ if finite} \wedge p : d \rightarrow 2]\} \quad \text{and} \quad p \leq q \text{ iff } p \supseteq q.$$

If

   i) $S_n = \{p \in P \mid p(n) \text{ is defined}\}, n < \omega$

   ii) $A \supseteq \{S_n \mid n \in \omega\}$, and

   iii) G is $\mathbf{P}$-generic over A,

then $\bigcup G$ is a function from ω into 2.

9. If $\mathbf{B}$ is a complete Boolean algebra, if A is a class and if F is an A-complete ultrafilter on B, then $F - \{\mathbf{0}\}$ is $\langle |B| - \{\mathbf{0}\}, \leq\rangle$-generic over A.

10. Find a condition on a partial order structure such that its Boolean algebra of regular open sets will satisfy the c.c.c. Can you find an equivalent condition?

Section 4

11. Show that the Boolean algebra of regular open sets for the partial order structure of Exercise 8 does not satisfy the $(\omega, 2)$-DL.

Section 5

12. Show that if $\mathbf{P}$ is a partial order structure, if A is a set and if G is $\mathbf{P}$-generic over A, then G is a filter for $\mathbf{P}$. Thus G is sometimes referred to as a $\mathbf{P}$-generic filter.

Section 7

13. We say that mathematical induction holds in a model M of Godel-Bernays set theory (GB) if for every formula φ of the language of GB,

$$[\varphi(0) \wedge (\forall n)[\varphi(n) \rightarrow \varphi(n + 1)]] \rightarrow (\forall n)[\varphi(n)]$$

is true in M. Show that if M is a standard model of GB, then M satisfies mathematical induction.

14. Let the strong Löwenheim-Skölen Theorem be the statement that every structure for a language $\mathscr{L}$ (which is a sequence of classes) has an elementary substructure of power $\overline{\overline{\mathscr{L}}}$. Show that if "$GB$ + there is a standard model of ZF which is a set" is consistent, then the strong Löwenheim-Skölem Theorem is not a Theorem of GB + mathematical induction. (Hint: Look at the minimal model.) We remark that the strong Löwenheim-Skölem Theorem may be proven in Morse-Kelly set theory, which has stronger comprehension axioms than does GB.

15. Give a construction for $M[K]$ that does not presuppose that $K \subseteq M$.

Section 8

16. What is the relationship between the sets in $L[K; F]$ of rank not more than α and $\{D_\alpha(t) \mid t \in T_\alpha\}$?

Section 9

17. In view of the fact that $V[F] \subseteq V$, discuss the statement "$V[F]$ is an extension of V."

Section 10

18. Show that if φ is any limited or unlimited formula, then $\{p \mid p \Vdash \varphi\}$ is a regular open set. Assuming forcing to be defined by the statements of Theorem 10.4, find a statement about $p \Vdash \varphi$ and $p \Vdash \neg\neg\varphi$ that is equivalent to the statement that $\{p \mid p \Vdash \varphi\}$ is regular open.

Section 11

19. Prove that the partial order structure **P** of Definition 11.1 is isomorphic to the partial order structure of Exercise 8. What relationship is there between the set a in Theorem 11.3 and the function $\bigcup G$ of Exercise 8?

20. Why is Corollary 11.4 not a proof that if ZF is consistent then so is $ZF + AC + GCH + V \neq L$?

21. Refer to Theorem 11.6. Show that $M[G_1]$ and $M[G_2]$ are not necessarily elementarily equivalent in the language $\mathscr{L}_0(C(M) \cup \{G(\ \ )\})$.

22. Show that the partial order structure of Theorem 11.10 is isomorphic to the structure $\langle Q, \leq \rangle$ where

$$Q = \{q \mid (\exists d)[d \subseteq \overline{\overline{\alpha \times \omega}} \wedge \overline{\overline{d}} < \lambda_0 \wedge q : d \to 2]\}$$

and $q_1 \leq q_2$ iff $q_1 \supseteq q_2$ (cf. Definition 11.1 and Exercises 8 and 19).

Section 12

23. Show that the partial order structure of Theorem 11.1 is isomorphic to the strong product of copies of the partial order structure of Exercise 8 or 19.

24. In courses in naive set theory a finite set is sometimes defined as a set that is equinumerous with none of its proper subsets. Show that this is a satisfactory definition only if the Axiom of Choice is assumed.

Section 13

25. In view of the fact that $V^{(\mathbf{B})} \subseteq V$, discuss the statement "$V^{(\mathbf{B})}$ is an extension of V." (cf. Exercise 16.)

26. Suppose $\langle K, \leq \rangle$ is a partial order structure and **B** is the Boolean algebra of regular open sets of **B**. Let $f_0 : K \to B$ be defined by $f_0(k) = [k]^{-0}$. Then $V[f_0]$ and $V^{(\mathbf{B})}$ are **B**-valued structures. Define a Boolean elementary embedding $I : V[f_0] \to V^{(\mathbf{B})}$ (i.e., such that for any formula φ and terms $t_1, \ldots, t_n \in V[f_0]$),

$$[\![\varphi(t_1, \ldots, t_n)]\!] = [\![\varphi(I(t_1), \ldots, I(t_n))]\!].$$

Is I one to one? Onto? $V[f_0]$ is a class of names for sets. In what sense can this also be said of $V^{(\mathbf{B})}$?

27. Suppose **B** is a complete Boolean algebra. For which $u \in V^{(\mathbf{B})}$ is

$$\{v \in V^{(\mathbf{B})} \mid [\![v \in u]\!] = \mathbf{1}\}$$

a set? For which $u \in V^{(\mathbf{B})}$ is

$$\{v \in V^{(\mathbf{B})} \mid [\![v \in u]\!] \neq \mathbf{0}\}$$

a set?

Section 17

28. If M and N are models of ZF and $M \subseteq N$, show that any cardinal in N is a cardinal in M (cf. Theorem 17.1).

29. If $\mathbf{P}$ is a partial order structure, then we say that $\mathbf{P}$ satisfies the γ-chain condition (where γ is a cardinal) if $\overline{\overline{S}} \leq \gamma$ for every $S \subseteq |\mathbf{P}|$ such that $(\forall p, q \in S)\,[\neg\,\text{Comp}\,(p, q)]$. Show that if M is a countable standard transitive model of ZFC, if $\mathbf{P} \in M$, if G is $\mathbf{P}$-generic over M, and if $\alpha > \gamma$ is a cardinal in M, then α is a cardinal in $M[G]$. Give two proofs of this, one using Theorem 17.4 (see Exercise 10), and the other based on Theorems 10.4 and 10.6.

Section 18

30. Suppose $\mathbf{P}$ is a partial order structure such that whenever $\langle p_i \mid i < \omega\rangle$ is a sequence with $p_0 \geq p_1 > \cdots$, then $(\exists p \in |\mathbf{P}|)(\forall i \in \omega)[p \leq p_i]$. Let $\mathbf{B}$ be the Boolean algebra of regular open sets of $\mathbf{P}$. Then $\mathbf{B}$ satisfies the $(\omega, 2)$-DL.

31. Use the result of Exercise 30 to show that if M is a countable standard transitive model of ZFC if $\mathbf{P} \in M$ has the property of Exercise 30, and if G is $\mathbf{P}$-generic over M, then $\mathscr{P}(\omega)$ is the same in M and $M[G]$.

32. Give a direct proof of Exercise 31 based on Theorems 10.4 and 10.6. Can the proof be generalized to give a stronger theorem?

Section 20

33. Give a proof for the remark after Definition 20.1.

34. Give a condition on a partial order structure such that its Boolean algebra of regular open sets satisfies the (ω, ω_α)-WDL. Can you find an equivalent condition?

Section 22

35. If $\mathbf{B}$ is a complete Boolean algebra and $\mathbf{B}_0$ is a dense subalgebra of $\mathbf{B}$ then is the completion of $\mathbf{B}_0$ necessarily isomorphic to $\mathbf{B}$?

Subject Index

Index of Symbols

Vol. 7 SERRE: A Course in Arithmetic. x, 115 pages approximately. Tentative publication date: February, 1973.

Vol. 10 COHEN: A Course in Simple-Homotopy Theory. xii, 112 pages approximately. Tentative publication date: February, 1973.

Vol. 11 CONWAY: Functions of One Complex Variable. x, 313 pages approximately. Tentative publication date: July, 1973.